W9-AHP-899

A series of student texts in

CONTEMPORARY BIOLOGY

Adaptation to Thermal Environment

man and his productive animals

Laurence E. Mount

M.B., B.S. (Lond.), M.D., C.M. (McGill)

Head of Department of Applied Biology, Agricultural Research Council
Institute of Animal Physiology, Babraham, Cambridge
Special Professor of Environmental Physiology
University of Nottingham School of Agriculture

University Park Press
Baltimore

First published in Great Britain 1979
by Edward Arnold (Publishers) Limited
First Published in the United States of America
by University Park Press, Baltimore, Maryland

Mount, Laurence Edward
Adaptation to Thermal Environment:
Man and his Productive Animals.
(A series of student texts in contemporary biology)
Includes index.
1. Veterinary physiology. 2. Body temperature—
Regulation. 3. Human physiology. I. Title.
[DNLM: 1. Body temperature regulation.
2. Environment. 3. Climate. 4. Adaptation,
Physiological. QT165 M928c] 79-44936
ISBN 0-8391-1420-6

Printed in Great Britain

Preface

This book deals with the heat exchanges that take place between animals and their thermal environments. The first six chapters give an account of the basic principles of the subject. There follow six chapters that are concerned specifically with man and those animals that are of agricultural importance.

The book is intended for several groups of readers who may have different interests. In universities and polytechnics, students of biological subjects, including students on agricultural, veterinary and medical courses, should find in it an introduction to climatic physiology. The species-specific chapters and parts of the earlier chapters should be useful to students in agricultural colleges and to those concerned with animal production, to designers and architects, and to environmental engineers. Parts of the book should have some value for senior classes in schools.

Much of the material has been developed in lectures that I have given as Special Professor of Environmental Physiology in the School of Agriculture of the University of Nottingham, and some parts of the work are derived from my earlier book, 'The climatic physiology of the pig', published by Edward Arnold in 1968.

It is a pleasure to acknowledge the help of colleagues who read different parts of the text and suggested many improvements:
Dr G. Alexander, C.S.I.R.O., Australia; Dr J. A. Clark, University of Nottingham; Professor O. G. Edholm, University College London; Dr M. F. Fuller, Rowett Research Institute, Aberdeen; Dr D. L. Ingram, Institute of Animal Physiology, Babraham; Dr A. J.

McArthur, University of Nottingham; Professor W. V. Macfarlane, University of Adelaide; Professor J. L. Monteith, F.R.S., University of Nottingham; Dr S. A. Richards, Wye College, University of London; Dr M. W. Stanier, Institute of Animal Physiology, Brabraham; Professor A. J. F. Webster, University of Bristol; Professor G. C. Whittow, University of Hawaii; and Professor E. J. W. Barrington, F.R.S., Editor of the Contemporary Biology Series published by Edward Arnold.

I am grateful to a number of authors and publishers, societies and journals for permission to reproduce tables and diagrams. Their names appear in the figure legends and in the list of references. The publishers include: Academic Press; American Physiological Society; American Society of Agricultural Engineers; Animal Production; Australian Journal of Agricultural Research; British Antarctic Survey Bulletin; British Medical Bulletin; British Medical Journal; Butterworths; Cambridge University Press; Charles C. Thomas; Condor; Copyright Agency of U.S.S.R.; Ecology; Elizabeth Licht; Elsevier Scientific Publishing Company; Ergonomics; European Association for Animal Production; Federation of American Societies for Experimental Biology; Her Majesty's Stationery Office; John Wiley; Journal of Experimental Zoology; Journal of Physiology; Journal of University of Newcastle-upon-Tyne Agricultural Society; Lancet; Lea and Febiger; Medical and Biological Engineering; Nature; Oxford University Press; Physiology and Behaviour; Pitman; Poultry Science; Research in Veterinary Science; Royal Society of Queensland; Science; Society for Experimental Biology; Springer Verlag; Swets and Zeitlinger; W. B. Saunders; and Zoological Society of London.

Finally, I should like to thank the staff of the Library and the Photographic Section of the Institute of Animal Physiology; the Director of the Institute, Dr B. A. Cross, F.R.S., for his permission to undertake the work; the staff of Edward Arnold, for their courteous assistance and co-operation; and my secretary, Jane Kilvert, for her invaluable help.

Babraham, Cambridge
1979

L. E. M.

Contents

PREFACE v

UNITS AND SYMBOLS xi

1 TEMPERATURE REGULATION AND HEAT BALANCE: A GENERAL SURVEY 1
Homeothermy and Poikilothermy 1
Mechanism of Temperature Regulation 4
Heat Production and Temperature 6
Heat Balance 7
Heat Exchange 9
Heat Storage 10
Energy Retention 12

2 HEAT PRODUCTION 14
Heat Production in Relation to Environmental Temperature 14
Thermal Neutrality 16
Critical Temperature 17
Limiting Temperatures 17
Body Size 21
Plane of Nutrition 26

Thermal Insulation 31
24-hourly Variation 32
Body Temperature 32
Sites of Heat Production 35
Non-shivering Thermogenesis 35
Other Factors 36

3 HEAT LOSS 40
Heat Transfer Coefficients 42
Evaporative Heat Loss 42
Cutaneous Evaporative Loss 46
Respiratory Evaporative Loss 48
Non-evaporative Heat Transfer 50
Radiation 50
Convection 64
Conduction 72
Calorimetry 73

4 THERMAL INSULATION 79
Tissue and External Insulations 79
Heat Flow and Insulation 81
Tissue Insulation 85
Coat Insulation 92
Air-ambient Insulation 96

5 CLIMATIC ZONES AND ASSESSMENT OF THERMAL ENVIRONMENT 99
Solar Radiation 99
Climatic Zones 101
The Assessment of Thermal Environment 105
Operative Temperature 106
Equivalent Effects of Environmental Variables 108
Micro-environment 113

6 ADAPTATION TO THE THERMAL ENVIRONMENT 116
Behavioural Adaptation 117
Physiological Adaptation 120
Morphological Adaptation 125
Genetic Adaptation 125

	Adaptation of the Newborn Animal to its Thermal Environment	129
	Adaptation to Low Temperatures	135
	Adaptation to High Temperatures	138
	Torpidity	141
	Poikilotherms	142
7	MAN	145
	Metabolic Rate	146
	Man in a Cold Environment	148
	Acclimatization to Cold	154
	Man in a Hot Environment	157
	Acclimatization to Heat	159
	Heat Stress	168
	Thermal Comfort	171
	Clothing	172
	The Newborn Infant	178
8	PIG	182
	General Features	183
	Responses to Thermal Environment	184
	Newborn Pig	185
	Growing Pig	190
	Implications for Husbandry	200
	Pigs in the Tropics	206
	Comparisons with Other Species	207
9	CATTLE	209
	Breeds	209
	Thermal Neutrality and Response to Cold	210
	Responses to High Temperatures	216
	Reproduction	222
	Implications for Husbandry	225
10	SHEEP, GOAT AND DEER	228
	Responses to Thermal Environment	229
	Adaptation to Heat	235
	Newborn Lamb	241
	Deer	245

11 COMPARATIVE ASPECTS OF UNGULATES IN HOT CLIMATES 247
Animal Production 248
Adaptation to Hot Climates 250
Evaporative Cooling 250
Water Economy 257
Adaptation to Hot Arid Conditions 259

12 BIRDS 269
Temperature Regulation and Metabolic Rate 269
Exposure to Cold 273
Exposure to Heat 277
Water Requirements 280
Flight 280
Husbandry 281

APPENDICES 287

BIBLIOGRAPHY 293

REFERENCES 295

SUBJECT INDEX 323

Units and Symbols

The units employed are in general those of the Système Internationale (SI), with variations as necessary for convenience or for material introduced from elsewhere. The Joule (J) is used as the unit of energy, and the watt (W) as the unit of metabolic rate or heat transfer. Factors for converting other units, such as the kilocalorie (kcal), to SI units are given in Appendix 1.

The negative exponent has been used instead of the solidus. This avoids the ambiguity that can arise when more than two quantities are involved in one expression. For example, a heat transfer coefficient is written in terms of $W\ m^{-2}\ {}^{\circ}C^{-1}$, instead of W/m^2 per °C; the use of the double solidus, $W/m^2/{}^{\circ}C$, is clearly inadmissible because the expression is then algebraically incorrect.

The symbols used in the text, apart from the recognized symbols for SI units, are as follows:

a	coefficient of thermal expansion of air
A	surface area
c	specific heat
c_p	specific heat at constant pressure
CT	critical temperature
d	diameter
D	insulation-wind-decrement
f	fleece length
F	radiative interchange factor
F	food intake

g	acceleration due to gravity
h	per cent wetted area
H	rate of heat transfer per unit area
H_c	convective heat transfer per unit area
H_d	conductive heat transfer per unit area
H_e	evaporative heat transfer per unit area
H_n	combined non-evaporative (sensible) heat transfer per unit area
H_r	radiant heat transfer per unit area
I	thermal insulation per unit area
I_e	external insulation per unit area
I_t	tissue insulation per unit area
k	thermal conductivity
k	partial efficiency
K	combined radiant-convective heat transfer coefficient
K_C	convective heat transfer coefficient
K_E	evaporative heat transfer coefficient
K_O	standard cooling rate
K_R	radiant heat transfer coefficient
L	thickness of medium
M	metabolic rate
P	pressure
Q	volume flow per unit time
r	thermal resistance
r	radius
R	evaporative impedance
R_F	effective radiant flux
S	rate of heat storage
t	time
T	temperature
T_a	air temperature
T_b	body temperature
T_c	core temperature
T_e	environmental temperature
T_g	globe thermometer temperature
T_o	operative temperature
$\bar{T}_r$	mean radiant temperature
T_{re}	rectal temperature
T_s	skin temperature
T_t	outer coat temperature
T_{wb}	wet bulb temperature
U	latent heat of vaporization of water
v	kinematic viscosity

V	windspeed
W	body weight
W^b	metabolic body size
X	fractional concentration difference
y	velocity of reaction
Y	quantity of water vapour per unit volume
γ	psychrometer constant (0.66 mbar $°C^{-1}$ at 20°C and 1013 mbar)
ε	emissivity
θ	equivalent temperature
ρ	density
σ	Stefan–Boltzmann constant 5.67×10^{-8} W m^{-2} K^{-4}

I

Temperature Regulation and Heat Balance: a general survey

HOMEOTHERMY AND POIKILOTHERMY

Animals can be described either as homeotherms or as poikilotherms. The term ***homeotherm*** is applied to an animal that usually maintains a stable deep body temperature within relatively narrow limits although the environmental temperature may fluctuate and although the animal's activity may vary greatly; this is characteristic of the mammals and the birds. The temperature in the homeotherm that is regulated is the deep body temperature, sometimes referred to as the core temperature. In the cold, the zone of core temperature shrinks and the limbs and peripheral tissues become colder; the resulting temperature distributions for man are shown in Fig. 1.1. When the organism is exposed to warm surroundings, very large parts of the tissues of the limbs and the peripheral tissues of the trunk have temperatures similar to that of the core.

The core temperature is not constant throughout the day. It shows a 24-hourly variation, with higher values in the day-time and lower values at night in diurnal mammals, such as man and pig, and the reverse in nocturnal mammals like the mouse. Some creatures, for example the humming bird, are part-time homeotherms; they become torpid for parts of the 24 hours with marked falls in their body temperatures. Other animals show changes in body temperature in different seasons. Hibernators, such as the ground squirrel, become dormant during the winter; aestivators become dormant during the summer.

Homeotherms therefore have body temperatures that vary in different parts of the body and at different times. However, the essential feature of the homeotherm is that core temperature is maintained around a level that is independent of environmental fluctuation. Marked departures from the zone of maintained core

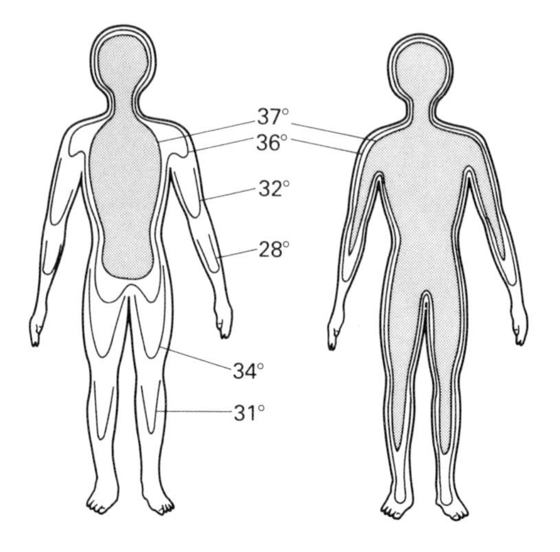

Fig. 1.1 Diagrammatic isotherms in the human body exposed to cold (left) and warm (right) conditions. The core of deep body temperature (stippled) shrinks in the cold, leaving a peripheral shell of cooler tissue (from Aschoff and Wever, 1958, by permission of Springer–Verlag).

temperature occur only when thermoregulation fails, resulting from too hot or too cold an environment, or due to drugs, disease or injury. The so-called 'primitive' mammals, including the monotremes and marsupials, have lower metabolic rates and lower body temperatures than eutherian (or higher) mammals.[134, 456] Nonetheless, as Fig. 1.2 indicates, homeothermy in the monotremes and marsupials is stable although at lower deep body temperatures of 30 to 35°C compared with 37 to 39°C in the eutherians. The birds, on the other hand, have deep body temperatures that are stabilized around 40°C or above.

The term ***poikilotherm*** refers to an animal whose body temperature tends to follow the environmental temperature, implying temperature variation rather than temperature stability. Poikilotherms include the reptiles, amphibia, fishes and invertebrates. However, the body temperatures of many poikilotherms are not simply environment-dependent; many of these animals can exercise considerable control over their body temperatures.[496] One example of this that is often quoted occurs in the desert iguana, which by moving in and out of shade can maintain a remarkably stable temperature.[140, 135] Another

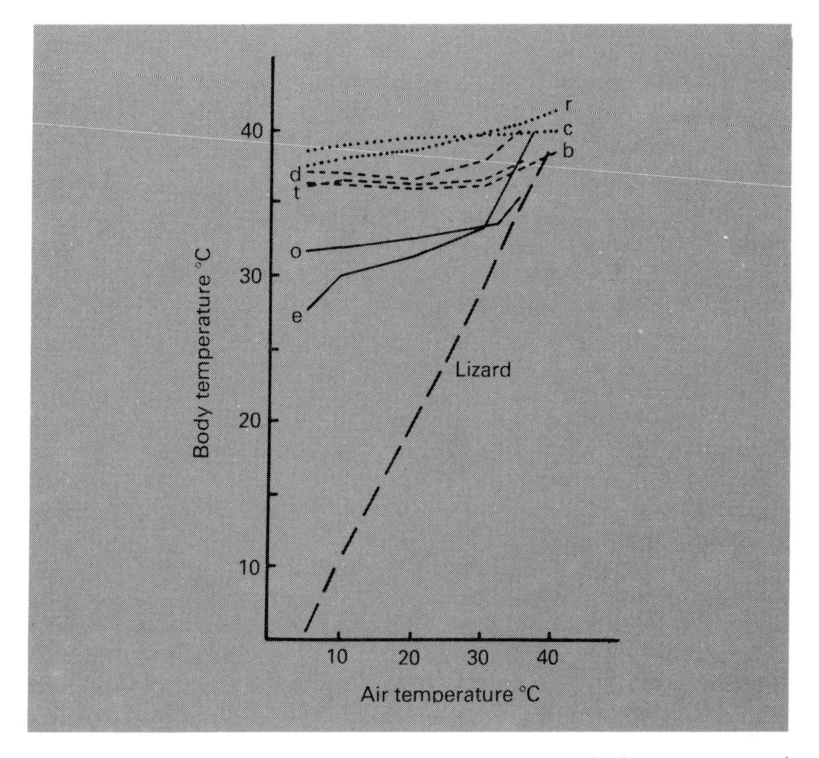

Fig. 1.2 The data of Martin (1903) on the relations between body temperature and air temperature in various groups of mammals and a lizard; lizard, *Cycolodus gigas*; r, rabbit; c, cat; d, *Dasyurus maculatus*; b, *Bettongia cuniculus*; t, *Trichosurus vulpecula*; o, *Ornithorhynchus anatinus*; e, *Tachyglossus aculeatus* (redrawn from Martin, 1903, by Dawson, 1973; by permission of Academic Press, copyright).

example occurs in some winged insects, which undergo a 'warming up' period prior to flight, raising the body temperature by several degrees through muscular effort,[33] and temperature regulation in the bumblebee is highly developed.[213] Through behavioural responses the iguana uses the diversity of its environment to regulate its temperature

Table 1.1 Set points of behavioural thermoregulation in some lizards and insects (from Heath, 1970, by permission of the American Physiological Society).

Animal	Set point (°C)	Motor pattern
ECTOTHERMS		
Iguanidae		
Phrynosoma coronatum	37.7	Shade-seeking
(California grassland)	34.2	Leave shade
Phrynosoma cornutum	37.5	Shade-seeking
(Texas grassland)	34.8	Leave shade
Phrynosoma m'calli	40.4	Shade-seeking
(Desert)	34.9	Leave shade
Cicadadae		
Magicicada cassini	31.8	Shade-seeking
	25.0	Leave shade
ENDOTHERMS		
Sphingidae		
Celerio lineata	38.0	Cessation of shivering
	37.7	Shade-seeking
	34.8	Resumption of shivering
Saturniidae		
Rothschildia jacobae	36	Cessation of shivering
	~32	Resumption of shivering

and for this reason can be described as ectothermic, whereas the insect uses metabolic heat and is endothermic. In both cases there is some temperature regulation with apparent set points for temperature control (Table 1.1). There is, though, no basically independent core temperature maintained in the poikilotherm as there is in the homeotherm, where regulation is controlled by the autonomic nervous system regardless of the animal's behaviour.

MECHANISM OF TEMPERATURE REGULATION

Temperature regulation in the homeotherm depends on nervous controls that are only imperfectly understood. There are peripheral temperature receptors in the skin, and central temperature receptors particularly in the hypothalamus in the midbrain, also in other parts of the brain and in the spinal cord. The peripheral receptors fall into two groups, those that initiate nerve impulses under cold conditions and those that do so under warm conditions. The measurements illustrated in Fig. 1.3 were obtained using a fine-pointed thermode (a temperature-controlled probe) to stimulate the receptor in the skin, and recording electrodes to pick up the impulses in the nerve fibre leading from the receptor (see Ingram and Mount[258] for further discussion). Information from both peripheral and central receptors influences the control of body temperature, which is centrally mediated in the hypothalamus. The hypothalamus probably works to a set-point temperature with its own receptors, and takes account of

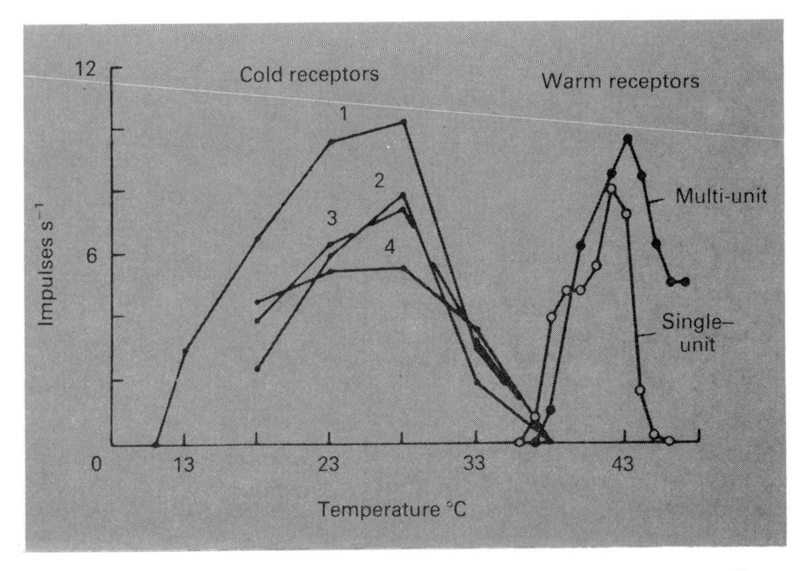

Fig. 1.3 Response of 'warm' and 'cold' peripheral temperature receptors (Iggo, 1969, by permission of Journal of Physiology).

information coming from other deep areas and from the skin. Deviations of body temperature lead to physiological and behavioural responses. When the set-point temperature is exceeded, heat dissipation is increased by sweating or panting, and heat conservation is decreased by peripheral vasodilatation, which is the dilatation of skin blood vessels under warm conditions. When the core temperature falls below the set-point temperature, heat dissipation is reduced, heat conservation is increased by pilo-erection (the fluffing up of the coat) and by peripheral vasoconstriction (the closure of skin blood vessels that occurs under cool conditions), and heat production is increased (by shivering or non-shivering thermogenesis).

These responses constitute the essence of those reactions to a fluctuating environment that distinguish the homeotherm from the poikilotherm. They are responses that involve the autonomic nervous system and they are involuntary. Behavioural responses, particularly changes in posture and changes of position in relation to micro-environment, are common to both homeotherm and poikilotherm; for example, both the pig and the desert iguana move in and out of shade depending on conditions. However, these responses are not beyond voluntary control, since they can be over-ridden by conscious control in man, and by other stimuli, such as the search for food and escape from predators, in animals.

The subject of temperature regulation has stimulated a great deal of work and given rise to much speculation and model building in attempts to explain how the regulation is achieved. There is a correspondingly large scientific literature on the subject, including many reviews.[44, 49, 68–9, 200, 216, 219, 258, 367, 498, 499]

HEAT PRODUCTION AND TEMPERATURE

Homeotherms have much higher rates of heat production than poikilotherms, even when the body temperatures are equal. In the hot desert at 37°C, a mammal weighing 10 g would produce 3.5 times as much heat as a 10 g lizard, and a 10 kg mammal would produce nearly seven times as much heat as a 10 kg lizard although in each case there is the same body temperature of 37°C.[496] When the environmental temperature falls, the homeotherm's heat production increases and so provides the capability for control of body temperature, whereas the poikilotherm's temperature tends to fall with that of the environment.[220]

In the cold, the homeotherm not only produces more heat, it also increases its overall thermal insulation by adopting a more compact posture and by pilo-erection and vasoconstriction. The combination of an increased heat production and an increased insulation allow the core temperature to be maintained although the core-environment temperature difference increases. If conditions become too cold, the demand for heat production exceeds the animal's metabolic capacity, body temperature begins to decline, and, unless it is rescued, the animal dies in hypothermia (low body temperature).

When the surroundings become warmer, the homeothermic animal's need is no longer to produce more heat and to conserve it by increasing insulation, but instead to dissipate heat and reduce insulation. This is achieved by adopting an extended posture with a raised skin temperature due to peripheral vasodilatation and an increased skin blood flow, and by losing heat by evaporation, either by sweating or by panting. The animal's heat production also falls as the surroundings warm up, but not below a certain level. This level is that of the least metabolism, or minimal metabolic rate. As the temperature rises further, heat production stays at this level at first, and then eventually begins to rise again, together with an increase in body temperature. The rise in body temperature occurs in spite of the animal's heat-dissipating systems, and if it continues the animal dies in hyperthermia (high body temperature).

The cold and hot limits for thermoregulation in the homeotherm are thus determined by heat-production capacity at low temperature and by heat-dissipating capacity at high temperatures. The range of environmental temperature in which the animal's metabolic rate is at a minimum, constant and independent of the environmental temperature is classically known as the ***zone of thermal neutrality***; other definitions may also be used, but these also include minimal metabolism[390, 393] (see Chapter 2). The temperature at the lower end of the zone is known as the ***critical temperature***; at temperatures below the critical level the homeothermic animal's metabolic rate must increase if the body temperature is to be maintained.

HEAT BALANCE

A balance can be set up between an animal's heat production on the one side and its heat loss on the other (Table 1.2). The result of the balance is that in the homeotherm the core temperature is maintained.

Feeding increases heat production. When the food intake is high, the level of heat production in the zone of thermal neutrality is higher than when food intake is low. The critical temperature falls when the food intake is increased, so that what were cool conditions when the subject was on a lower food intake are now included in the thermoneutral zone (see Chapter 2).

Non-shivering thermogenesis is a term applied particularly to heat production from brown fat in the new-born animal and in the hibernator (see Chapter 2). Heat production apart from that in brown fat also increases without noticeable shivering as the environmental temperature falls below thermal neutrality; shivering becomes apparent as conditions become cooler. Muscular activity increases heat production, sometimes very considerably.

Heat loss is affected by physiological, morphological and behavioural adaptations and by factors in the environment (see Chapters 3 and 6). The emphasis is on conservation or dissipation of heat, depending on environmental conditions being either cold or hot. Peripheral vasoconstriction and pilo-erection increase the peripheral insulation. Under warm conditions peripheral vasodilatation leads to dissipation of heat from the surface in spite of the subcutaneous fat layer because blood flow through the blood vessels of the fat layer effectively 'short circuits' the insulation of the fat. Counter-current heat exchange occurs in animals tissues; it depends essentially on arterial blood giving up some of its heat to venous blood returning from the

Table 1.2 The factors that are involved in heat balance in the homeotherm.

Heat production	Heat loss
Minimal metabolism	*Decreasing Heat Loss: conservation of heat:*
Feeding (resulting in heat increment)	Peripheral vasoconstriction
Additional non-shivering thermogenesis	Coat (pilo-erection in the cold)
Shivering	Subcutaneous fat
Muscular activity	Counter-current heat exchange
	Compact posture
	Increasing Heat Loss: dissipation of heat:
	Peripheral vasodilatation
	Sweating
	Panting
	Extended posture

periphery. The blood that reaches the periphery is then already cooled as a result of heat conservation and less heat is lost to the environment (Chapter 4).

For heat conservation the animal's behavioural responses lead it to adopt a compact posture with the minimum exposure of body surface to the environment. For heat dissipation the posture is extended, with limbs spread out; coupled with peripheral vasodilatation, and a resulting rise in skin temperature, this increases heat transfer to the surroundings. Evaporative loss is more important than non-evaporative in hot conditions, because the environmental temperature approaches skin temperature so that the driving force for non-evaporative heat loss is progressively reduced. Evaporative heat loss takes place mainly through sweating or panting, the dominance of one mechanism or the other depending on the particular species (see Chapter 3).

The homeothermic animal uses its environment in achieving a heat balance. For example, animals may huddle with their fellows, build nests, or seek out micro-environments with relatively favourable climatic characteristics. These actions allow core temperature to be maintained on lower metabolic rates than would otherwise be the case and so reduce the food requirement. In warm regions, animals seek shade and the water requirement for sweating or panting is then reduced as a result of the decreased solar heat load. Man can survive under both very cold and very hot conditions by providing himself with suitable micro-environments in clothing, vehicles and buildings (see Chapter 7).

HEAT EXCHANGE

Heat is exchanged with the environment through the four channels of evaporation (H_e), radiation (H_r), convection (H_c) and conduction (H_d), so that

$$H = H_e + H_r + H_c + H_d$$

where H = the total heat transfer, which may be a net loss or gain.

The modes of heat transfer fall into two groups: the evaporative, which is particularly important for maintaining heat balance under warm conditions, and the non-evaporative, which is dominant under cold conditions. Non-evaporative heat exchanges include the radiative, convective and conductive components, sometimes termed sensible heat transfers because they depend on temperature differences. The temperature difference may be in either direction, that is either the animal or the environment may be the hotter. In the latter case, the animal receives a heat load from the environment in addition to the heat it produces by its metabolism, and if thermal balance is to be maintained the sum of these two heats must be dissipated by evaporative means. Evaporative transfer does not depend on temperature differences but on differences in water vapour pressure between animal and environment.

Non-evaporative (or sensible) heat transfer may therefore be a heat loss from the animal in cool conditions or in the zone of thermal neutrality, or a heat gain by the animal (a heat load) under hot conditions. There may be a loss in one channel, and a gain in another; for example, an animal may receive radiant and convective heat loads coupled with a conductive loss to a cool floor, resulting in either a net loss or gain in non-evaporative heat exchange. Evaporative transfer is always considered for practical purposes as a net heat loss, although under some conditions it is possible to expose an animal so that net condensation occurs on part of it, indicating a gain of heat. The modes of heat transfer, and the factors of particular importance for each mode, are treated in more detail in Chapter 3. The effects of thermal insulation in reducing heat transfer are discussed in Chapter 4 and assessment of the thermal environment in terms of the different forms of heat transfer is considered in Chapter 5.

HEAT STORAGE

The rate of heat production or metabolic rate, M, is affected by a number of factors including body size, thermal insulation, level of food intake, activity and environmental conditions (see Chapter 2). When the animal's mean body temperature remains unchanged, the animals is in thermal equilibrium, heat is not being stored, and M = H. This no longer holds when the mean body temperature varies.

The mean body temperature ($\bar{T}_b$) cannot be directly measured. It can be thought of as the sum of the products of the masses of each part of the body and their respective temperatures, divided by the total body mass (W):

$$\bar{T}_b = \frac{W_1 T_1 + W_2 T_2 + \ldots W_n T_n}{W}$$

This is an approximation because it is assumed that the specific heat is constant over the whole body, whereas in fact it varies depending on the water content of each part.

The mean body temperature is sometimes estimated in practice as

$$\bar{T}_b = 0.7 T_{re} + 0.3 \bar{T}_s$$

where T_{re} = rectal temperature and $\bar{T}_s$ = mean skin temperature.

If the animal's mean body temperature rises by $\Delta\bar{T}$ per unit time, some of the metabolic heat that would otherwise be lost to the surroundings is instead stored in the body:

$$\begin{aligned} M &= H + S \\ \text{or } H &= M - S \end{aligned}$$

where S = rate of heat storage

$$= W \Delta \bar{T} c$$

where c = mean specific heat of the body; a value of about 3.5 J °C^{-1} g^{-1} for man has often been used.

When the mean body temperature falls, the total heat lost to the surroundings exceeds the rate of heat production by the amount of heat lost form the body store:

$$M = H - S$$
$$\text{or } H = M + S$$

so that heat storage is negative.

In general, then:

$$M = H + S$$

where S may be positive or negative. Reference to Fig. 1.1 shows that $\bar{T}$, and therefore S, can change without change in the core temperature, so that temperature regulation in the homeotherm applies to the body core although changes in stored heat may occur peripherally.

However, under some conditions, particularly during muscular activity at high environmental temperatures, the rate of dissipation of heat cannot match the rate of its production, and this leads to heat storage with a rise in the core temperature. The extent to which this occurs is illustrated by measurements made on a cheetah, goat and dog when they were running at 10 km h^{-1} for 15–30 minutes at 22°C. The cheetah stored about 70% of the heat it produced during the run, the goat about 35%, and the domestic dog about 4%. The cheetah is the world's fastest land animal, capable of reaching speeds of 110 km h^{-1} for short bursts, but it does not continue running after its body temperature has risen to 40.5–41°C. This accords with the sprint distance of rather less than 1 km over which the cheetah chases its prey, and suggests that the duration of its sprint is limited by its tendency to become hyperthermic. The cheetah does not use evaporative heat loss to a greatly increased extent when it runs; the goat and dog both pant during running (panting is a particularly effective means of heat dissipation in the dog) and the increase in their evaporative heat loss concurrent with exercise results in decreased heat storage compared with the cheetah.[492]

The comments that have been made on heat storage as the difference between heat production and heat loss apply to an animal that is either resting or active, but which through its activity is not transferring any of its metabolic energy to work done on the environment. If a man hoists a weight through a vertical distance, he does some external work that is stored as potential energy in the lifted weight; in addition, some of the external work he does is dissipated as frictional heat in the hoisting machinery, because no machine is 100% efficient. Both the potential energy due to the weight's raised position and the frictional heat have the same dimensions, that is those of energy, and they may be added together to give the total external work done. By the law of conservation of energy, the equation of energy balance then becomes:

$$M = H + S + \text{rate of doing external work.}$$

For considering the ways in which metabolic heat is lost to the environment, and to consider only a heat balance, it is simpler, and permissible, to assume that the animal is doing no external work. No external work is done, for example, when the stationary animal grooms itself; all the work involved in such activity is liberated as heat from the animal. However, if external work is done in a particular case it must be included as part of the animal's metabolic rate budget, and the animal's metabolic rate then exceeds its rate of heat production by an amount equal to the rate at which external work is done. Confusion sometimes arises from statements that the measurement of metabolic rate is an estimate of free energy. This is not so; it is the net heat production, and not the free energy, that is measured in calorimeters.[543]

ENERGY RETENTION

One of the consequences of its high metabolic rate at the resting minimal level is that the homeotherm needs a high food intake if it is to maintain its overall energy balance. Additional to meeting the requirements of maintenance, food must be supplied to meet the demands of activity, growth and thermoregulation in the cold.

When the body is in thermal equilibrium, so that the mean body temperature remains unchanged, the equation for energy derived from the food and energy either retained in the body as new tissue or lost to the environment as heat is

$$ME = ER + H$$

where ME = metabolizable energy intake, ER = energy retained in the body in the tissues, mainly as fat and protein, H = heat lost to the surroundings.

Metabolizable energy (ME) is referred to instead of food energy (GE, gross energy) because some of the food energy is lost in the faeces, urine and (particularly in the ruminant animal) as methane. The breakdown is as follows:

$$GE = DE + \text{faecal energy}$$

where DE = digestible energy,

and DE = ME + energy losses in urine and methane,

so that GE = ME + energy losses in faeces, urine and methane.

These various quantities can be measured by collecting faeces, urine and methane from animals fed at known rates. The total heats of combustion of samples of the food, faeces and urine are measured in a bomb calorimeter and, using the known heat of combustion of methane, the animal's ME intake can be calculated by difference from the gross energy of the food. In man, pig and poultry, the energy loss as methane is usually so small that it can be ignored; it is important in cattle, sheep and other ruminants, where it amounts to about 8% of GE, with considerable variation depending on feed.[58]

2

Heat Production

HEAT PRODUCTION IN RELATION TO ENVIRONMENTAL TEMPERATURE

Animals produce heat as a result of their metabolic activity. The general relation between a homeotherm's heat production and the environmental temperature, critical temperature and thermoneutral zone has been discussed in Chapter 1; this relation can now be considered in more detail. An animal's rate of heat production is usually determined from its rate of consumption of oxygen, whereas heat loss can be estimated by a number of direct means. The various methods employed in animal calorimetery are discussed in Chapter 3.

In Fig. 2.1, the evaporative and non-evaporative (sensible) components of heat loss are represented as well as heat production. When the temperature rises above zone CD, evaporative loss begins to increase rapidly, whereas under cool conditions most of the heat produced is dissipated by non-evaporative means. It is assumed that the rate of change of stored heat is zero, so that the rate of total heat loss (H) equals the rate of heat production at any given environmental temperature. The deep body or core temperature is included in the diagram; its rise above E and fall below B correspond to the hot and cold limits mentioned in Chapter 1. There is a hyperthermic rise in heat production at E, and a hypothermic collapse of heat production at B.

The curve for non-evaporative heat loss consists primarily of two straight lines. That part which lies at temperatures below C (the

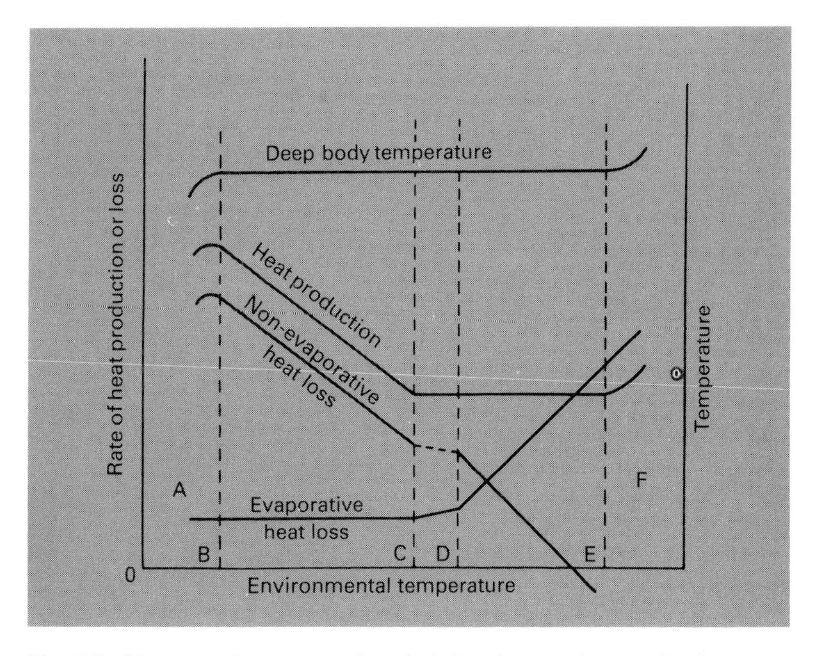

Fig. 2.1 Diagrammatic representation of relations between heat production, evaporative and non-evaporative heat loss and deep-body temperature in a homeothermic animal.
A, zone of hypothermia; B, temperature of summit metabolism and incipient hypothermia; C, critical temperature, CT; D, temperature of marked increase in evaporative loss; E, temperature of incipient hyperthermal rise; F, zone of hyperthermia; CD, zone of least thermoregulatory effort; CE, zone of minimal metabolism; BE, thermoregulatory range (Mount, 1974a).

critical temperature, CT) refers to the animal with peripheral vasoconstriction. The slope of the line, $\Delta H/\Delta T$, gives a measure of the thermal conductance between the animal and its environment; this is the reciprocal of the thermal insulation. Above D, under warm conditions, the slope of the line is steeper, indicating a higher thermal conductance, that is a greater heat transfer per degree temperature difference. This is due to peripheral vasodiliatation occurring when the temperature reaches zone CD immediately above the critical temperature.

In a completely dry system with no evaporative loss, the line for heat loss at temperatures above D would cut the temperature axis at the deep body temperature, because at that point no heat production would be required to maintain body temperature. For the homeothermic animal, extrapolation of the line for non-evaporative heat loss (not the line for total heat loss including evaporative) in Fig. 2.1 cuts the temperature axis just below the deep body temperature. The intercept is the same for the non-evaporative heat loss line corresponding to either the low conductance at temperatures below C or the high conductance at temperatures above D. The difference between the intercept and the deep body temperature, which occurs because all homeotherms lose some heat by evaporation, is explained in Appendix 2; it does not affect the present discussion.

THERMAL NEUTRALITY

The zone of thermal neutrality has been referred to in Chapter 1 as the range of environmental temperature bounded at its lower end by the critical temperature and at its upper end by the hyperthermic point. It is defined most commonly as the range of environmental temperature within which metabolic rate is minimum, constant and independent of temperature. However, if an animal is at its minimal metabolic rate but is maintaining its deep body temperature only by greatly increased sweating or panting it can hardly be considered to be in a neutral condition.[390] For this reason, a number of alternative definitions of thermal neutrality have been suggested:

(1) Minimal metabolism, as just mentioned, bounded on each side by rising metabolic rate, CE in Fig. 2.1.
(2) Least thermoregulatory effort, coinciding with minimal material demand either for food to provide energy in the cold or for water for the evaporative dissipation of heat under hot conditions; this zone is bounded at the colder limit by rising metabolic rate and at the warmer limit by a marked rate of increase in evaporative loss, CD in Fig. 2.1.
(3) Zones defined for particular purposes, such as the preferred thermal environment (comfort zone), and zones that are optimal for given requirements such as growth rate or the development of thermoregulation in the young animal; these zones do not necessarily coincide with either minimal metabolism or least thermoregulatory effort.

To avoid confusion, the particular definition to be applied to the thermoneutral zone should be indicated when the term is first introduced in any particular connection.

Fig. 2.1 is a diagram that does not necessarily portray the pattern that results from plotting actual measurements of heat production or heat loss, but it is useful in showing relations between different quantities and different zones. It is a generalized diagram in that the scale values depend on species, age, plane of nutrition, adaptation history and factors in the environment. All these conditions must be defined before results from different animals can be compared. In some animals, the zone of minimal metabolism is very wide, as in man and cattle, whereas in others it is narrower, as in pig and mouse. Below the critical temperature the heat production rises steeply for the newborn baby or newborn pig, whereas for the adult furred animal, which has much greater thermal insulation, the slope is very shallow.

CRITICAL TEMPERATURE

In Fig. 2.1 the critical temperature, C, is sharply defined, with a rising heat production to the left as environmental temperature falls from C, and a constant rate of heat production in the thermoneutral zone to the right as the temperature rises. It is sometimes possible to obtain such sharp definition with an immobile subject, for example a new-born kitten or puppy, or with man, but usually the relation between heat production and environmental temperature is more curvilinear in the region of the critical temperature, in the manner shown by the lower curve in Fig. 2.2a. It is then necessary to estimate an 'effective critical temperature'. One way of doing this is to take the point of intersection of extrapolations from the slope of rising heat production in the cold and from the constant heat production of thermal neutrality.

LIMITING TEMPERATURES

The animal makes a response to the thermal demand of the environment, a concept that applies particularly in the cold where heat production must match requirements for the maintenance of body temperature.[85] When the demand exceeds the capability of the metabolic response, hypothermia ensues. The thermal demand of the environment is zero in the thermoneutral zone, at least in respect of heat production. Above the hyperthermic point, the animal's powers of heat dissipation are not adequate to prevent a rise in deep body temperature and this leads in turn to a rise in metabolic rate (see Fig. 2.1).

The lower temperature limits in still air that can be met by metabolic response and the upper limits that can be met by heat dissipation, together with critical temperatures, are given in Table 2.1 for man, pig and sheep, both newborn and mature. These species illustrate a remarkable range in cold limit, from +27°C for the human infant to a calculated value of −200°C (or even lower) for the mature

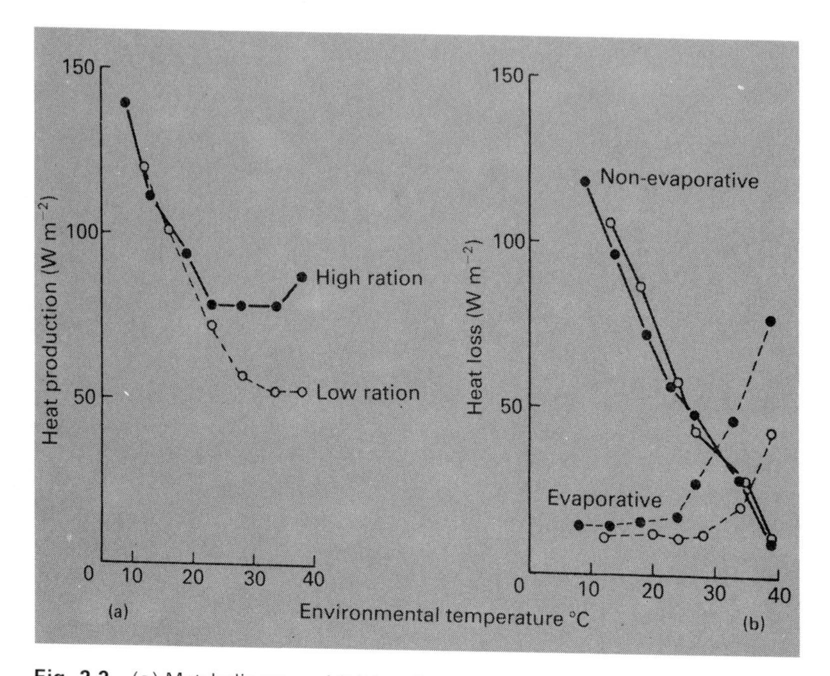

Fig. 2.2 (**a**) Metabolic rate and (**b**) heat loss at various environmental temperatures in one closely clipped sheep during periods on a high ration (●) and a low ration (○) (Graham, Wainman, Blaxter and Armstrong, 1959; Blaxter, Graham, Wainman and Armstrong, 1959; reproduced from Alexander, 1974).

long-haired sheep; even the almost hairless new born pig has a cold limit as low as 0°C. The thermoneutral zone above 28°C for nude man typifies him as a tropical animal, whereas thermal neutrality for the sheep includes the temperatures of most habitable areas of the earth, in accordance with the widespread distribution of that species. However, these temperatures refer to still-air conditions; wind and precipitation raise critical temperatures considerably. The hot limits for the three species are much closer together; the exceptional value is the relatively low hot limit for the mature pig, which is due to that animal's inability to sweat or pant effectively.

The relation of cold limits to the maximum metabolic rate can be demonstrated in the case of the newborn baby, piglet and lamb, where the maximum rates have been measured. The maximum rate multiplied by the thermal insulation (see Chapter 4) gives an approximate estimate of the body-environment temperature difference

Table 2.1 Critical temperatures, °C, and approximate cold and hot environmental limits for thermoregulation, under still-air conditions, for the newborn and mature of unclothed man, pig and long-haired sheep (values from Hey, 1974; Mount, 1968a; Alexander, 1974).

	Critical temperature	Cold limit	Hot limit
Man			
newborn	33	27	37
mature	28	14	43
Pig			
newborn	34	0	36
mature	10	−50	30
Sheep			
newborn	30	−100	40
mature	−20	−200	40

that can be sustained (Table 2.2): for a more accurate assessment, the evaporative loss should be subtracted first. The calculations yield cold limits in the region of those given in Table 2.1.

So far, the thermal attributes of the environment have been characterized only in terms of the environmental temperature. The measurement that is most commonly made to represent the thermal environment is the air temperature, and this equals the environmental temperature in a standardized environment where (i) the air and mean radiant temperatures are equal to each other, (ii) there is a regimen of free convection without forced draughts of air and (iii), in application to livestock, the floor is insulated. Such standardized conditions can be produced for experimental purposes, but they are not usual in general experience, either for man or for animals. Complex thermal environments will be considered in Chapter 5.

In addition to the environmental temperature, the factors that have the greatest effect on an animal's rate of heat production are body size,

Table 2.2 The calculation of the cold limits for thermoregulation from the measured maximum metabolic rate and thermal insulation in the newborn human infant, piglet and lamb.

	A: Maximum metabolic rate, $W m^{-2}$	*B*: Thermal insulation, $°C m^2 W^{-1}$	*C*: Maximum sustainable difference between core and environmental temperatures, °C, from $A \times B$	*D*: Predicted cold limits, °C, from core temperature minus *C*: compare Table 2.1
Baby	70	0.17	12	24
Piglet	160	0.23	37	2
Lamb	290	0.40	116	−77

plane of nutrition and thermal insulation. These and a number of other factors will now be considered.

BODY SIZE

A large homeothermic animal produces more heat than a small animal. This could reasonably be expected, although if the animals were not far apart in size and the larger had much more thermal insulation, the smaller might be producing more heat. Taking two animals that are highly disparate in size, say the mouse and cow,[302] it is obvious that the mouse can produce only a very small fraction of the heat produced by the cow; in fact the 20 g mouse produces heat at the rate of about 0.2 W (watt), whereas the resting 600 kg cow produces about 380 W. If now the heat production is calculated per unit of body weight, the mouse produces 10 $W\,kg^{-1}$, and the cow only 0.6 $W\,kg^{-1}$. Each unit mass of mouse is clearly very much more active metabolically than each unit mass of cow. How then can the metabolic rates of mouse and cow be compared with each other?

This is not simply an academic question. In the first place, it is necessary to find some common basis for comparing the metabolism of animals different in size, although not necessarily as different as mouse and cow, so that the effect on metabolism of influences other than body size can be studied. In the second place, the relation of metabolic rate to body size has a highly practical corollary in its application to the determination of feeding standards for the maintenance requirements of farm animals.

One approach to the problem might be to relate metabolism to some function of body size other than body weight, and, because an animal's heat loss might be thought to be proportional to its surface area, this particular function of body size and shape has been extensively used. Kleiber[299] refers to the 'surface-area law' applied to homeotherms as stating that 'the basal metabolism of animals differing in size is nearly proportional to their respective body surfaces'. This rule was first put forward by Sarrus and Rameaux[454] and for many years subsequently workers in the metabolic field tended to regard their results as being at fault if they did not show proportionality to surface area. Table 2.3 gives metabolic rates from various animals in terms both of body weight and of surface area. When compared with body weight, surface area offers a much more uniform basis for comparison between species.

Another more general approach has been to relate metabolic rate to empirically determined functions of body weight. From large numbers of measurements of metabolic rates of animals of different weights it is clear that the logarithm of the metabolism tends to vary directly with the logarithm of the body weight, according to the expression:

$$\log \mathrm{M} = \log \mathrm{a} + \mathrm{b} \log W$$

where M = metabolic rate, W = body weight, and a and b are

Table 2.3 Voit's table on surface law for fasted animals (from Kleiber, 1975, by permission of John Wiley and Sons Inc.).

Animal	Number of determinations	Average weight (kg)	Metabolism	
			$\mathrm{W\,kg^{-1}}$	$\mathrm{W\,m^{-2}}$
Horse	8	441	0.55	45.9
Pig	2	128	0.92	52.2
Man	5	64.3	1.55	50.5
Dog	15	15.2	2.49	50.3
Rabbit	5	2.3	3.64	37.6
Goose	6	3.5	3.23	49.3
Hen	2	2.0	3.44	48.8

constants; this is the celebrated 'mouse to the elephant' curve from Benedict.[41] This is equivalent to:

$$\mathrm{M} = \mathrm{a}W^{\mathrm{b}}$$

The double-log plot provides a linear relation, and so allows the exponent b to be calculated from the slope, which is given by the regression coefficient. W^{b} is then termed the metabolic body size.

Many different values have been determined for b under different experimental conditions. Alternatively, b can be fixed, and the quantity a varied to fit the results. One example, where b = 1, makes metabolic rate proportional to body weight. The exponent b may also be fixed in relation to some other parameter, such as surface area; in the special case of the surface areas of bodies similar in shape and density b is 0.67.

Brody[77] and Kleiber[299, 300, 301, 302] have discussed the exponent extensively. Brody found an interspecific value of 0.734 to fit determinations of metabolic rate over a wide range of species from various investigations. He subsequently approximated this, first to 0.73 and then to 0.7. Kleiber, however, found a value of 0.756 as the best regression coefficient, and rounded this to 0.75 as a convenient value for calculation. Kleiber disapproved of Brody's figure of 0.7 as being too close to the two-thirds power; Kleiber considered a three-quarters power rule to be more appropriate to the relation between metabolic rate and body weight. Brody's 0.73 was the value adopted by a National Research Council (U.S.) Conference in 1935, in preference to 0.75. Such fine discrimination is, however, questionable in view of the considerable variation in the measurements on which the exponent is based. Kleiber[301] commented that a significant difference between proportionality to the two-thirds power of body

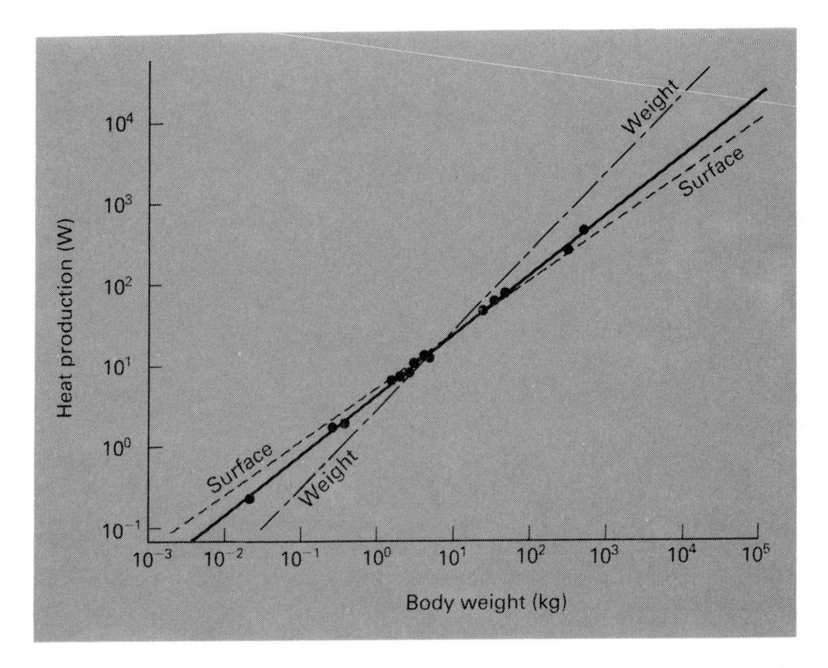

Fig. 2.3 Relation of the logarithms of metabolic rate and body weight in mammals (from Kleiber, 1947, by permission of American Physiological Society).

weight and the three-quarters power could be established only with series of animals showing a weight range of ninefold or more in extent; to establish a significant difference between proportionality to body weight or to the three-quarters power requires a threefold weight difference (Fig. 2.3). At a meeting of the European Energy Metabolism Symposium, in 1964, the three-quarters power of body weight was adopted as the ***interspecific*** reference base (see Appendix 3).

Some approximation of this sort is convenient for comparison between animals of different species. Table 2.4 collects the values for the mouse and cow example mentioned earlier. It is clear that as an interspecific reference base neither the rate of heat production per whole animal nor per kg body weight is satisfactory, but using either $kg^{0.75}$ or surface area gives quantities for mouse and cow that are more nearly commensurate, as the last column of the table shows. For experimental purposes it is often advantageous to determine the relation between metabolic rate and body weight on a within-experiment basis for the exclusion of variations in body weight from treatment effects.[187]

Table 2.4 Comparison of the metabolic rates of mouse and cow, in terms of body weight, metabolic body size, and body surface area (values from Kleiber, 1975).

	Mouse	Cow	Ratio of values, mouse/cow
Body weight, kg	0.02	600	
$kg^{0.75}$	0.053	121	
Surface area, $m^2 = 0.09\ kg^{0.67}$	0.0065	6.5	
Heat production, W	0.2	380	0.0005
$W\ kg^{-1}$	10	0.6	17
$W\ kg^{-0.75}$	3.8	3.2	1.2
$W\ m^{-2}$	31	59	0.5

Kleiber[302] gives $3.4\ W\ kg^{-0.75}$ as the mean value for fasting metabolic rate derived from a number of species. Fasting metabolism has been measured as $2.7\ W\ kg^{-0.75}$ in wether sheep,[58] $4.2\ W\ kg^{-0.75}$ in cattle[66], and $4.4\ W\ kg^{-0.75}$ in pigs.[112] In the pig this value rises to $7–12\ W\ kg^{-0.75}$ in fed animals, depending on the plane of nutrition,[113] and feeding also increases the metabolic rate in the other species. As a general approximation for fed animals, metabolic rates are $7\ W\ kg^{-0.75}$ in the pig, $6\ W\ kg^{-0.75}$ in cattle, and $5\ W\ kg^{-0.75}$ in sheep.

Hemmingsen[217] has discussed the subject of energy metabolism and body size at some length, ranging from unicellular organisms to large mammals and even to beech trees. From interspecific calculations involving considerable ranges of weight, he found in general that metabolism can be related to weight raised to 0.75 power for poikilotherms as well as homeotherms; the slope of the three lines in the log-log plot of Fig. 2.4 is 0.75 in each case, a smoothed value obtained by the method of the least sum of squares assuming the exponent to be the same for each of the three sets of values.[544]

Another form of metabolic body size has sometimes been used, the ***active body mass*** or ***lean body mass***. This may be applicable when metabolic rate is either at a maximum or at a minimum, when it

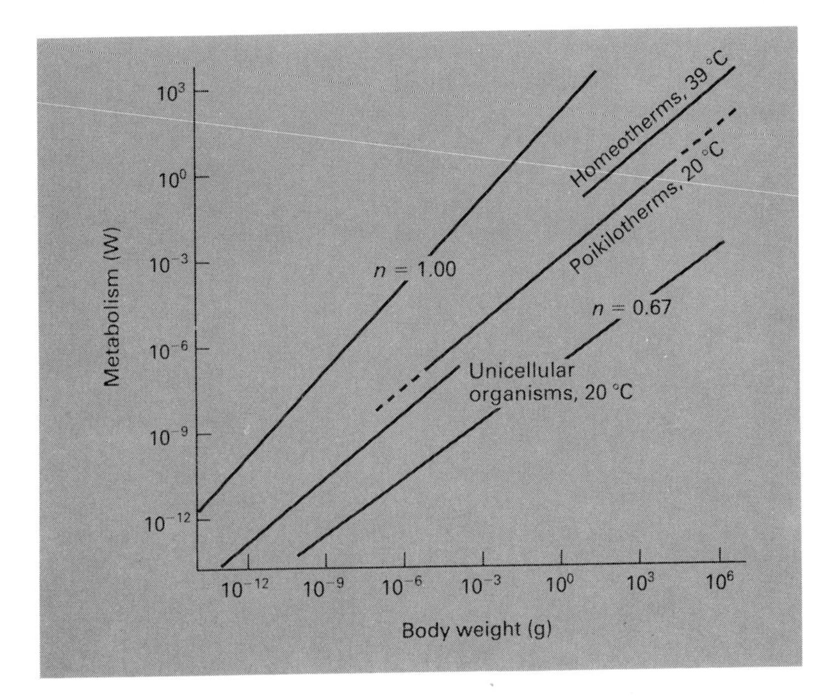

Fig. 2.4 Double logarithmic plots of metabolism against body weight for unicellular organisms and larger poikilotherms at 20°C, and for homeotherms. The two outer lines are drawn to demonstrate proportionality of metabolism to body weight ($n = 1.00$) and to body surface area ($n = 0.67$) (from Hemmingsen, 1960, by permission of the author).

represents an integration of the several products of tissue mass and metabolic rate per unit mass for the different tissues of which the organism is composed. The environmental conditions that produce the upper and lower rates are the cold limit and thermal neutrality, respectively, so that the rates are expressions of the organism's potential metabolic activity. However, for intermediate conditions it is the environment that determines the metabolic rate, and not the active body mass, with the result that this concept is not suitable as a basis for metabolic body size unless its use is restricted to limiting conditions.[386]

PLANE OF NUTRITION

When an animal is resting in a thermally neutral environment, its rates of heat production and loss depend on its mean rate of food consumption. This effect is evident in Fig. 2.2, and it is shown

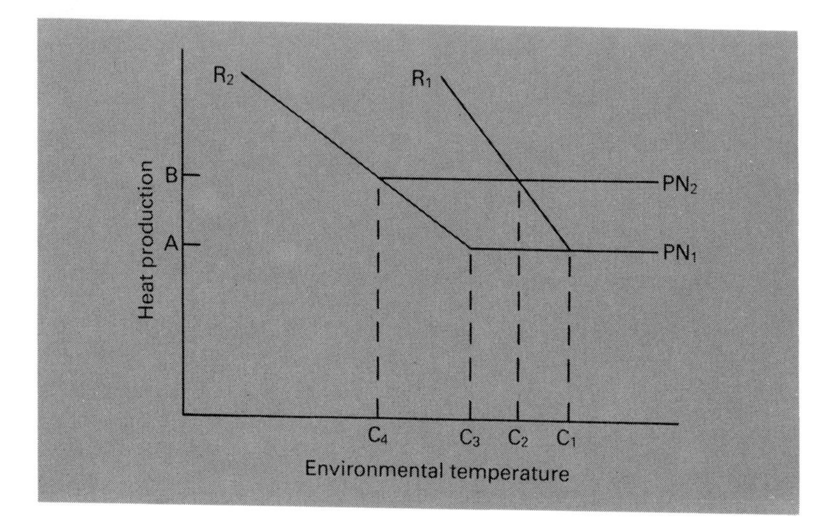

Fig. 2.5 Generalized diagram of heat production against environmental temperature for a homeothermic animal of low (R_1) or high (R_2) thermal insulation, on a low (PN_1) or high (PN_2) plane of nutrition. C_1, C_2, C_3 and C_4 are the corresponding critical temperatures (Mount, 1976c, by permission of Pitman, Ltd.).

diagrammatically in Fig. 2.5. For a poorly insulated animal with thermal insulation R_1, the critical temperature is C_1 on a plane of nutrition PN_1. When PN_1 is increased to PN_2, the critical temperature falls to C_2. When the animal's insulation increases from R_1 to R_2, and the animal is fed at PN_1, the critical temperature falls from C_1 to C_3. When PN_1 is then increased to PN_2, the critical temperature falls still further to C_4.

The particular combination of plane of nutrition and thermal insulation thus determines the temperature at which the resting animal begins to be affected by cold. Raising the food intake from PN_1 to PN_2 gives an increment of heat production equal to AB which substitutes for what would otherwise be necessary thermoregulatory heat production if the food intake remained at PN_1 and the environmental temperature lowered from C_1 to C_2. Similarly, heat

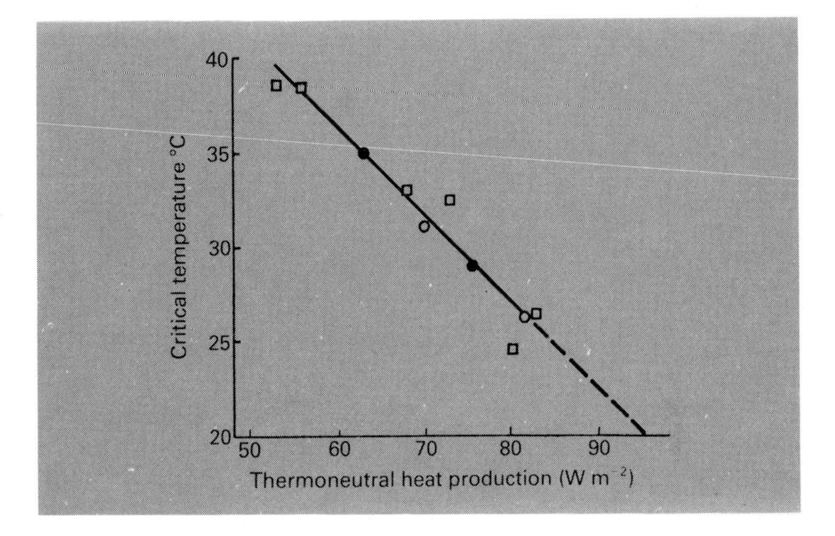

Fig. 2.6 Critical temperature of shorn sheep and thermoneutral heat production. The points refer to wethers (□), non-pregnant ewes (●), and pregnant ewes (○) (Graham, 1964, by permission of Australian Journal of Agricultural Science).

produced by exercise substitutes for thermoregulatory heat in the cold, although exercise decreases thermal insulation through movement and consequently affects the slope R. The converse is that in a warm environment these increments in heat production must be dissipated as extra heat loss if hyperthermia is not to ensue.[258, 393] Fig. 2.6 relates the fall in critical temperature to the rise in thermoneutral heat production in sheep.

Raising the plane of nutrition also lowers the temperature of the hyperthermic point (E in Fig. 2.1) so that the zone of minimal metabolism itself is shifted downwards along the temperature axis as the plane of nutrition increases. This means that a temperature that is within the thermoneutral zone for an animal on a given level of feeding may be either above or below the zone when the level is changed. One result of this, for animal production, is that animals on higher planes of nutrition approach maximum productivity at temperatures below the optimal temperatures needed by animals on lower planes. Keeping animals on high food intakes at high temperatures makes it difficult for the extra heat produced to be dissipated and may mean that the animals become hyperthermic.

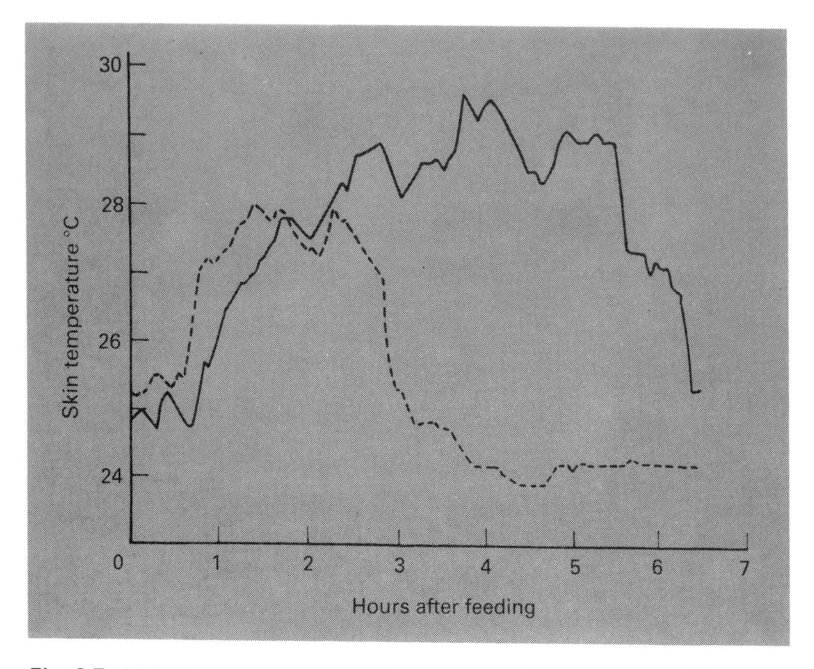

Fig. 2.7 Variations in skin temperature following food intake in pigs of 26 kg body weight at an air temperature of 3°C. Intake of 0.5 kg barley and 100 g milk powder (———); intake of 0.25 kg barley and 50 g milk powder (------). Sørensen, 1962, by permission of the author).

The animal indicates its preference for a lower environmental temperature when the plane of nutrition is high. This has been shown in behaviour experiments in which pigs kept at 10°C were trained to switch on a bank of infra-red heaters for a short period at a time, so that they could control their thermal environment. When the animals were given 400 g food per day they chose to receive 300 periods of heating per hour, but when they were fed 900 g food per day they took only 200 periods of heat per hour[25] (see Chapter 8).

Feeding raises the skin temperature, the extent and duration of the

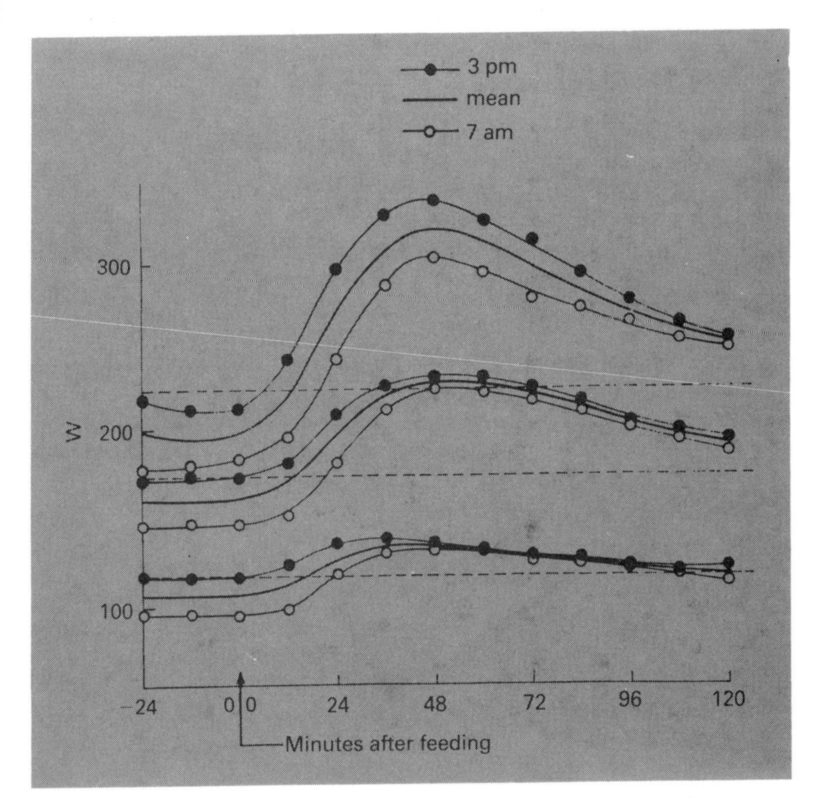

Fig. 2.8 Heat production of growing pigs fed at 3 pm and 7 am compared with the mean 24-hour heat production rate (------) at 18–20°C. Mean body weights of pigs, with reference to the curves, from above downwards: 85, 54 and 30 kg (Neergaard and Thorbek, 1967, by permission of European Association for Animal Production).

rise depending on the amount of food taken (Fig. 2.7). The warming is related to the heat increment of feeding, which is demonstrated in Fig. 2.8.

Energy retention

An animal's energy requirement for maintenance is met when its intake of metabolizable energy (ME) equals its rate of heat production, so that in the expression:

$$ME = H + ER$$

where H = rate of heat loss which, at thermal equilibrium, is the same as the rate of heat production (see Chapter 1), and ER = the rate of energy retention, ER = 0 and ME = H.

When an animal's food intake is in excess of its requirements for maintenance and thermoregulation, there is a net increase in the energy content of the organism which is in the form of both protein and fat in the growing animal but largely fat in the mature animal.

When the intake of ME is increased by an amount ΔME, and ER increases by ΔER, the ratio ΔER/ΔME is termed the ***partial efficiency*** (k), which indicates the proportion of the increment of ME above the maintenance level that becomes ER, as distinct from the gross efficiency ER/ME. In the thermoneutral zone, k is often about 0.7, indicating that 70% of an increment of ME becomes ER, and 30% is added to heat loss. In Fig. 2.5 an increase in intake from PN_1 to PN_2 (corresponding to ME_1 to ME_2) is accompanied by an increase in heat production from A to B at thermal neutrality, with a corresponding increase in heat loss, H, and if $k = 0.7$ then $(B - A)$ is 30% of $(ME_2 - ME_1)$.

Under cold conditions, k tends towards unity because there is substitution of the heat increment of feeding for the extra thermoregulatory heat that is required to satisfy the increased environmental demand. This effect is shown in Fig. 2.2a, where at the higher temperatures the plane of nutrition determines the rate of heat production, whereas at the lower temperatures the environment determines heat production independently of the plane of nutrition. Under cold conditions the whole of the increment in intake, ΔME, then tends to become ΔER.[393] Below the maintenance level of food intake, when ER is negative, the partial efficiency with which ME is used for maintenance purposes, k_m, is often about 0.8 under thermoneutral conditions. k_m would also be expected to rise towards unity in the cold, like k.[111]

The relations just outlined between ME, H and ER do not on the whole move sharply from one state to another. The curvilinearity of the relation between heat production and environmental temperature in the region of the critical temperature arises because animals adapt to changing conditions by changes in posture, variations in activity, huddling with other animals in the cold and making variable use of bedding and other physical attributes of their surroundings. Similarly, a partial efficiency of unity is by no means always found under cool conditions, but the general tendency is for the partial efficiency to increase in the cold compared with the warm. If the partial efficiency at thermal neutrality is known, then since $\Delta ER = k \cdot \Delta ME$ it is possible to determine the additional heat production to be expected from additional food intake above the maintenance level (ME_m). If the animal's total heat production now exceeds the sum of this additional heat and the heat production of the animal on a maintenance level of food intake (when ER = 0), the excess is equal to the extra thermoregulatory heat (ETH). The effective critical temperature (CT) is then ETH/G degrees above the temperature of the measurements (T), where G = thermal conductance, that is the slope of heat production against environmental temperature at sub-critical levels:[393]

$$ME = ME_m + \frac{ER}{k} + ETH$$

$$CT = T + \frac{ETH}{G}$$

This follows from the geometry of the plot of heat production in the interval BC of Fig. 2.1. Effective critical temperatures calculated in this way for groups of pigs gave 17°C for a food intake of 45 g per kg body weight per day, and 11°C for a food intake of 52 g per kg per day.[398]

Interactions of this nature between the plane of nutrition and the environmental temperature are clearly important in determining the use that an animal makes of its food for maintenance and growth purposes.

THERMAL INSULATION

It is evident from the earlier discussion that differences in thermal insulation between animals affect the relation between heat loss and environmental temperature. The effect is shown in Fig. 2.5. The slope for the poorly insulated animal is steeper than that for the well insulated. The steeper line indicates a higher thermal conductance, where a higher rate of heat production is needed to maintain a given difference between the deep body and environmental temperatures, which is what obtains in the shorn sheep compared with the sheep with a fleece. The intersections of the minimal metabolic rate lines with the rising heat loss lines give the critical temperatures for two feeding levels.

Two effects of thermal insulation are illustrated in this diagram. The first is that when the thermal insulation is increased on food intake PN_1, the critical temperature falls from C_1 to C_3 although the metabolic rate remains unchanged; environmental temperature must drop below C_3 before the animal has to make a metabolic response to cold. The second effect is that the critical temperature can be lowered by increasing either the thermal insulation or the plane of nutrition.

The subject of thermal insulation will be considered in more detail in Chapter 4.

24-HOURLY VARIATION

A diurnal mammal like a pig or a man has a higher metabolic rate, body temperature and activity during daylight than during the night (Fig. 2.9). For nocturnal mammals, such as the mouse, these quantities reach their maxima during the night (Fig. 2.10). 24-hourly variations are important in the estimation of an animal's heat production, because short-term measurements of metabolic rate made over one or two hours may not only be unrepresentative of the mean 24-hourly value but may also, if made at different times of day, suggest differences that may be thought of as due to experimental treatment when they are instead due to 24-hour variations.

In experiments on energy balance full 24-hourly determinations of heat production are essential if the equation

$$\mathrm{ME} = \mathrm{H} + \mathrm{ER}$$

is to be balanced satisfactorily. Heat storage effects (see Chapter 1) are minimized by measurements of 24 hours' duration, not only because any given heat storage, when expressed as a rate over a period, becomes progressively smaller as the duration of measurement increases, but also because both the beginning and end of the measurement period are in the same phase of the 24-hourly cycle of the animal's metabolism and body temperature.

It has been a common practice, particularly in medicine, to refer to 'basal metabolism', defined as the metabolic rate of the resting subject in the post-absorptive state at thermal neutrality. Although this is useful clinically, its value for energy balance purposes is doubtful on account of the restricted conditions. It is more useful to measure metabolic rate under a series of defined conditions of environment and food intake.

BODY TEMPERATURE

Normally the deep body temperature is regulated in the homeotherm; it remains stable while metabolic rate varies. However, at the extremes of incipient hypothermia and hyperthermia, when

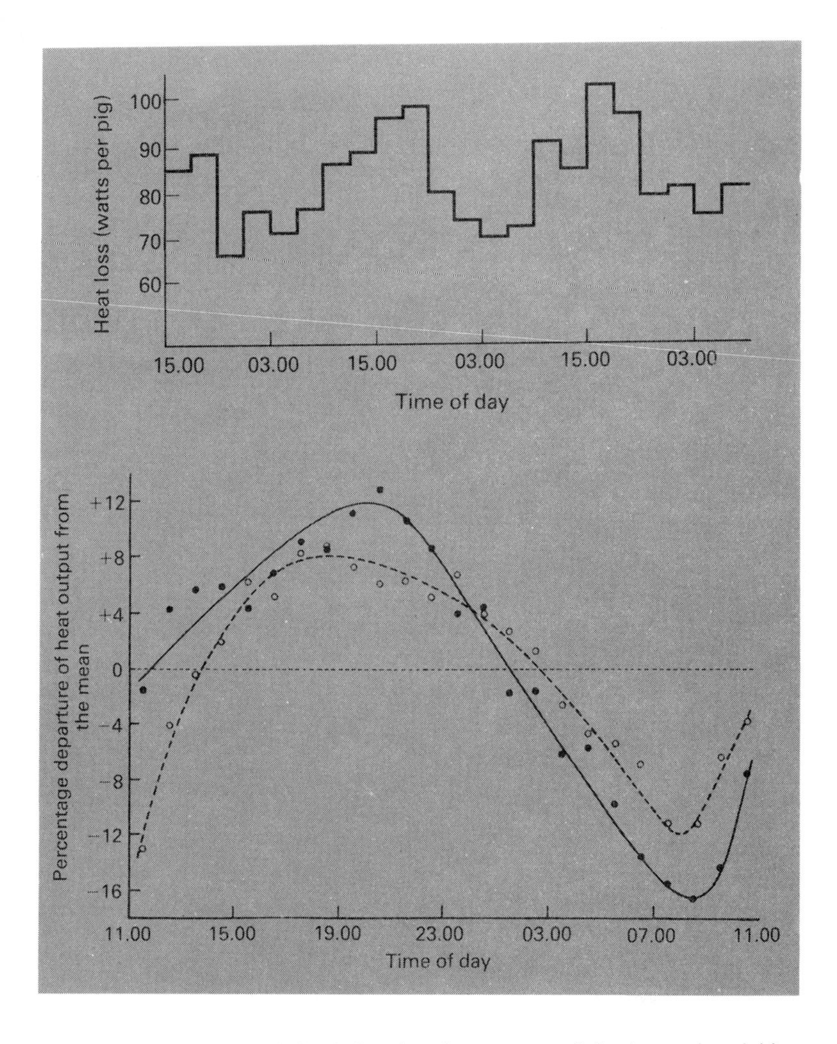

Fig. 2.9 (**a**) 24-hour variation in heat loss from a group of six pigs, each weighing approximately 20 kg, at 20°C in the Babraham pen calorimeter (Mount, 1968b, from Adaptation of domestic animals, ed. E.S.E. Hafez, Lea and Febiger, 1968). (**b**) Mean cycles of heat output for eight pigs subject to a regimen with 24-hour periodicity. Weight range 3 to 11.5 kg. ○—○, runs beginning at 11.00 hrs; ●—●, runs beginning at 23.00 hrs (Cairnie and Pullar, 1959, by permission of Cambridge University Press).

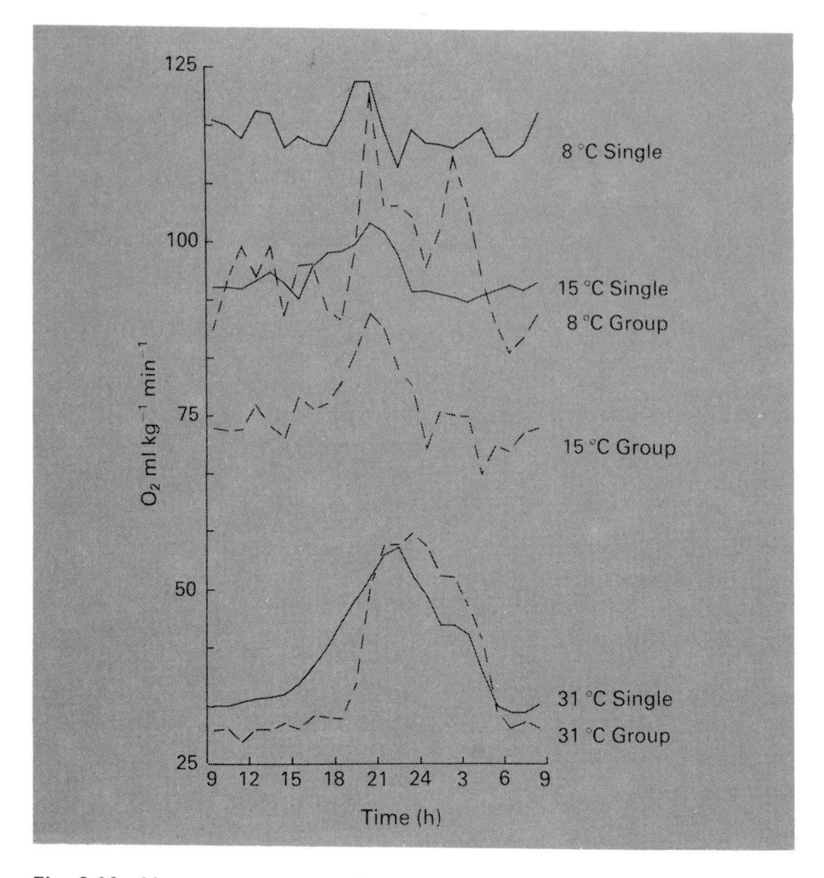

Fig. 2.10 Mean oxygen consumption rates over the 24-hour period for both single mice and groups at environmental temperature of 8, 15 and 31°C. Single mice (————), groups of mice (– – – – –). Note that the ordinate scale does not extend to zero (Mount and Willmott, 1967, by permission of Journal of Physiology).

metabolic rate is at a cold-induced maximum or a warm-induced minimum, further movement of environmental temperature away from the zone of regulation results in a change in body temperature which itself then determines the metabolism. In the cold, a fall in body temperature limits the heat-producing power of the tissues; in the warm, a rise in body temperature leads to an increased heat production. These are sometimes referred to as Q_{10} effects: Q_{10} is the factor by which the velocity of a reaction (y) is multiplied for a rise in temperature of 10°C. In the equation

$$Q_{10} = \left(\frac{y_2}{y_1}\right)^{\frac{10}{T_2 - T_1}}$$

Q_{10} is commonly about 2 for many chemical reactions, and values close to this have sometimes been observed for the rate of heat production by the whole animal.[386]

Age

As an animal grows from birth its metabolic rate in early life increases rapidly in relation to body weight and then more slowly later. In man, for example, metabolic rate is proportional to body weight (W) during the first year after birth, then to approximately $W^{0.6}$ subsequently.[234] In cattle, sheep and pigs the early period of rapid development is relatively much shorter than in man.[77, 386]

SITES OF HEAT PRODUCTION

The various tissues of the organism contribute to the total heat production to different degrees. Measurements suggest that the gastro-intestinal tract contributes about 25% of the total, the liver 10–15%, muscle about 20% (resting) and the remainder arises chiefly from the heart, kidneys, nervous system and skin. There is a high correlation of heat production with protein synthesis.[40, 502, 528, 529]

NON-SHIVERING THERMOGENESIS

The regulation of body temperature in the cold requires heat production additional to that from resting metabolism in a thermo-neutral environment. The additional heat can be derived from muscular work, shivering or non-shivering thermogenesis (NST).[82] NST is the liberation of metabolic heat by processes that do not involve muscular contraction; it is of great significance for the newborn mammal, reaching a level of 2–3 times the minimal metabolism; it declines as the animal grows older. It is associated with the calorigenic action of noradrenaline, and is more pronounced in smaller animals than in larger.[215]

Brown fat is recognized as an important source of NST, although NST also occurs in other tissues, particularly skeletal muscle.[272, 273]

About two-thirds of NST can arise in brown fat in the newborn rabbit, for example, where the temperature of the site of heat production in brown adipose tissue may be 2–3°C higher than colonic temperature. The only newborn animals that have been investigated and found not to contain brown fat or to show a calorigenic response to noradrenaline are the pig and fowl.[163, 315]

Brown adipose tissue grows after birth at different rates in different species; it occurs characteristically in the inguinal, axillary and subscapular regions, and around the deep blood vesels. It can be calculated that 35 g of brown adipose tissue could use all the extra oxygen and produce all the extra heat associated with the human infant's response to cold exposure. Heat production in brown adipose tissue depends on a large supply of oxygen, and for this reason it is susceptible to hypoxia. Fat is the principal fuel, and the sympathetic nervous system appears to control thermogenesis in brown adipose tissue.[245] Brown fat is important in the hibernating animal, where during the initial phase of awakening it may contribute 80% of the total heat production.

Brown fat is present in numerous vacuoles in the cell, as compared with the single vacuole of the white fat cell, and brown fat cells contain many mitochondria, indicative of the potentially high level of metabolic activity (Fig. 2.11). When brown fat cells are empty, they appear brown, although when full of fat droplets they are not nearly so easily distinguishable from white adipose tissue.

Adaptation to cold in small mammals appears to depend considerably on NST,[272] requiring increased food consumption. Larger animals adapt to cold by increasing the insulation of the coat, an adaptation that is not so effective in the smaller animal. Shivering probably depends on temperature changes both at the periphery and in the body's core; the higher the core temperature, the lower must be the skin temperature if shivering is to occur. Heat production during muscular work is a substitute for shivering, but it is additional to NST. During cold adaptation, the threshold body temperature for shivering falls.

OTHER FACTORS

Endocrines and muscular activity

The endocrines, particularly the thyroid gland, are highly significant in influencing metabolic rate; their role in adaptation is discussed in Chapter 6, also the sparing of heat production that occurs in the cold

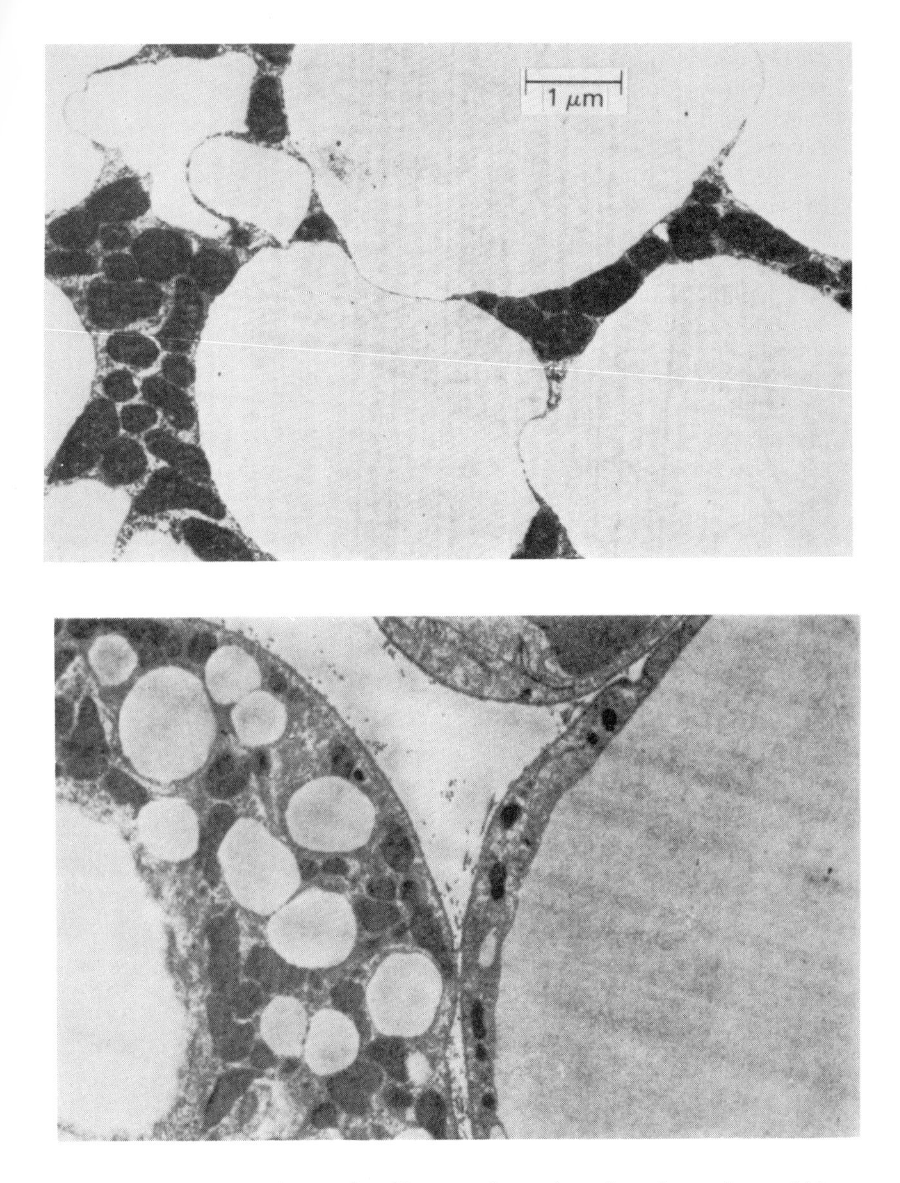

Fig. 2.11 Electron micrographs of brown adipose tissue from the newborn rabbit. (**a**) only a part of one cell is shown: large mitochondria surround the numerous vacuoles of fat. (**b**) a developing white adipose cell (right) next to a developing brown adipose cell (left) (Hull, 1966, by permission of British Medical Bulletin).

in social thermoregulation by groups of animals. Drugs (including anaesthetics) and emotional responses can have marked effects on metabolic rate.

Very high levels of metabolic rate are achieved by activity, as in running mammals and flying birds, sometimes over short periods. Taylor[492] has calculated the metabolic rates of animals running at their top speeds. He finds that, for the short periods that the maximum effort is made, both small animals (such as mice) and large animals (horses and elephants) can increase their metabolism up to 20 times, whereas animals of intermediate size (dog, cheetah and gazelle) can increase their resting metabolic rate by a remarkable 40–60-fold. The cheetah (see Chapter 1) achieves a calculated 54-fold increase in heat production, albeit for a very short period when pursuing its prey. The energy cost of running in man is high compared with the cost in mammals of corresponding body sizes.

Metabolic rate and heart rate

Various degrees of correlation have been reported between metabolic rate and heart rate by workers with different species. A reliable correlation would provide a useful method for assessing an animal's rate of heat production without close restraint, heart rate being recorded by telemetry.

However, the possibilities of obtaining useful quantitative estimates by this means appear to be limited. Webster[522] found good correlation in individual sheep, either resting or exposed to cold or feeding, but there was considerable variation between individual animals.[281] Similar results were reported for ducks by Wooley and Owen;[555] other measurements have been made on chickens.[289] Holmes, Stephens and Toner[240] found a significant regression of carbon dioxide production on heart rate in Jersey calves. In man, Dauncey and James[127] observed that large errors could result from estimating energy expenditure from heart rate; the errors were reduced considerably when the individual-specific relations between metabolic rate and heart rate were taken into account.

AEROBIC AND ANAEROBIC METABOLISM

Normally the pattern of metabolism in mammals and birds is aerobic, with bouts of anaerobic metabolism when the oxygen supply is limited relative to the activity that is undertaken. Examples of this in homeotherms occur in diving mammals and birds. When these animals are under water their heart rates are slowed, and lactic acid accumulates in the muscles as a result of anaerobic metabolism. When the animal surfaces and begins to breathe again, the circulation quickens, the lactic acid from the muscles is metabolized in the liver and the aerobic pattern of metabolism is re-established.[266] Per unit amount of substrate in the tissues, anaerobic metabolism is much less productive of energy than aerobic metabolism because the reaction is not taken to completion but is arrested at the lactic acid stage of the metabolic cycle. In general, the embryo and the newborn animal tolerate anaerobic conditions more readily than the adult.

3

Heat Loss

Two modes of heat transfer between the animal and its environment, the non-evaporative (or sensible) and the evaporative, have already been mentioned in earlier chapters. Non-evaporative heat transfer depends on a temperature difference, whereas evaporative heat transfer depends on a difference in water vapour pressure.

Non-evaporative heat transfer takes place through the channels of radiation, convection and conduction. Each of these has a particular temperature in the environment with which it is associated. For radiative transfer this is the mean radiant temperature of the surroundings, for convective transfer it is the air temperature, and for conductive transfer it is the temperature of the surface that is in contact. In addition to the temperatures, other factors influence heat transfer through the three channels and these are given in Table 3.1.

When the environmental temperature rises, non-evaporative heat loss is progressively reduced until it becomes zero when the environmental and body temperatures are equal. Under hot conditions an animal's ability to regulate its body temperature therefore depends on effective evaporative heat loss. Indeed, when the environmental temperature exceeds the body temperature, the body gains heat from the surroundings (where the non-evaporative heat loss line above D in Fig. 2.1 extends below the temperature axis), and then if a rise in the deep body temperature is not to take place both the animal's metabolic heat production and the heat load that it receives from its surroundings must be dissipated through the evaporative channel.

Table 3.1 Factors that influence the different modes of heat transfer between organism and environment (Ingram and Mount, 1975a, by permission of Springer–Verlag).

Mode of transfer	Animal characteristics	Environmental characteristics
Radiant	Mean radiant temperature of surface; effective radiating area; reflectivity and emissivity	Mean radiant temperature; solar radiation and reflectivity of surroundings
Convective	Surface temperature; effective convective area; radius of curvature and surface type	Air temperature; air velocity and direction
Conductive	Surface temperature; effective contact area	Floor temperature; thermal conductivity and thermal capacity of solid material
Evaporative	Surface temperature; percentage wetted area; site of evaporation relative to skin surface	Humidity; air velocity and direction

Evaporative heat transfer takes place at two sites on the animal: the skin surface and the upper respiratory tract. Evaporative heat transfer involves essentially the vaporization of water; in addition to the difference between the water vapour pressures at the animal's surfaces and in the environment, other factors are involved in evaporative heat transfer; these are also given in Table 3.1.

HEAT TRANSFER COEFFICIENTS

For estimating heat exchange, it is convenient to use coefficients to link the animal's surface parameters with those of the environment. Such a coefficient for non-evaporative heat transfer has the units of $W\,m^{-2}\,{}^{\circ}C^{-1}$ (W = watt), which implies that if the coefficient, the animal's effective surface area for the particular mode of heat exchange, and the surface-environment temperature difference are known, the rate of heat exchange can be calculated as the product of the three quantities.

For evaporative heat transfer, the coefficient is in terms of $W\,m^{-2}\,mbar^{-1}$, referring to the vapour pressure difference between the skin surface and the environment. The vapour pressure at the skin surface is determined by the rate of sweating and the ambient humidity; it can be obtained by assuming a 'per cent wetted area' of the skin[85,258] (see p. 84). Evaporative loss from the respiratory tract can be estimated from the respiratory minute volume by assuming the expired air to be saturated at the temperature at which it leaves the body.[341]

EVAPORATIVE HEAT LOSS

The loss of heat by evaporation depends on the change of state of water from liquid to gas, a process that takes up heat from the surroundings and so lowers the temperature at the place where the vaporization occurs. It follows that the site of evaporation is important in determining what is going to be cooled in this process. Water evaporating on the surface of an animal's coat takes up heat largely from the surrounding air, and cools the animal to only a very limited degree. In a bare-skinned animal, like pig or man, the evaporation of water on the body surface takes up most of the heat required for the process from the body itself, and so constitutes an efficient cooling mechanism. The quantity of heat that is taken up is the latent heat of vaporization of water, which is 2501 J g^{-1} at 0°C, decreasing to 2406 J g^{-1} at 40°C.[341, 369]

The vapour pressure difference that is the driving force for vaporization is a difference between absolute humidities and not between relative humidities. Absolute humidity is measured by the vapour pressure or by the mass of water vapour per unit volume; relative humidity is the proportion of the saturation level of vapour pressure or vapour-holding capacity per unit volume that is represented by the water vapour actually present. When the temperature rises, the saturation vapour pressure rises, because the holding capacity for water vapour per unit volume increases. Consequently, if the actual vapour pressure is fixed then the relative humidity falls. For example, at 20°C the saturation vapour pressure is 23.3 mbar, and in a saturated atmosphere the relative humidity would be 100%. At 50% relative humidity at 20°C the vapour pressure is 11.7 mbar. However at 30°C the saturation vapour pressure is 42.0 mbar, and then a vapour pressure of 23.3 mbar gives a relative humidity of only 55% and 11.7 mbar gives only 28%.

The psychrometric chart (Fig. 3.1) and the inset diagram give the relations between water vapour pressure, dry bulb (DB) and wet bulb (WB) air temperatures, the dew point temperature and relative humidity.[102] DB is air temperature obtained with a dry thermometer; WB is the air temperature obtained when the thermometer bulb is covered by a film of water, which is achieved in practice by surrounding the bulb by a wet wick. The WB reading is less than DB due to evaporative cooling.

The depression of WB below DB depends on DB and the ambient humidity, because these control the rate of evaporation from the WB. If the absolute humidity of the environment, that is the water vapour pressure, is held constant, and DB is increased, WB increases. The water vapour pressure at the WB surface is the saturation vapour pressure at WB temperature, and consequently as WB rises, and the saturation vapour pressure at the WB also rises, the driving force for vaporization (WB vapour pressure minus ambient vapour pressure) increases, producing increased vaporization and increased cooling of the WB. The new equilibrium value of WB is then determined by the balance between the warming effect of a rise in DB and the cooling effect of increased vaporization from the WB. The result is that WB rises less than DB, so that the depression of WB below DB is increased. If the ambient humidity is now increased, with DB held constant, the vaporization from the WB is decreased with the result that WB rises, so reducing the WB depression and indicating a rise in the relative humidity.

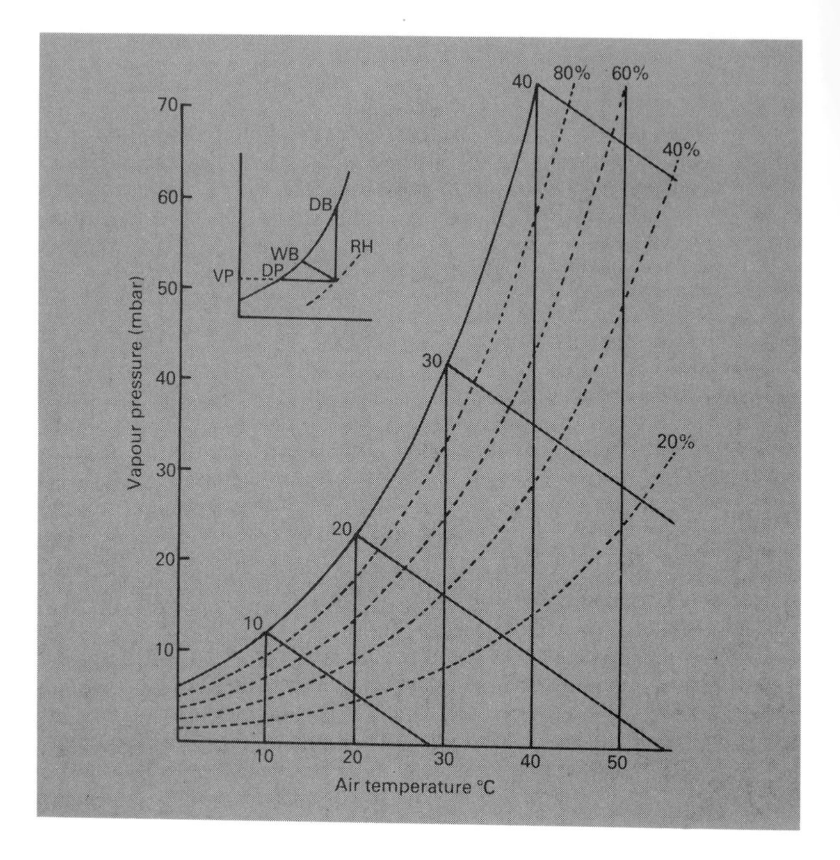

Fig. 3.1 Psychrometric chart relating dry (DB) and wet bulb (WB) temperatures to vapour pressure (VP), dew point (DP), and relative humidity (RH) (Ingram and Mount, 1975a, by permission of Springer–Verlag).

These effects are reflected in the psychrometric chart (Fig. 3.1). As an example, a DB of 20°C and a WB of 12°C correspond to a relative humidity of 37%, whereas a DB of 20°C and a WB of 15°C, following a rise in the absolute humidity, correspond to a relative humidity of 60%. The DB/WB combination of 20/12°C corresponds to a vapour pressure of 9 mbar, with a dew point of 5°C; the 20/15°C combination gives a vapour pressure of nearly 14 mbar and a dew point of 11.7°C.

The dew point is the DB temperature that corresponds to 100% relative humidity, that is saturation with water vapour, for the particular ambient humidity considered. As the DB falls, the relative humidity rises until it reaches 100% when DB equals the dew point; when the DB falls below this point, water condenses. The dew point is directly related to the vapour pressure (DP and VP in Fig. 3.1) and consequently gives a measure of the absolute humidity; it can therefore be used in practice instead of the DB/WB combination for determining the ambient humidity. The difference between the saturation vapour pressure and the actual absolute vapour pressure is termed the ***saturation deficit***, which is sometimes referred to as a percentage, when it is equal to (100 − relative humidity %). This is a measure of the drying power that is available.

An animal's surface temperature and the percentage wetted area of the surface (explained in Chapter 4) together determine the mean vapour pressure at the surface (Table 3.1). Air movement is an important factor that leads to the replacement of vapour-laden air near the skin surface by drier air, which results in increased evaporation from the surface.[258]

Evaporative cooling takes place either on the skin or on the mucosal surface of the upper respiratory tract; in both cases, vaporization of water on the surface leads to cooling of the blood flowing in the immediately underlying tissue. Depending on the species, evaporative cooling occurs mainly on the skin or mainly in the upper respiratory tract. The water that evaporates on the skin comes from sweat secreted by the sweat glands in the skin, or sometimes, as in the pig, from the surroundings.

Vaporization in the respiratory tract takes place in the upper respiratory tract, because it is here that variations in the temperature of the air and the water vapour pressure are apparent. In the lungs these conditions are nearly invariant, even under extreme conditions,[85] so that the statement that evaporative loss takes place in the lungs is incorrect. The boundary between the external environment and the animal's internal environment is situated more towards the oral end of the respiratory tract.

The effectiveness of evaporative cooling is high both on the skin and in the upper respiratory tract. Monteith[369] estimates that for man, who can produce more sweat than any other species and who can have a totally wetted skin surface (see Chapter 7), the diffusion resistance in the respiratory tract exceeds the boundary layer resistance by an order of magnitude, so that the rate of water loss by evaporation in the respiratory tract is much smaller than the cutaneous evaporation rate when the skin is covered with sweat. A high rate of sweating is consequently superior to panting in bringing about evaporative heat loss, provided that the transfer of water vapour is not impeded by coat or clothing.[99]

CUTANEOUS EVAPORATIVE LOSS

All animals lose water through the skin by diffusion, but the chief control of evaporative loss from the skin is through the sweat glands. Some animals can achieve very high rates of sweating, and the high evaporative loss that accompanies this allows them to survive in hot environments without becoming hyperthermic. Other animals have very little sweating capability, so that if they are able to control their body temperatures in hot environments they need to wet their surfaces with water from the environment in the form of wallows or sprays. The pig falls into this category. Some animals, such as mice, wet their coats by licking when conditions are very hot, but the excessive use of body water that is involved, particularly in small mammals, makes this procedure appropriate only when the animal is in an extreme situation.

The rates of cutaneous water loss for several species are given in Table 3.2. Man can sweat copiously, and it is due to this that he can survive in dry heat at an environmental temperature of 50°C or more with a normal body temperature (see, for example, Blagden[53, 54], quoted by Ingram and Mount[258]).

Sweat glands

Sweat glands were formerly considered to be either *eccrine* or *apocrine* in type, depending on their mode of secretion. The glands that occur in large numbers in the skin of man are the characteristically eccrine glands. They produce the watery sweat secretion that is important for thermoregulation under hot conditions. The secretion of apocrine glands was originally described as composed of material derived from cells lining the glands, as opposed to the secretion of sweat from intact cells that occurs in eccrine glands. It is now apparent that there are many different types of sweat gland, so that the simple division between eccrine and apocrine is not realistic.[274]

The numbers of sweat glands in man range from 2000 per cm^2 in the palm of the hand and sole of the foot down to 100–200 per cm^2 in the limbs and trunk. Cattle have an average of 1800 per cm^2, and sheep about 240–340 per cm^2.

The chief substance in sweat is sodium chloride, at a concentration between 0.2 and 0.4 g per 100 ml. When the rate of sweating increases the concentration rises. Acclimatization to hot environments leads to a fall in sodium chloride concentration in sweat, but this does not

Table 3.2 Water loss through the skin of various species below and above their critical temperatures (Ingram, 1974, by permission of Butterworths).

Animal	Below critical temperature ($g\,m^{-2}\,h^{-1}$)	Above critical temperature ($g\,m^{-2}\,h^{-1}$)	Reference
Pig (low humidity)	+7 to +16	+24 to +32	Ingram[248]
Pig (high humidity)		−15 to −23	Ingram[249]
European cattle	+12 to +16	+67 to +144	McLean[338]
Sheep with sweat glands	+12	+63	Brook and Short[77a]
Sheep without sweat glands	+12	+32	Brook and Short[77a]
Man	+6 to +10		Hertzman *et al.*[222a]
Man (at rest)		+150	Winslow and Herrington[546a]
Man (at work)		+1200	Robinson[447]

At 20°C, 1 $g\,m^{-2}\,h^{-1}$ is equivalent to a latent heat loss of 0.68 $W\,m^{-2}$

occur if extra salt is taken by the subject. The concentrations of urea and lactic acid are higher in sweat than in blood[258] (see Chapter 7).

RESPIRATORY EVAPORATIVE LOSS

The continuous movement of air to and fro over the upper respiratory surfaces leads to evaporation in a manner similar to the effect of wind on a moist skin surface. The conditioning of the inspired air is highly efficient, both for equilibration to body temperature and for saturation with water vapour, processes that occur before the inspired air reaches the alveoli. The mucosal lining of the respiratory tract gives up water vapour to the inspired air and so undergoes evaporative cooling.

When air is expired, some water is condensed on the mucosa, which is below the dew point, with the release of latent heat, some non-evaporative heat is returned to the mucosa and the air leaves at a temperature lower than deep body temperature. This is a form of counter-current heat exchange (Fig. 3.2) that results in savings of heat and water in the organism. Water is saved because the expired air is not carrying as much water vapour as it would if it were saturated at the deep body temperature. This counter-current process is of considerable significance in conserving water in desert animals. The kangaroo rat, for example, can live without water intake; it derives water from its metabolism, and its respiratory counter-current exchange is very efficient, with an expired air temperature as low as 24°C. Sometimes the temperature of air expired by birds is closer to the environmental temperature than to the deep body temperature, and the air expired by small rodents is on occasion even cooler than the inspired air.[455] In man, the temperature in the pulmonary alveoli is 37°C and in quiet respiration the temperature of the expired air is about 32°C. Reference to Fig. 3.1 shows the considerable fall in saturation vapour pressure between these two temperatures, with a corresponding decrease in the water content of the expired air.[520]

The component of non-evaporative heat transfer involved in conditioning the respired air is small when compared with the evaporative heat transfer. In man at an ambient temperature of 20°C, the total conditioning of the respired air, non-evaporative and evaporative combined, requires about 11 W, a little more than 10% of the resting metabolism. At 0°C, a man breathing dry air loses nearly 20% of his heat production in heating and humidifying his respired air, a loss that is smaller than would be incurred if counter-current exchange did not take place.

Panting

Panting occurs characteristically in the dog. It consists of shallow, rapid respiratory movements that greatly increase the movement of air to and fro in the upper respiratory tract, leading to a corresponding increase in evaporative cooling. The tidal air is small;

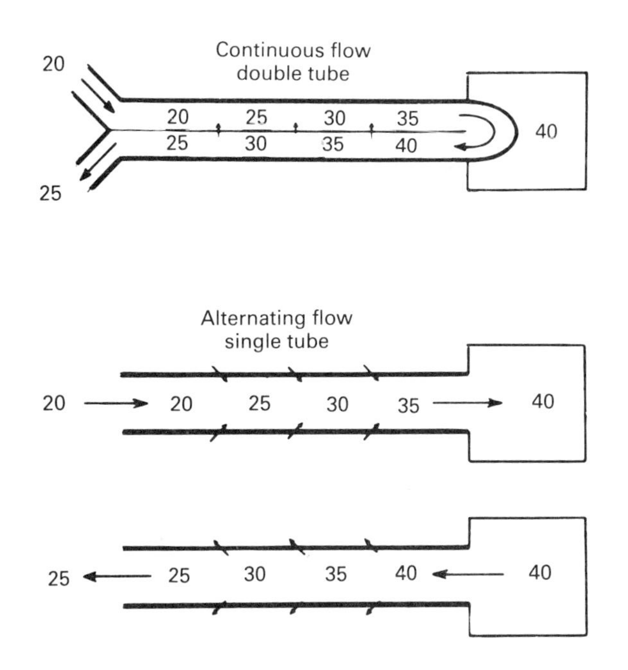

Fig. 3.2 Two types of heat exchange system. The lower system represents that in the respiratory tract, in which inspired air draws heat and water from the walls and gives both back again at expiration; numbers are °C (Jackson and Schmidt–Nielsen, 1964).

the importance of this lies in the small effect that panting has on pulmonary ventilation, so that there is no excessive loss of carbon dioxide and consequently no disturbance of the acid-base equilibrium in the form of a respiratory alkalosis[195] (see Chapter 9). Other animals, such as pigs and sheep, show increased respiratory frequency under hot conditions; cattle also show this response although they are more capable of sweating than sheep. The tendency is for panting to occur in animals that have relatively low sweating capabilities.[49]

Rapid panting moves air quickly to and fro over the turbinates, leading to evaporation from the mucosa and cooling of the blood flowing through the region. The higher the body weight, the lower is the panting frequency. In the adult Merino the rate rarely exceeds 350 min^{-1}, whereas in the lamb the rate reaches 450 min^{-1}.[347]

The increases in the respiratory minute volume during panting range from 10-fold in the ox to 12-fold in the sheep, 15-fold in the rabbit, and 23-fold in the dog. Some birds also pant and flutter the gular pouch at resonant frequencies (see Chapter 12). In birds, as in mammals, the increase in respiratory rate during panting is accompanied by a reduction in tidal volume.[435]

NON-EVAPORATIVE HEAT TRANSFER

Whereas evaporative heat loss involves essentially the vaporization of water, non-evaporative heat exchange takes place through the three channels of radiation, convection and conduction as the result of temperature differences.

When the environmental temperature is reduced, an animal reduces its effective radiating surface area, so that although the radiant heat flux density is increased, the total flux is not as great as might be expected from localized measurements on the animal's surface. The same applies to convective and conductive heat loss. If the animal is resting on a floor when the temperature is reduced, it changes its posture, rising from a relaxed extended position to one in which it is supported on its limbs in a flexed posture; this change is associated with a marked reduction in conductive heat loss. Under hot conditions, the surface area that is available for heat exchange is increased by the animal adopting an extended posture.

RADIATION

The precise computation of the net exchange of heat by radiation between an animal and its surroundings would be an extremely difficult if not impossible task, but a high degree of precision would be out of keeping with the considerable variation that always attends work on living animals. The physical and mathematical basis of the subject of heat exchange by radiation is considered in many textbooks; a comprehensive treatment, which permits the selection of the level of approximation appropriate to animal experiments, is given by Jakob.[270, 271] The subject is presented here, derived from Jakob, so that the relative importance of the factors concerned in an animal's radiant heat exchange can be made explicit.

The Stefan–Boltzmann Law for total radiation from a perfectly black body is given by

$$H_r = \sigma A T^4$$

where H_r = heat transfer rate by radiation; σ = the Stefan–Boltzmann constant, $5.67 \times 10^{-8}\ W\,m^{-2}\,K^{-4}$; A = effective radiating area of body and T = absolute temperature of the radiating surface. This is the one-way radiation from a given body, depending on the fourth power of the absolute temperature T. What is required in practice is the net radiant exchange, which is the difference between the radiant energy leaving the body and the radiant energy entering it from the environment.

Emissivity

Before formulating an equation to give the net radiant exchange, account must be taken of the deviation of radiating surfaces from the perfectly black body referred to in stating the Stefan–Boltzmann Law. This is done by introducing into the equation values for the emissivities of the exchanging surfaces. Emissivity has a maximum value of unity, and this is the value for a perfectly black opaque body, which reflects none of the incident radiation but absorbs all of it. A perfectly opaque reflector, on the other hand, has an emissivity of zero; it reflects all incident radiation and absorbs none of it.

Such bodies do not exist in nature; a matt black surface may have an emissivity between 0.95 and 1.0. A nearly perfectly black surface is produced by a hollow enclosure with only a small opening. The opening itself then acts as a surface with absorptivity = emissivity = 1. The proof of this is given by Jakob[270], and rests on the facts that for an opaque body the sum of absorbed and reflected radiation equals unity, and for a partially transparent body the sum of absorbed, reflected, and transmitted radiation equals unity; and the absorptivity equals the emissivity. Thus an opaque body that absorbs much radiation reflects little and *vice versa*. The term 'black body' can lead to confusion, because, as Monteith[369] points out, even fresh snow, like most natural objects, behaves like a black body in the waveband 3–100 μm, as distinct from the solar radiation reflected by snow. He therefore refers to a 'full radiator' instead of a black body.

Emissivity varies with the wavelength of the radiation used; Fig. 3.3 shows the variation in reflection power (and consequently the

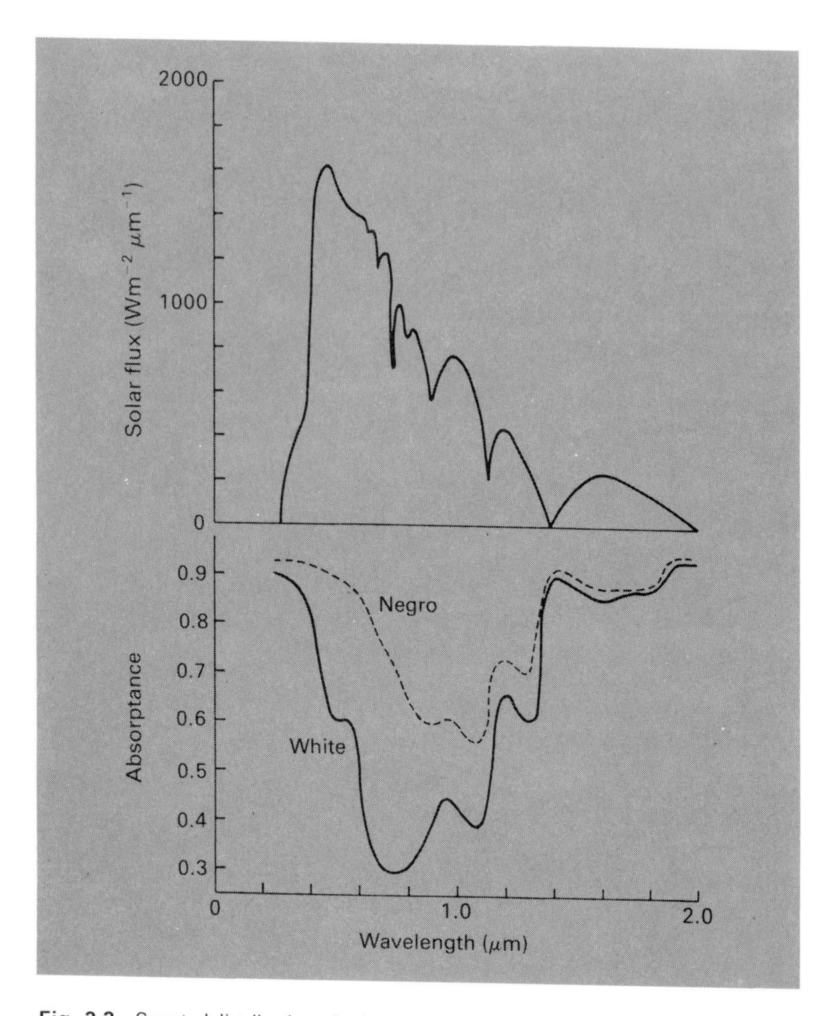

Fig. 3.3 Spectral distribution of solar radiation (upper curve) and of absorptances of white and Negro skin (lower curves) (Redrawn from Gates, 1962, Jacquez, Huss, McKeehan, Dimitroff and Kuppenheim, 1955, and Jacquez, Kuppenheim, Dimitroff, McKeehan and Huss, 1955, by Kerslake, 1972, by permission of Cambridge University Press).

emissivity) of human skin when the wavelength of radiation is varied. For short-wave solar radiation many surfaces, including white paint and pale-coloured animal coats, have a high reflectance. For long-wave radiation, however, of wavelength greater than 3 μm, most surfaces, including white paint, white coats and pale skin, have an emissivity that approaches unity. Thus for short-wave radiation the reflectance is dependent on colour, whereas for long-wave it is independent. For wavelengths longer than 5 μm the emissivity of skin is close to one, and this is true for most surfaces at this wavelength. Some highly polished metals, such as silver, aluminium and copper, have low emissivities at long wavelengths.

When the emissivity (ε) is taken into account, the statement for total radiation from a body becomes

$$H_r = \varepsilon\sigma AT^4$$

and $H_{r\,net}$ (the radiant heat exchange resulting from the algebraic summation of incoming and outgoing radiation) is given by

$$H_{r\,net} = \sigma A(\varepsilon_1 T_1^4 - \varepsilon_2 T_2^4)$$

where ε_1 and ε_2 = the emissivities of the surfaces of the body and the surroundings and T_1 and T_2 = the absolute temperatures of the body and its surroundings, respectively.

In addition to emissivity, the ratio of the surface area of a given body to that of its surroundings is important, as for example when measurements are made on a man or animal in an enclosed chamber. These requirements are met by Christiansen's equation:[271]

$$H_{r\,net} = \frac{1}{1+\varepsilon_1(1/\varepsilon_2 - 1)A_1/A_2}\,\varepsilon_1\sigma A_1(T_1^4 - T_2^4) \qquad (3.1)$$

where A_1 and A_2 = the effective radiation areas of the inner body and the enclosure, respectively.
This equation may be re-written:

$$H_{r\,net} = F\sigma A_1(T_1^4 - T_2^4) \qquad (3.2)$$

where F is termed the radiative interchange factor, and is defined by equations (*3.1*) and (*3.2*); this factor takes account of both the configuration and the emissivities of the surfaces.

What is important in experiments in which the heat exchange of an animal within an enclosure is considered is that F, the radiative interchange factor, can be determined *for a given situation* without reference to the consitution of the factor. The chief disadvantage of this method is that any change in emissivity of the surface of either the animal or of the enclosure would not be recognized, and so would cause error. However, the most probable change in long-wave emissivity would be due to dust or other particulate matter, which would tend to raise the emissivity towards unity. At the long infrared wavelengths of emission of heat from an animal's skin, the emissivity is already close to unity, so that any error due to dust would arise only in a polished enclosure. Where the emissivity, ε_2, of the enclosure surface approaches unity, the value for F given by equations (*3.1*) and (*3.2*) tends towards ε_1.

Wavelength

The wavelength of maximum emission of radiation from a surface is given by Wien's displacement law[270, 369] as 2897/T μm, where T is the absolute temperature (degrees K). For a skin temperature of 30°C, the value is 2897/303 = 9.6 μm; for 35°C it is 9.4 μm.

Heat exchange by radiation in animals is conveniently considered in two parts. The first of these deals with exchange when radiation from the surroundings is all long-wave, that is derived from surfaces at a range of temperatures extending from several hundred degrees centigrade downwards (a wavelength of maximum emission of radiation of 5 μm corresponds to a surface at about 300°C). The second part includes the effects of shorter wavelengths, including the visible spectrum (0.4–0.7 μm) and ultraviolet. Fig. 3.4 shows the natural division that occurs between long-wave and short-wave radiant flux in the wavelength region of 3 μm.

Long-wave radiant exchange

An important feature of long-wave radiation is that it does not pass through the majority of substances that are transparent to the shorter wavelengths of the visible spectrum, such as glass and most plastics. For example, the Perspex (polymethyl methacrylate) walls of a hospital incubator for babies is opaque to long-wave radiation, so that the baby loses heat by long-wave radiant exchange with the surface of the Perspex wall and the mattress on which it lies. The temperature of the wall is therefore important in determining the baby's radiant heat loss, because it occupies the major part of the solid angle subtended at the baby. The wall temperature lies nearly midway between the temperatures of the incubator air and the room air, so that in a cool room the baby's radiant heat loss can be considerable.

A thin Perspex shield placed between the baby and the wall reduces this loss. The shield tends towards the temperature of the incubator air and, since it is opaque to long-wave radiation from the baby, it acts as a warmer radiant environment than the incubator wall (Fig. 3.5). The use of such a radiant shield significantly lowers the baby's oxygen consumption rate from about 9.0 to 7.6 $ml\,kg^{-1}\,min^{-1}$ (mean result from measurements on six babies), and its effect on skin temperature is shown in Fig. 3.6.[229]

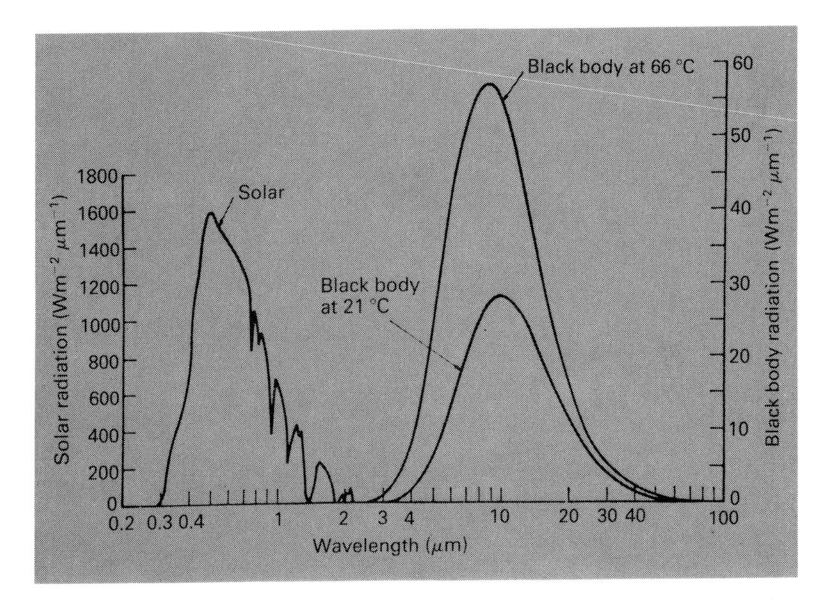

Fig. 3.4 Solar radiation at the ground for a zenith sun, and radiation from black bodies at 21 and 66°C (Bond, Kelly, Morrison and Pereira, 1967, by permission of American Society of Agricultural Engineers).

Polyethylene is one of the few substances that transmits long-wave radiation, and in Fig. 3.7 the percentage transmittances of glass, Perspex and polyethylene sheet are given in relation to wavelength. The transmittances of both glass and Perspex fall towards zero as the wavelength increases above 2–5 μm, but the transmittance of polyethylene remains high, with sharp troughs recurring at regular wavelength intervals due to the interference pattern associated with the CH_2 grouping in the molecule. If polyethylene is used instead of glass in making a horticultural greenhouse, some of the advantage that glass confers of preventing direct radiant heat loss from inside is not retained. The value of the polyethylene, in common with glass, would then lie in reducing convective heat loss due to wind.

For radiant heat emitted by the skin, transmittance is at a minimum

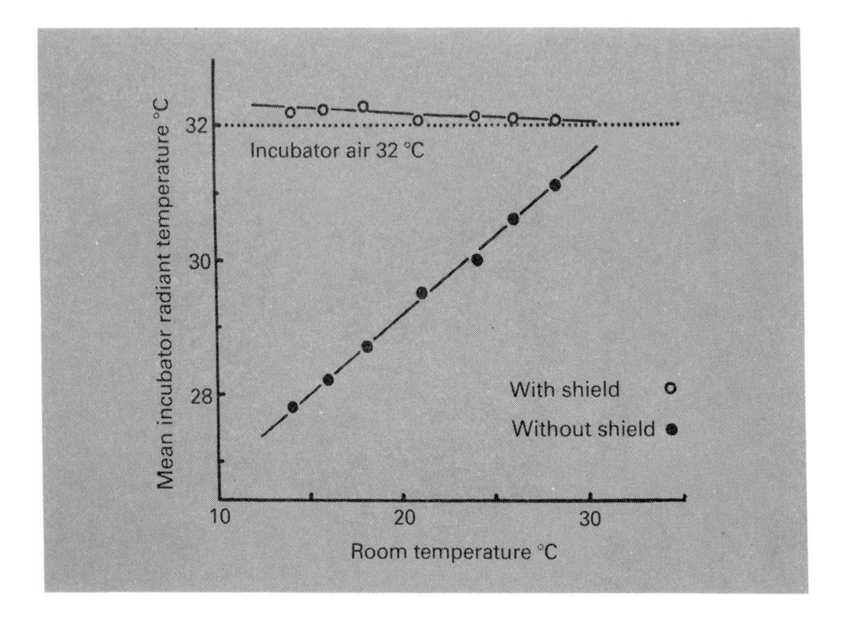

Fig. 3.5 The effect of room temperature on the mean radiant temperature within an incubator, with and without an internal shield at the temperature of the incubator air (Hey and Mount, 1966, by permission of The Lancet).

through glass and Perspex, but high through polyethylene. The property of polyethylene of transmitting such long-wave radiation has been used in shielding a radiometer,[170] in experiments on the radiation cooling of dairy cattle[464] and in the determination of the radiant heat exchange of the newborn pig.[381, 386]

Another approach to the separation of radiant from convective heat transfer was adopted by Joyce, Blaxter and Park[283] in their work on sheep. They exposed two sheep in two respiration chambers in which the air velocities and their distributions were similar, but in each chamber there were different wall emissivities in the infrared. Varying the emissivity changed the effective black-body temperature of the walls to which the animals were losing heat, and in each set of circumstances an appropriate radiative interchange factor could be determined. The results of these and other experiments gave good agreement between measurements made in the laboratory and other measurements made under a variety of conditions out of doors.

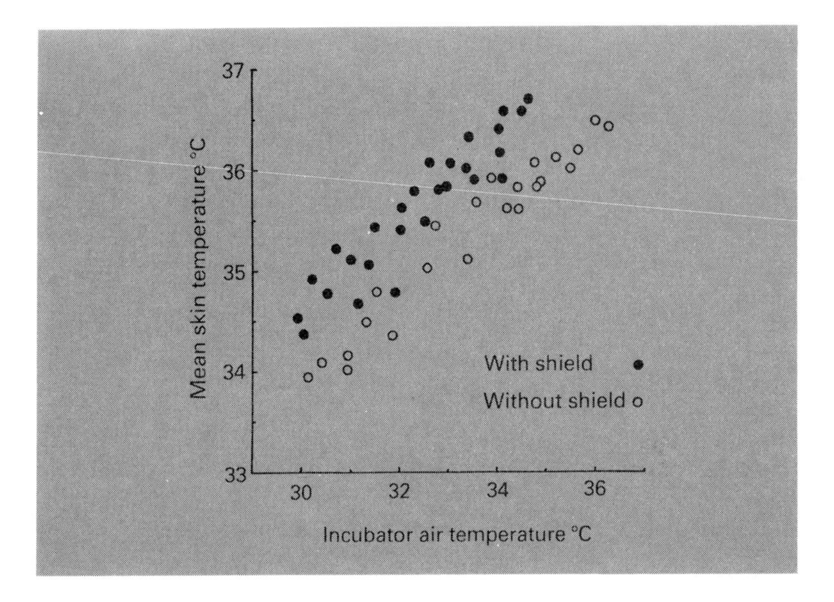

Fig. 3.6 The effect of a radiant shield on the mean skin temperature of 13 healthy human infants in an incubator, over a range of incubator air temperatures. Room temperature: 20 to 22°C (Hey and Mount, 1967, with acknowledgement to the Editor of Archives of Disease in Children).

Full-spectrum radiant exchange

The long-wave radiant energy exchange of an animal out of doors is with the ground, surrounding objects and the atmosphere, and the animal receives short-wave radiant energy from the sun directly, as a diffuse solar radiation due to scattering in the atmosphere, and by reflection from the ground surface.

The behaviour of man and animals in orientating themselves in relation to the sun, and in adopting various postures, produces a considerable variation in the incident load of solar radiation. For a sphere of radius r, the profile exposed to the sun is a disc of radius r with an area of πr^2. This is exactly one quarter of the surface area of the sphere. For an irregularly shaped body such a precise relation would not be expected to apply, although as shown in Fig. 3.8 for a human subject facing the sun at zero altitude the radiation area is close to 25% of the total body surface area, the percentage being low at high azimuth angles and when the sun is in an overhead position. Radiation geometry is discussed extensively elsewhere.[336, 369]

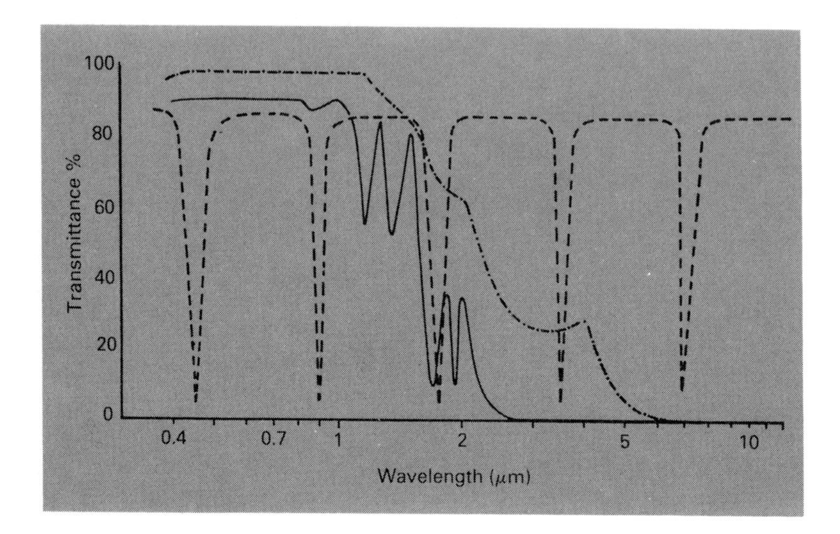

Fig. 3.7 The percentage transmittance of polyethylene, glass and Perspex for radiation extending from the visible spectrum into the infra-red (partly diagrammatic, for clarity) (Mount, 1968a).

In bright sunlight, the solar radiation intercepted by an animal may be several times the metabolic heat production. White coats reflect more than half the incident solar radiation, and dark coats absorb most of it, so that differences in reflectance can have large effects on the size of the environmental heat load that an animal has to bear. Radiation falling on the coat may be reflected, absorbed, or transmitted by forward scatter towards the skin; radiation penetration is greater in white than in dark coats, and depends on the density of the coat.[153, 246, 259]

The conductivity of coats for long-wave radiation transfer ranges from $15\,\mathrm{mW\,m^{-1}\,{}^{\circ}C^{-1}}$ for fleece to $4\,\mathrm{mW\,m^{-1}\,{}^{\circ}C^{-1}}$ for cattle coats.[97, 100] The method of thermography, which presents a thermal image of the body instead of a light image, gives a detailed

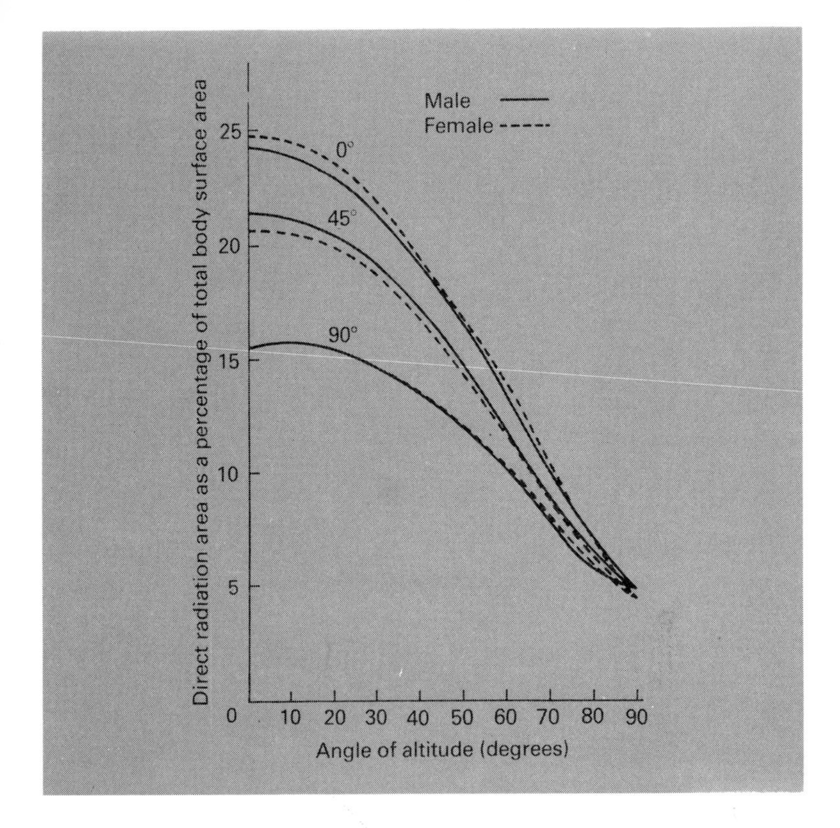

Fig. 3.8 Projected area, expressed as proportion of total area, of male and female subjects as a function of solar altitude. Curves are shown for three angles of azimuth (0° corresponds to facing the sun) (Underwood and Ward, 1966, by permission of Ergonomics).

comparative picture of temperature distribution over the surface, and amongst other things can be used to study the relation between coat colour and coat temperature under different conditions. Cena[95] discusses the characteristics of the method (Fig. 3.9).

For long-wave radiation, the depth of coat that is needed to absorb 95% of the incident radiation ranges from about 1 mm for cattle to about 4 mm for sheep.[97, 100] For short-wave radiation, the relation between the incident radiation and its absorption and reflection is complicated by the scattering of radiation in the coat and the effects of a secondary flux of radiation reflected by the skin. Cena[95] discusses

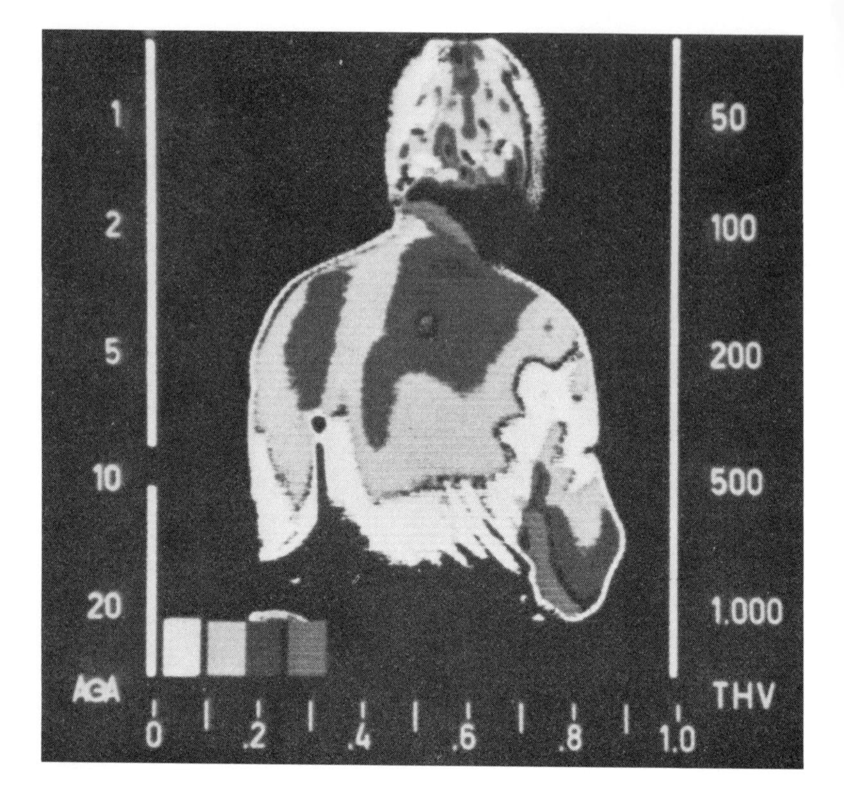

Fig. 3.9 Thermograms of a human subject wearing a white shirt with a black '3' on the back. In the thermogram on this page, the subject is exposed to weak radiation. In the thermogram on p. 61, there is an irradiance of 800 W m^{-2}, and the black '3' is now 3°C hotter than the shirt surface. In each thermogram, the scale of shades on the abscissa progresses from colder on the left to warmer on the right (from Cena, 1974, by permission of Butterworths, London).

the resulting equations, and concludes that when the reflection coefficient for the coat as a whole is 0.1, the absorption coefficient of the coat is about 0.7; when the reflection coefficient exceeds 0.7, the absorption coefficient tends to zero. Values for the cleaned coats of several species are given in Table 3.3; in practice, the bulk of reflection of 'white' coats rarely exceeds 0.5.

Although for long-wave radiant exchange many surfaces behave like black bodies in having emissivities close to unity with correspondingly low reflectances, exposure to short-wave or full-spectrum radiation is accompanied by much greater variations in

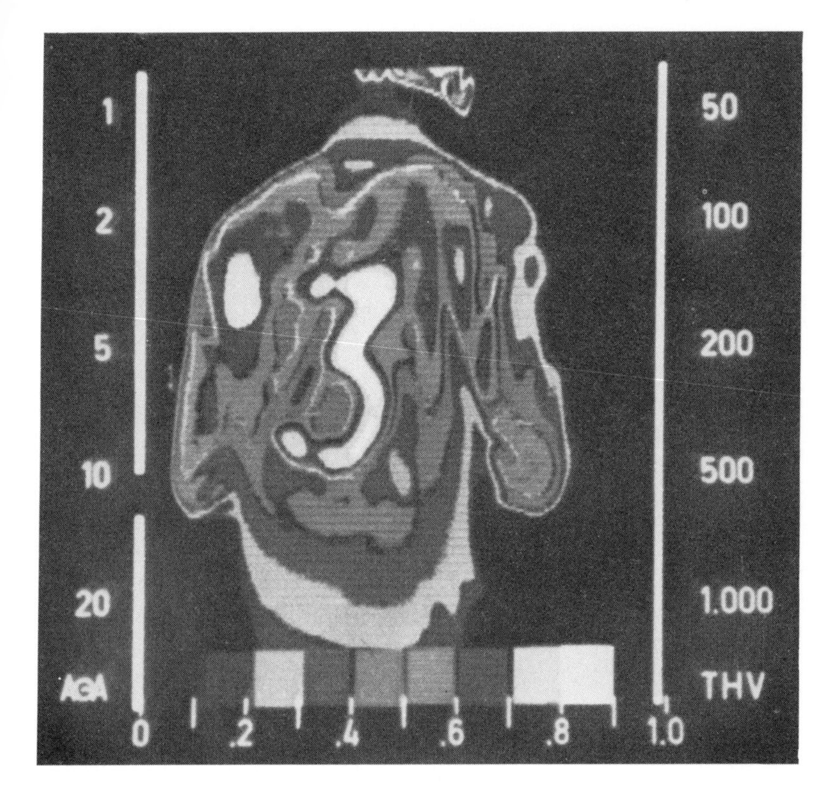

reflectance. These differences are apparent in Table 3.4, which gives the excess temperature over air temperature reached by various materials when they are exposed to bright sunlight.

Radiative heat transfer coefficient

For small differences of temperature, the fourth-power relation of the Stefan–Boltzmann Law can be replaced by a first-power law as an approximation. The coefficient for long-wave exchange, K_R, is then given by $4\sigma T^3$, where σ is the Stefan–Boltzmann constant (5.67×10^{-8} W m^{-2} K^{-4}) and T is the mean absolute temperature.[258, 302] From this, when T = 295 K (22°C), $K_R = 5.8$ W m^{-2} °C^{-1}; and at

Table 3.3 Values for the physical parameters of cleaned coats (from Cena, 1974).

	Number of hairs per unit surface area of skin (cm^{-2})	Hair diameter (μm)	depth (cm)	Coat reflection coefficient	Absorption coefficient
Dorset Down sheep	1430	48	8.0	0.79	0.001
Clun Forest sheep	1460	42	5.0	0.60	0.05
Welsh Mountain sheep	850	65	3.5	0.30	0.30
Rabbit	4200	31	2.0	0.81	0.001
Badger	240	71	1.8	0.48	0.11
Cow	1260	44	0.6	0.63	0.04
Goat	110	83	2.3	0.42	0.16
Fox	3600	20	1.4	0.34	0.26
Deer	520	150	1.5	0.69	0.025

Table 3.4 The temperatures of various surfaces exposed to bright sunshine, expressed as excess above air temperature to nearest 0.5°C (from Landsberg, 1964, by permission of Elizabeth Licht).

Natural surfaces		**Structural materials and parts**	
Soil surface under 5 cm grass	0.1	Brick steps	6.5
Intermittently shaded soil surface	1.0	Brick wall, unpainted	10.5
Tree leaves	2.0	Wooden siding, red paint	8.5
Short grass tips	6.0	Window surface	7.0
Bare sand	9.5	Wood block	
Bare rock	11.0	white lead paint	11.0
Bare dark soil	15.5	pink paint	11.0
Road surfaces		yellow ochre paint	14.5
Gravel drive	8.0	red oil paint	15.5
Concrete	17.5	soot covered	17.0
Asphalt	22.5	Cork	
Roofing materials		white oil paint	7.5
Asphalt shingles	17.0	black oil paint	28.0
Dark slate	18.0	Stone	
Asbestos shingles	19.0	white oil paint	6.5
Galvanized metal		black oil paint	18.0
fresh aluminium paint	6.5		
unpainted	16.0		
red paint	19.5		
black paint	26.0		

305 K, $K_R = 6.4\ W\ m^{-2}\ {}^{\circ}C^{-1}$. For a mean radiant environmental temperature of 15–20°C, and a surface (skin or coat) temperature of 30–35°C, K_R is thus close to $6\ W\ m^{-2}\ {}^{\circ}C^{-1}$. The temperatures for this purpose are the black-body temperatures.

CONVECTION

Convective heat exchange is a form of heat transfer that depends on the redistribution of molecules within a fluid, as distinct from the conduction of heat in which there is no actual translocation of molecules. Heat exchange by convection depends on the surface temperature of the body, its shape, surface characteristics and size, and on the air temperature and the air movement rate that impinges on the body (Table 3.1).

Forced and natural convection

The convective movement of molecules in a fluid (gas or liquid) is due primarily either to an external force acting on the fluid or to a buoyancy force resulting from differences of temperature in the fluid. The first type is termed forced convection; an example of this is the movement of air by a fan. The second type is termed natural (or free) convection and an example is the way in which air rises when it is warmed by a hot body.

Forced convection dominates when wind impinges on a body; natural convection dominates in still air or at low windspeeds. Kerslake[296] comments that for man forced convective heat transfer predominates at windspeeds above $0.2\ m\ s^{-1}$, and natural convection at lower windspeeds. The position is similar for the pig, where, for animals of 4–60 kg, forced convection dominates above $0.2\ m\ s^{-1}$ and natural convection below $0.1\ m\ s^{-1}$, with a regimen of mixed convection between $0.1–0.2\ m\ s^{-1}$.[396, 397]

The pattern of air flow that results around a body may be either laminar (stream-lined) at lower velocities, or turbulent, at high velocities, for both forms of convection. Wind velocities can be very high, but the high velocities that are sometimes associated with natural convection are not always so apparent. The natural convection boundary layer of air around a warm body is the zone through which convective heat exchange takes place in still air. In a standing man the air flow has a velocity exceeding $0.3\ m\ s^{-1}$ (Fig. 3.10). By contrast, the layer is thinner around the head of the lying subject and the velocity only $0.05\ m\ s^{-1}$.[109] Over a standing man, air flow due to natural convection is rapid and turbulent in the region of the head, but laminar around the lower limbs (Fig. 3.11).

The movement of air by natural convection over the surface of man and animals and other warm objects, and the associated boundary layer, can be made visible by Schlieren photography (Fig. 3.12).[323] The rate of heat loss by natural convection from a man whose skin surface temperature is 10°C above air temperature may be 30–40 W m^{-2}, which is a large part of the resting metabolism.[365,366]

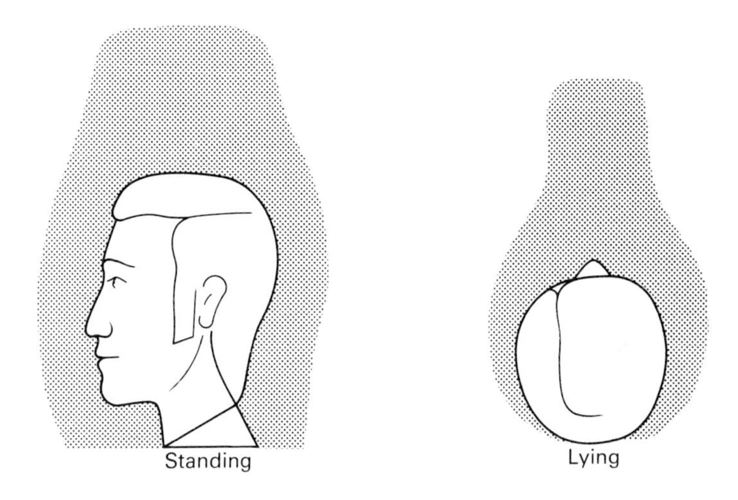

Fig. 3.10 The dotted area indicates the extent and shape of the convective boundary layer for the two postures, standing and lying (Clark and Toy, 1975a, by permission of Journal of Physiology).

Convective heat transfer coefficients

As with evaporative and radiative heat transfers, coefficients can be used to express convective heat transfer. With forced convection, the magnitude of the coefficient depends on the windspeed, whereas with natural convection the size of the coefficient depends on the temperature difference between body and air. So that a single equation can be written to cover a wide range of conditions, it has been found useful to employ dimensionless quantities[146,309] that include the coefficients. The coefficients can be calculated from empirical relations between dimensionless numbers: Nusselt and Reynolds numbers for

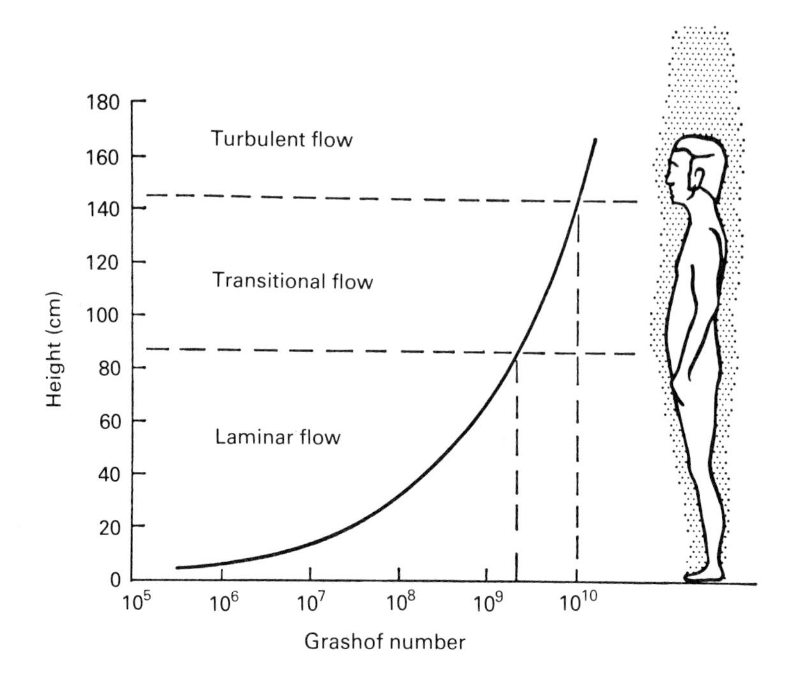

Fig. 3.11 Variation of Grashof number with vertical height over the body surface for a skin temperature of 33°C and an air temperature of 20°C, showing the regions of laminar, transitional and turbulent flow. The dotted area around the outline of the human figure indicates the convective boundary layer flow. The maximum flow velocity is 0.5 m s^{-1}; the plume extends 1.5 m above the head and the volume of air moving over the head can be 600 l min^{-1} (Clark and Toy, 1975a, by permission of Journal of Physiology).

forced convection, and Nusselt and Grashof numbers for natural convection:[369, 396]

$$\text{Nusselt number (Nu)} = \frac{K_C d}{k}$$

where K_C = convection coefficient, W m^{-2} °C^{-1}; d = characteristic dimension of body, m (taken as trunk diameter); and k = thermal conductivity of dry air, W m^{-1} °C^{-1}.

$$\text{Reynolds number (Re)} = \frac{Vd}{v}$$

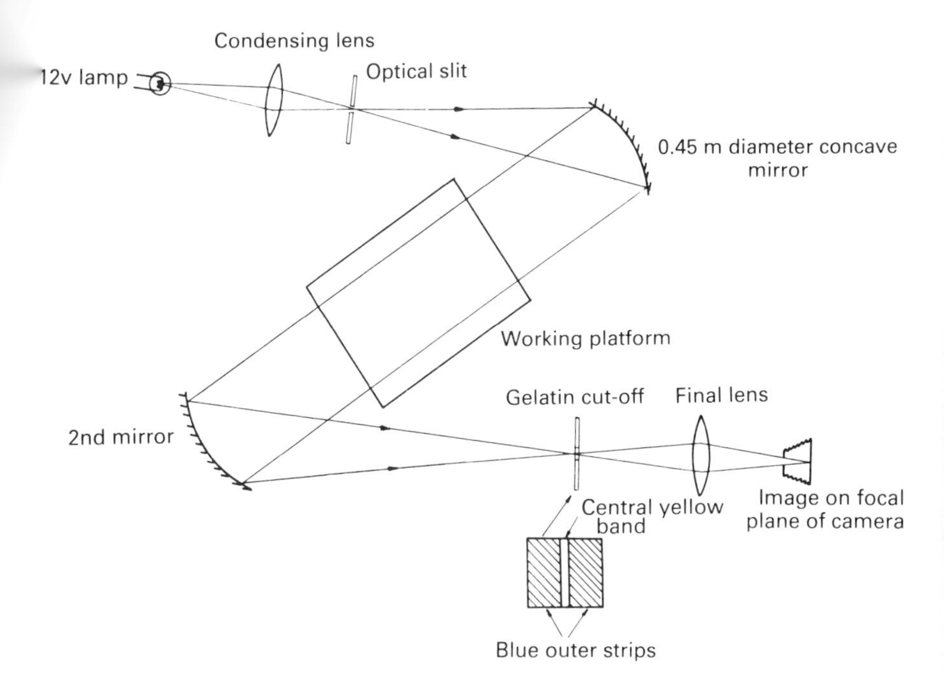

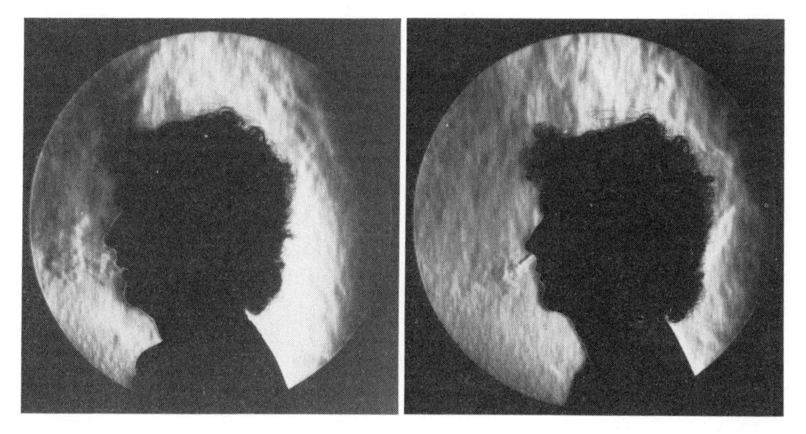

Fig. 3.12 Upper: diagram of the Schlieren optical system that is used to detect warmed air through its change in refractive index.
Lower: Schlieren patterns around the head; in both cases, warmed air is rising around and above the head, with warmed air issuing from the mouth (left) and from the nose (right). (Clark and Cox, 1974, by permission of Medical and Biological Engineering; photographs by coutesy of Dr R. P. Clark.)

where V = air velocity, $m\,s^{-1}$; and v = kinematic viscosity of dry air, $m^2\,s^{-1}$.

$$\text{Grashof number (Gr)} = \frac{a g d^3 (T_s - T_a)}{v^2}$$

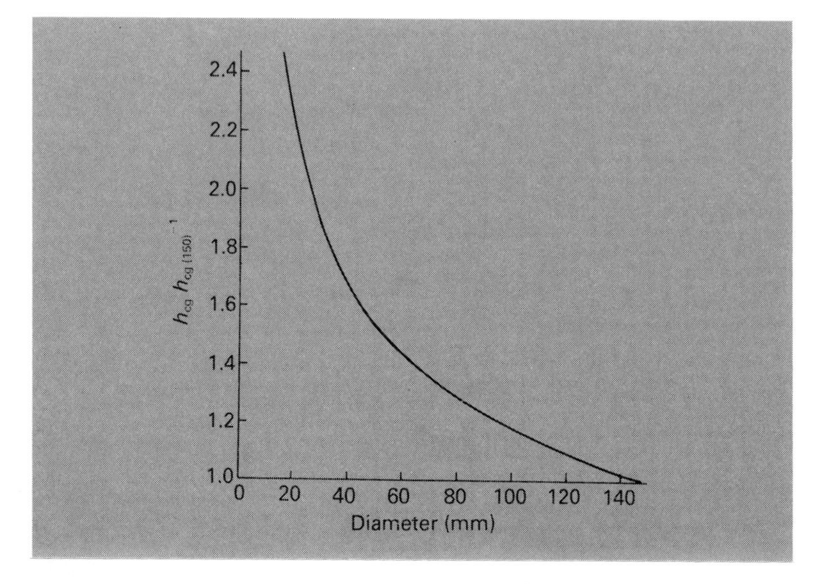

Fig. 3.13 The effect of size on convection coefficient for a globe. The abscissa shows the diameter of the globe, the ordinate the ratio of its convection coefficient, h_{cg}, to that of a standard 150 mm diameter globe, $h_{cg\,(150)}$ (Kerslake, 1972, by permission of Cambridge University Press).

where a = coefficient of thermal expansion of air, $(1/T_{a\,(abs)})$; g = acceleration due to gravity, $m\,s^{-2}$; T_s = skin temperature, °C; and T_a = air temperature, °C.

A decrease in the size of the body leads to a greater convective heat loss per unit area for a given body-ambient temperature difference,[223] with the result that smaller animals and parts of smaller radius are more susceptible to convective cooling (Fig. 3.13). This accords with the results of calculations using the Nusselt–Reynolds and Nusselt–Grashof relations:

forced convection $\quad Nu = A\,Re^n$

natural convection $\quad Nu = B\,Gr^m$

where the values of A, B, n and m depend on whether the flow is laminar or turbulent and on the shape of the body and its orientation to the flow.[369] n is of the order of 0.5 and m about 0.25, so that in both cases K_C increases as d decreases.

It is commonly stated that the forced convection coefficient is

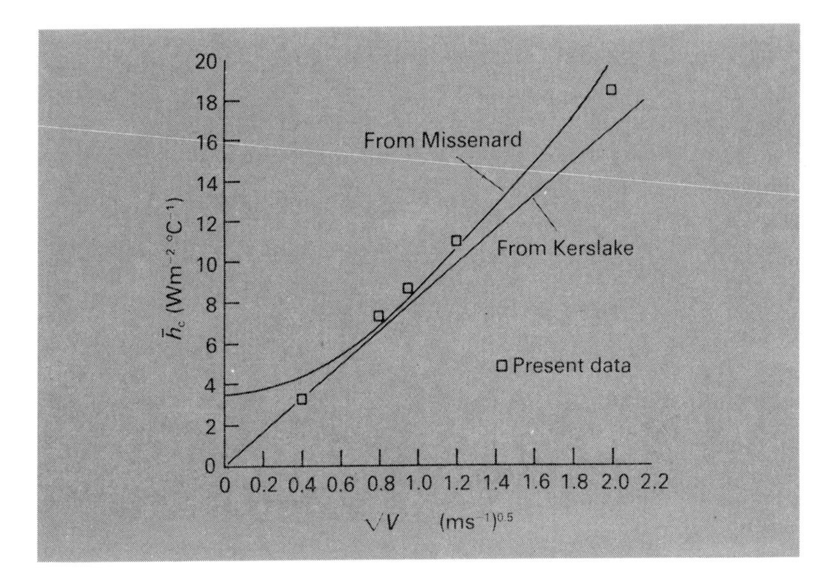

Fig. 3.14 Relation between the overall convection coefficient, $\bar{h}_c$, from the head and the square root of the air velocity (Clark and Toy, 1975b, by permission of Journal of Physiology).

proportional to the square root of the windspeed, that is to $V^{0.5}$.[52, 116, 410, 411] This holds for values of Reynolds number between 1000 and 10 000, although over a wide range of Reynolds number n varies from below 0.4 to above 0.8.[365] The proportionality of convective heat loss to $\sqrt{V}$ in man is a good approximation for windspeeds up to about 4 m s^{-1} (Fig. 3.14), corresponding to n = 0.5. At a windspeed of 1 m s^{-1}, the convective heat transfer coefficient for nude man is close to 8 W m^{-2} °C^{-1}; at 2 m s^{-1} it is about 12 W m^{-2} °C^{-1} and at 3 m s^{-1} it is about 15 W m^{-2} °C^{-1}. The distribution of forced convective heat transfer around a heated cylinder and the human head is illustrated in Fig. 3.15.

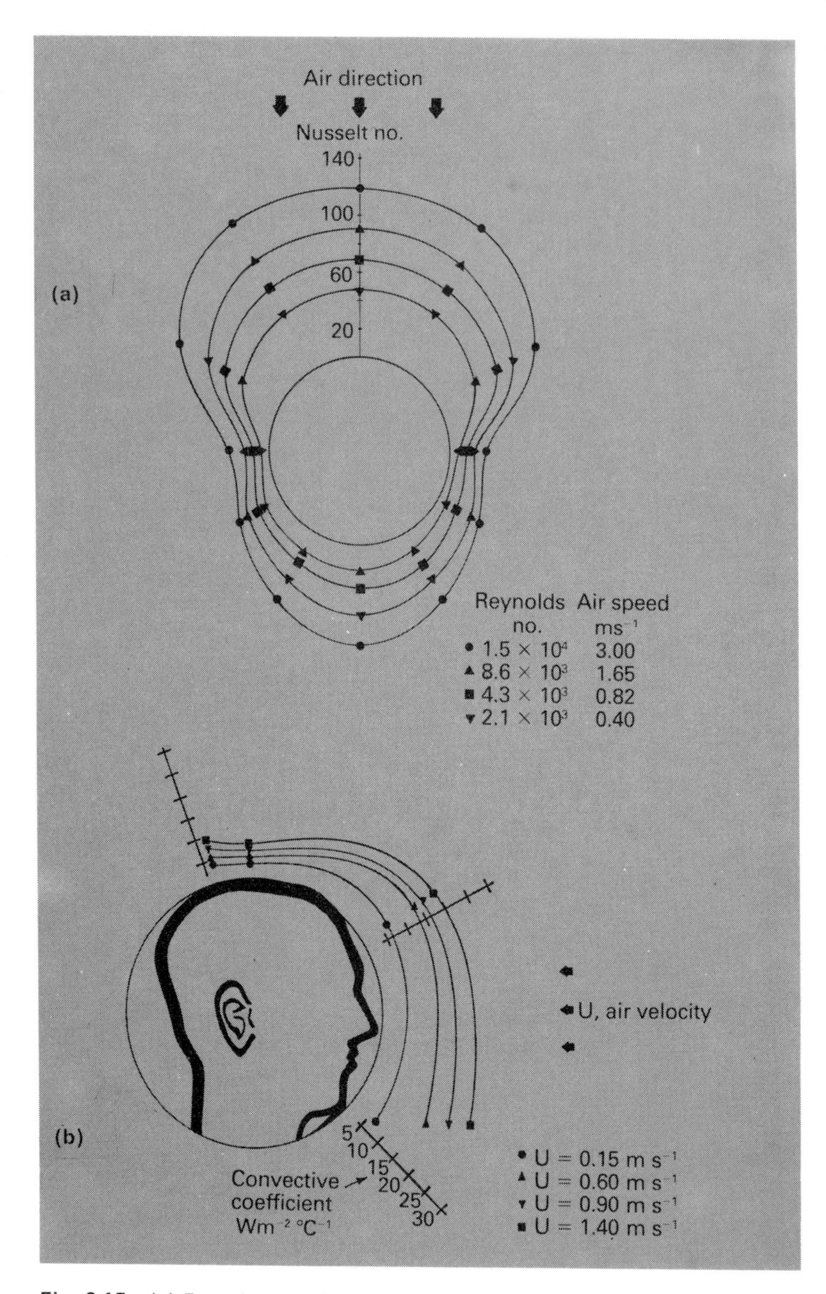

Fig. 3.15 (**a**) Forced convective heat transfer distribution around a vertical heated cylinder (Clark and Toy, 1975b, by permission of Journal of Physiology). (**b**) Local convective heat loss coefficient from the head in forced convection (Clark and Toy, 1975b, by permission of Journal of Physiology).

At low rates of air movement, the dependence of convective heat transfer on the temperature difference $(T_s - T_a)$ can be determined by comparing the Grashof (Gr) number with the square of the Reynolds (Re) number. Both Grashof and Reynolds numbers are dimensionless; the Grashof number is the ratio of a buoyancy force multiplied by an inertial force to the square of a viscous force, and the Reynolds

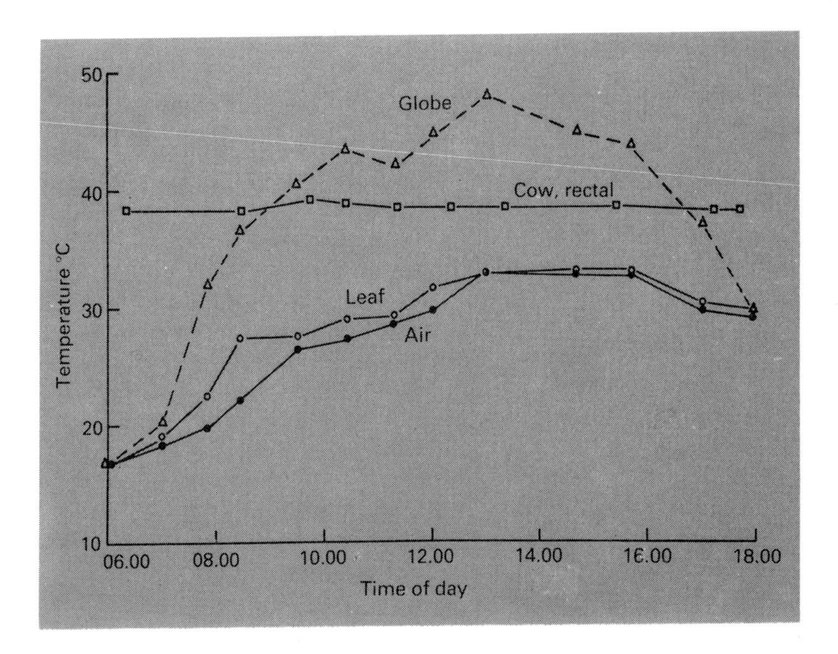

Fig. 3.16 The temperatures of a 6-inch black globe thermometer, English walnut leaves and a Jersey cow, exposed to full sunlight (on 28 July 1953, at Davis, California), compared with air temperature (Kelly, Bond and Heitman, 1954).

number is the ratio of inertial forces to viscous forces. The ratio Gr/Re^2 is then proportional to the ratio of buoyancy to inertial forces. When Gr/Re^2 is large, buoyancy forces are relatively large, and heat transfer is governed by free convection; when Gr/Re^2 is small, buoyancy forces are negligible and forced convection is dominant.[369]

In calculations made on convective heat transfer from the pig, the relations $Nu = 0.6\,Re^{0.5}$ and $Nu = 0.5\,Gr^{0.25}$ were used.[396] In this instance it was found that the forced convection coefficient at a windspeed of $0.2\,m\,s^{-1}$ was $3\,W\,m^{-2}\,°C^{-1}$ for the 60 kg pig and

$5\,W\,m^{-2}\,°C^{-1}$ for the 4 kg pig, increasing to 5 and $8\,W\,m^{-2}\,°C^{-1}$ at a windspeed of $0.5\,m\,s^{-1}$. The natural convection coefficients were approximately $3–4\,W\,m^{-2}\,°C^{-1}$ at environmental temperatures of 20–30°C.

The convection coefficients are therefore of the same order of magnitude as the $6\,W\,m^{-2}\,°C^{-1}$ often taken as a mean value for the radiation coefficient. For an environment where the air and mean radiant temperatures are equal to each other, this implies that the heat transfer rates by convection and radiation are approximately equal, when conditions are those of still air or low windspeed. Measurements made on pigs bear this out.[72,381]

The comparative effects of radiant and convective heat exchanges on three different types of body are illustrated in Fig. 3.16. The globe thermometer usually takes the form of a thermometer or thermocouple at the centre of a black-painted sphere that is commonly 15 cm in diameter.[38,85] Its temperature results from the combined effects of radiation and convection, and it shows a marked rise in temperature in full sunlight. The leaf's temperature closely follows that of the surrounding air as a result of the propotionately large convective loss due to its high surface-area/volume ratio, its movement and transpiration. The cow's deep body temperature remains at a steady level in consequence of thermoregulatory mechanisms involving increased evaporative heat loss and increased non-evaporative losses due to a high skin temperature.

CONDUCTION

Although heat transfer by conduction through contact with solid surfaces is not usually important for man, it can assume considerable proportions for animals that lie on the ground or on the floors of buildings and enclosures.[80] The thermal conductivity of the floor affects the rate of heat transfer, but another factor of considerable importance is its thermal capacity. The ratio of the thermal conductivity to the thermal capacity gives the thermal diffusivity which, as its name implies, indicates the rate at which heat diffuses throughout a given material.

When an animal lies on a cold floor, heat flows from its body and raises the temperature of the floor until an equilibrium is reached. The higher the thermal capacity, the more heat is required before equilibrium is determined by the thermal conductivity. However, if the thermal capacity is large the animal may have moved before equilibrium is reached, with the result that any insulation built into the floor is relatively ineffective. This situation commonly occurs with concrete floors that have underlying layers of insulation.

Heat loss from the pig to an insulated floor is often about equivalent per unit area to that lost from the free surface by radiation and convection.[386] A sheep living on cold, poorly-insulated ground may dissipate up to 30% of its minimum heat production by conduction,[178] which is comparable with heat loss per unit area from the free surface even under these conditions.

Small amounts of heat are also transferred to ingested food and drink. This heat of warming often amounts to about 3% of the animal's resting heat production. For example, if a pig weighing 20 kg drinks 2.5 litres of water at 20°C in one day, it imparts 200 kJ to raise the water through 19°C to the deep body temperature; this is 3% of the animal's heat production of 6700 kJ per day. At higher environmental temperatures, although the water temperature is higher and therefore requires less warming to body temperature, the pig's water consumption is increased.[386] In the sheep, the heat of warming both food and water to body temperature has been found to amount to 2–3 W, which is about 3% of the total heat loss needed to balance a heat production of about 100 W.[63]

CALORIMETRY

Calorimetry is the term applied to the measurement of heat transfer. Its significance for man and animals lies in the methods that have been devised for measuring heat production in the organism and the associated heat loss to the environment. Calorimetric methods as they are applied to the whole animal therefore have to take account of the various aspects of metabolism and heat transfer that have been discussed in the preceding and present chapters. The practical use of animal calorimetry lies in drawing up an energy balance that allows estimates to be made of energy requirements for maintenance, activity and growth.

The first useful measurements of animal heat were made in the second half of the eighteenth century, in conjunction with the development of calorimetry in the study of chemical combustion. The early development of animal calorimetry is interesting in illustrating the relation between theory and experiment in the evolution of a new approach. This can be best appreciated by comparing the early work of Crawford in Scotland with that of Lavoisier in France.

Adair Crawford[120, 121] carried out experiments in calorimetry in Glasgow in the summer of 1777, and described his work at a meeting of the Royal Medical Society in Edinburgh the following winter. He was a pupil of Joseph Black, and he made his calorimeter at Priestley's suggestion. The calorimeter depended in principle on the measurement of the rise in temperature of a jacket of water surrounding the reaction chamber, or animal chamber. Since heat loss from the calorimeter was largely prevented by insulation, the rise resulted from heat flow from the chamber. Crawford demonstrated that:

> 'the quantity of heat produced when a given quantity of pure air is altered by the respiration of an animal is nearly equal to that which is produced when the same quantity of air is altered by the combustion of wax or charcoal, and that when an animal is placed in a cold medium, it phlogisticates a greater quantity of air in a given time than when it is placed in a warm medium.'

At this time the phlogiston theory was still accepted, and the theoretical treatment of experimental results suffered accordingly. It was a great theoretical advance, therefore, when Lavoisier and Laplace[313] translated combustion into terms of oxygen consumption, and soon after, with their ice calorimeter, established a sound basis for animal calorimetry in relation to chemical combustion. The ice calorimeter depended on the collection of water from melting ice in a jacket surrounding the animal chamber. Haldane[192] compared Lavoisier's calorimeter unfavourably with Crawford's apparatus on two counts: first, the error involved in collecting water from melting ice was greater than that in measuring a rise in temperature in a water jacket; and, second, the animal was necessarily exposed to a cold environment. Haldane commented:

> 'It would thus appear that Crawford's method was in his own hands a very exact one. It seems exceedingly doubtful whether any observer up to the present (1889) has obtained more accurate results in animal calorimetry.'

However, whereas Crawford lived on with the phlogiston theory and its attendant obstruction to progress, Lavoisier advanced to 'oxygène' and the liberation of heat by chemical reaction. Crawford had come to the conclusion that the liberation of heat was due to changes in heat capacity of the reacting substances, and wrote (Crawford[121] pp. 370–1): 'We may conclude, therefore, that the sensible heat, which is excited in combustion, depends upon the separation of absolute heat from the air.' This publication was subsequent to that of Lavoisier and Laplace; Crawford discusses their concept of animal heat being due to 'chemical decomposition'. He disagrees with their view, and goes on to say (p. 380 of his book): 'Hence we may conclude, in general, that the evolution of heat from the air, in consequence of an alteration in its capacity, is the true cause of animal heat, and of that which is produced by the inflammation of combustible bodies.'

Although Crawford failed to understand the true significance of what was taking place, he demonstrated the increase in heat production that takes place in a guinea-pig in the cold. In spite of this achievement, however, he was unable to take the necessary step for the advancement of his subject because he could think only in terms of entrenched theory. It was left to Lavoisier to establish the basis for future metabolic work. Thus although Crawford's observations were the more accurate, the development of metabolic studies in fact followed Lavoisier's recognition of oxygen and the realization of the similarity between chemical combustion and animal metabolism.[386]

These early measurements depended on direct calorimetry, that is the measurement of heat loss. Two main types of direct calorimeters have emerged. The one depends on absorbing the animal's heat loss and measuring it as a rise in temperature in the absorbing medium, as was the case with Crawford's calorimeter; the other depends on the measurement of the temperature difference produced across a layer surrounding the animal as the result of heat flow from the animal to its surroundings.

The most celebrated instrument of the first kind was the Atwater–Rosa–Benedict apparatus, in which heat was removed in circulating water.[18, 328] Similar methods were used by Armsby and Fries[14] for cattle, and by Capstick[91], Capstick and Wood[92], and Deighton[139] for pigs; the pig calorimeter was developed initially by Hill and Hill.[233] Also belonging to this class is the 'air' calorimeter of Kelly, Bond and Heitman[295], which is similar to that of Auguet and Lefèvre[19] in which the animals' heat loss is measured from the temperature rise in the ventilating air stream. Heat sink calorimeters of the circulating water variety have been developed for animals both singly and in groups,[112, 400] and subsequently for birds[421] and man.[128]

The other method of direct calorimetry was first used by Richet[440] and Rubner[452], who estimated the rate of heat flow from the temperature difference in concentrically arranged air spaces around the calorimeter. Day and Hardy[138] used the same principle for the measurement of heat loss from babies, and Prouty, Barrett and Hardy[425] made an animal calorimeter on similar lines. Their apparatus consisted of inner and outer copper cylinders, fixed in relation to each other. Heat flow from inside the inner cylinder produced a temperature difference across the layer of air between the two cylinders, so that in practice heat flow was determined as directly proportional to this temperature difference, which was measured by thermocouples. An electric fan blowing air on the outer cylinder maintained the outer shell at room temperature, and in this way the shell and fan constituted a heat absorber of large capacity.

The concentric shell type of apparatus was succeeded by the accurate gradient layer calorimeter of Benzinger and Kitzinger[45]; Pullar[428] has also described a gradient layer calorimeter, and, more recently, McLean[339]. Hammel and Hardy[198] have used a gradient layer calorimeter for a dog, coupled with oxygen and carbon dioxide measurement, for the assessment of thermoregulatory responses. In these calorimeters the temperature difference across a thin layer of material lining the inside of the chamber is measured by numerous thermocouples distributed over its whole area, so that the animal's heat loss is accurately integrated.

Another form of direct calorimetry involves the assessment of the partition of heat exchange. In an extensive series of papers, Winslow, Herrington and Gagge developed the basis of partitional calorimetry in its application to man.[171–2, 175, 177, 547–51, 553] Their method was basically to expose the subject to an environment in which the components responsible for heat exchange by convection and radiation could be varied independently one of the other. They were then able to write heat balance equations for different sets of conditions, and so to arrive at the heat exchange taking place through each of the channels of convection, radiation, and evaporation (conduction played only a small part in their experiments). Heat transfer coefficients calculated from the surface temperature and dimensions of the organism can also give approximate estimates of heat exchange.[397]

Distinct from these direct calorimetry approaches, which may be termed heat sink and thermal gradient methods, are the systems of indirect calorimetry, in which the material exchanges accompanying metabolism are used to compute an animal's heat production. These again fall into two main groups: one is respiration calorimetry; the second is dependent on the carbon and nitrogen analysis of food intake and may often be combined either with respiration calorimetry or with carcass analysis.[56]

Respiration calorimetry in the form of measurement of oxygen consumption is very widely used to determine heat production either in closed-circuit or open-circuit systems. Developed from the early apparatus of Reignault and Reiset,[432] the closed-circuit method in which the animal is totally enclosed in a chamber forming part of the gas circuit has led to the development and use of apparatus of high accuracy.[5, 58, 77] Open-circuit systems have evolved from designs originally used by Haldane and by Pettenkofer (see Kleiber[302]); the use of the diaferometer and other devices for measuring concentrations of oxygen and carbon dioxide is now making the open-circuit method both easier and more accurate in use. An 'open and shut' calorimeter has been designed and used for large animals.[61] The fall in oxygen content in the chamber is measured during a given period, and the chamber is then opened for a period during which the chamber is flushed out with fresh air. The cycles of opening and shutting are repeated continuously.

Heat production can be calculated from oxygen consumption using energy equivalents for the oxygen consumed, depending on the respiratory quotient.[531] The Brouwer[78] equation for calculating heat production is

$$M = (16.2 \times O_2) + (5.02 \times CO_2) - (2.17 \times CH_4) - (5.99 \times N)$$

where M = heat production (kJ); O_2, CO_2 and CH_4 represent volumes of oxygen consumed and of carbon dioxide and methane produced (litres) and N is the quantity of urinary nitrogen excreted (g). McLean[340] has shown that in open-circuit calorimetry the heat production of ruminants can be predicted to within $\pm 2\%$ solely from the measurement of oxygen concentration and ventilation rate:

$$M = 20.47\,QX$$

where Q = flow rate of outlet air, litres per unit time, and X = inlet-outlet fractional concentration difference for oxygen.

One of the difficulties with the indirect calorimeter is that the capacity of the chamber is usually large relative to the circulation of gas, so that there is a long equilibration period following the introduction of an animal into the chamber. This can be overcome using a predictive equation to give the expected asymptotic level of oxygen consumption (or carbon dioxide production) from the early changes in gas concentrations.[343]

The various methods of calorimetry used in the past, and those in current use, have been described adequately in many text-books and other publications; reference may be made to Lusk,[328] Brody,[77] Kleiber,[302] and Blaxter.[58] Attention is drawn to them here to indicate the range of approach to the measurement of heat production and heat loss in the whole animal, because this is obviously of the greatest importance in assessing the effects of environment on energy exchange.

4

Thermal Insulation

Thermal insulation impedes heat flow. Consequently, animals that are poorly insulated lose heat readily. For example, the shorn sheep has much less thermal insulation than the sheep with a fleece, so that to maintain its body temperature in a cool environment the shorn sheep needs to produce heat faster than the unshorn. The effects of this are to increase the food intake that is required for maintenance, to modify the animal's behaviour, and to increase its susceptibility to climatic variation and to disease.

TISSUE AND EXTERNAL THERMAL INSULATIONS

An animal's thermal insulation consists of two components in series: the tissue insulation deep to the skin surface and the external insulation superficial to the skin (Fig. 4.1). The tissue insulation is the resistance offered to the flow of heat between the heat-producing tissues and the skin surface; it largely depends on the thickness of subcutaneous fat. In addition, peripheral vasoconstriction and vasodilatation change the blood flow and lead respectively to increased and decreased tissue insulation. The external insulation is due principally to an animal's coat or man's clothing, and to the air-ambient insulation between the surface of the coat (or bare skin) and the environment. The air-ambient insulation is the resistance offered to heat exchange by radiation and convection between the organism and its surroundings.

The tissue insulation and external insulation together constitute the animal's core-environment insulation. This can be described in two ways, either as insulation per unit of surface area or as overall insulation per animal. The insulation per unit area, sometimes called the specific insulation,[302] is different for different parts of the body; the limbs of cattle and sheep, for example, are more poorly insulated

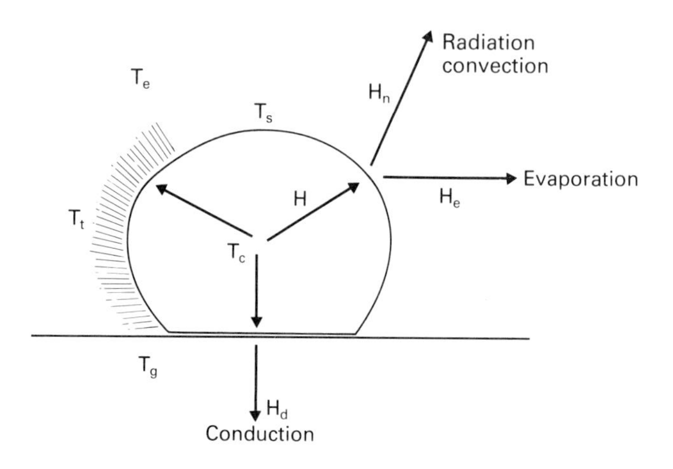

Fig. 4.1 Heat loss from animal to environment, passing through the internal tissue insulation between T_c (core temperature) and T_s (skin temperature), and the external insulation between T_s and T_e (environmental temperature). The external insulation includes the coat insulation between T_s and T_t (the outer coat temperature) and the air-ambient insulation between T_t and T_e, or T_s and T_e for a bare-skinned animal. $H = H_n + H_e$. Non-evaporative heat transfer (H_n) is by radiation and convection and by conduction to the ground (H_d) at temperature T_g. Heat transfer through the non-evaporative channels may constitute a heat gain to the animal under hot conditions. Evaporative heat loss (H_e), from the skin and the upper respiratory tract, depends on a difference in water vapour pressure and not on a temperature difference.

externally than the animal's trunk. Specific insulation also varies considerably between species, but, because it is related to unit area, comparisons can be made independently of body size.

This is not so with the overall insulation of the animal, which depends on body size. Posture also affects the overall insulation by modifying heat flow, depending on whether the attitude is flexed or extended. When an animal becomes active it exposes to the environment a larger effective area for heat exchange and increases air movement over its surface, so decreasing the insulation. The effects of activity in reducing overall insulation are accentuated by convective heat loss taking place more readily from those parts of the organism that have relatively small radii of curvature (see Chapter 3) so that the thermal insulation of the appendages is less per unit area than it is on the abdomen and thorax. For comparative purposes, insulation per unit area is the more useful quantity.

Insulation per unit area

Thermal insulation per unit area can be defined as a temperature difference divided by a heat flux density; in the animal, this might be taken as the core-environment temperature difference divided by the non-evaporative (or sensible) heat loss per unit of skin surface area. However, a difficulty now arises in that although heat is transferred necessarily as non-evaporative heat inside the body, dissipation of heat from the skin occurs through both the non-evaporative and evaporative channels. The non-evaporative heat flow per unit area that approaches the skin internally is therefore greater than the non-evaporative heat flow per unit area that leaves the skin externally. The change in heat flow at the skin surface (or at the mucosal surface in the respiratory tract) does not allow the core-environment insulation to be calculated in one step; it must be calculated in two steps, one for the internal insulation and one for the external, each with its own temperature difference and non-evaporative heat flux. The total insulation per unit area is then obtained as the sum of the internal and external specific insulations.

Simultaneous modes of heat transfer through a medium operate in parallel and are additive; for example, the non-evaporative and evaporative heat transfers from the body surface are additive to give the total heat loss. Similarly, thermal conductances and heat transfer coefficients (see Chapter 3) that act in parallel, such as those for radiation and convection, are additive to give combined values. Thermal insulation values, as the reciprocals of conductances, are additive when they are in series, as with the addition of tissue and external insulations to give the total core-environment insulation. If therefore radiative and convective insulation are used they must be combined as resistances in parallel, that is the sum of their reciprocals gives the reciprocal of the combined value.

HEAT FLOW AND INSULATION

The exchange of heat between the animal and its environment occurs in two modes, the non-evaporative and the evaporative. Non-evaporative heat loss predominates in the homeotherm in a cold environment, and evaporative heat loss under warm conditions. The dissipation of heat through the two modes can lead to the calculation of two corresponding resistances to heat flow, which may be termed the non-evaporative thermal insulation and the evaporative impedance.

Non-evaporative insulation

The core-environment insulation per unit area for non-evaporative heat flow, I, is given by

$$I = I_t + I_e$$

where I_t = tissue insulation; and I_e = external insulation. I_e includes the insulation due to the coat or clothing, and the air-ambient insulation.

The rate of non-evaporative heat flow is proportional to the thermal conductance of the medium through which heat is passing and to the temperature difference across it. This is a statement of Fourier's Law, which may be expressed as:

$$H = \frac{k}{L}(T_1 - T_2)$$

where H = rate of heat flow per unit area; L = thickness of medium through which heat is passing; k = thermal conductivity of medium; and $T_1 - T_2$ = temperature difference across the medium.

This can be re-written in a form corresponding to Ohm's Law for the flow of electricity, with electric current as the analogue of heat flow, voltage difference for temperature difference, and resistance for thermal insulation:

$$H = \frac{(T_1 - T_2)}{I} \qquad (4.1)$$

where I = L/k is the resistance to heat flow per unit cross-section area. Therefore,

$$I_t = \frac{T_c - T_s}{H}$$

$$\text{and } I_e = \frac{T_s - T_e}{H_n}$$

where T_c = core temperature; T_s = skin temperature; T_e = environmental temperature; H = total heat loss; and H_n = non-evaporative heat loss per unit area.

In the cold, when H_n is a large part of the total H, as a first approximation

$$(T_c - T_e) = HI$$

This was the approximation used in Table 2.2 to calculate the lowest environmental temperatures that could be withstood by the newborn piglet, lamb and baby when each was exerting the maximum cold-induced metabolic rate.

Newton's Law of cooling is sometimes invoked to describe heat loss from animals. However, this law is concerned with the cooling of a body, and it is an approximation in the form of

$$\frac{\Delta T}{\Delta t} \propto (T_s - T_e)$$

which states that the rate of fall of temperature ($\Delta T/\Delta t$) of a body is proportional to the temperature difference between the body's surface and the environment. This law is consequently not applicable to the dynamic steady state of an animal, where the body temperature does not fall progressively.[302]

Units of insulation

Thermal insulation per unit area can be expressed in terms of $°C\,m^2\,W^{-1}$. Corresponding terms in other units are $°C\,m^2\,hr\,kcal^{-1}$, $°C\,m^2\,d\,Mcal^{-1}$, $°F\,ft^2\,hr\,BTU^{-1}$, and so on; however, the use of SI units makes for easier calculation. Another unit of insulation that is sometimes used is the 'clo',[85] a unit chosen to represent approximately the insulation of normal indoor clothing worn by sedentary workers in comfortable surroundings. After allowing for an evaporative heat loss of 25%, such a person might lose 44 $W\,m^{-2}$ as non-evaporative heat. If the skin temperature is to be maintained at 33°C at an environmental still air temperature of 21°C, then the combined clothing plus air-ambient insulation is to be (33-21)/44 = 0.27 $°C\,m^2\,W^{-1}$. Of this, the air-ambient insulation is 0.11 to 0.12 $°C\,m^2\,W^{-1}$ and the clo unit is defined as 0.155 $°C\,m^2\,W^{-1}$.

Monteith[369, 370] has applied the concept of diffusion resistance to both non-evaporative and evaporative heat transfers. This application allows the heterogeneous collection of insulation units that has developed to be replaced by units of resistance, r. Insulation is a temperature difference per unit heat flux density and is equal to a resistance divided by a volumetric specific heat. For comparison with the resistance of the boundary layer around a body, the volumetric specific heat used is that of air, ρc_p, where ρ is density and c_p is specific heat at constant pressure; at 0°C this is 1.29 kJ m^{-3} $°C^{-1}$. From I $= r/\rho c_p$, for an insulation of 1 °C m^2 W^{-1}, $r = 1290$ s m^{-1} or 12.9 s cm^{-1}; the clo unit of 0.155 °C m^2 W^{-1} is then equivalent to a resistance of 2.0 s cm^{-1}. Resistance in terms of s m^{-1} is the reciprocal of velocity. Units of resistance are elegant and their use is unifying between different scientific fields; however, °C m^2 W^{-1} has a self-explanatory advantage that cannot be denied. The values in Tables 4.1 and 4.2 are given in both insulation and resistance units.

Evaporative impedance

Under hot conditions, evaporation from the skin surface and upper respiratory tract may be the only channel of heat transfer that is effective for heat loss and thermoregulatory control. Just as the non-evaporative insulation provides the resistance to non-evaporative heat flow, so an evaporative impedance can be calculated for latent heat transfer.

For evaporative heat transfer to take place, there must be a vapour pressure difference between animal and environment. If a high water vapour pressure is allowed to build up at the skin surface, vaporization of water on the skin is diminished and evaporative cooling decreases. An impermeable garment has this effect, thereby increasing the evaporative impedance and producing discomfort.

At the skin surface, the evaporative impedance, R_s, is given by an equation corresponding to equation (*4.1*), but with a vapour pressure difference in place of a temperature difference, and a loss of heat by vaporization per unit area, H_e, in a place of a non-evaporative heat flux:

$$R_s = \frac{P_s - P_a}{H_e} \tag{4.2}$$

where P_s = mean vapour pressure at the skin surface; and P_a = mean environmental vapour pressure, so that the units of evaporative impedance are mbar m^2 W^{-1}.

The impedance R_s is decreased by air movement and increased by coat or clothing. P_s depends on the 'percent wetted area', a concept which postulates that the vapour pressure over the wet area associated with a sweat gland is the saturated vapour pressure, and that over the intervening dry parts of the skin it is the environmental vapour pressure. On the basis of proportions it can then be shown that

$$P_s - P_a = \frac{h(P_{sat\,s} - P_a)}{100} \qquad (4.3)$$

where h = percent wetted area of skin, and $P_{sat\,s}$ = saturated vapour pressure at skin temperature over the wetted areas, so that from equations (*4.2*) and (*4.3*)

$$R_s = \frac{h(P_{sat\,s} - P_a)}{100\,H_e}$$

For respiratory evaporation, the impedance, R_r, can also be defined by an equation corresponding to equation (*4.1*)

$$R_r = \frac{P_{sat\,r} - P_a}{QU(Y_{sat} - Y_a)} \qquad (4.4)$$

where $P_{sat\,r}$ = saturated vapour pressure of the expired air; Q = respiratory ventilation per unit time; U = latent heat of vaporization of water; Y_{sat} = quantity of water vapour in saturated expired air, and Y_a = quantity of water vapour in ambient air.
From equation (*4.4*), R_r decreases as Q increases; in general terms this can be expected because the larger air volumes carry more water vapour.[85, 258]

TISSUE INSULATION

Turning now to the internal and external insulations associated with non-evaporative heat flow, the internal or tissue insulation will be considered first.

When a pig with a considerable layer of subcutaneous fat is exposed to cold, peripheral vasoconstriction is very effective in producing a high specific insulation. This was clearly shown by Irving's pigs[263–4] in Alaska, when tissue temperatures at various depths below the skin surface were measured using thermocouples sealed in hypodermic needles. At low environmental temperatures, the skin surface temperatures in these animals fell considerably compared with furred arctic mammals (Fig. 4.2) and the core temperature retreated deep to the skin surface (see Fig. 1.1).[1] However, in the furred polar bear, peripheral tissues also act as a significant part of the polar bear's insulation in a cold environment[414] (see p. 136). When there is no fat layer, as in the newborn pig, the increase in insulation is small on exposure to

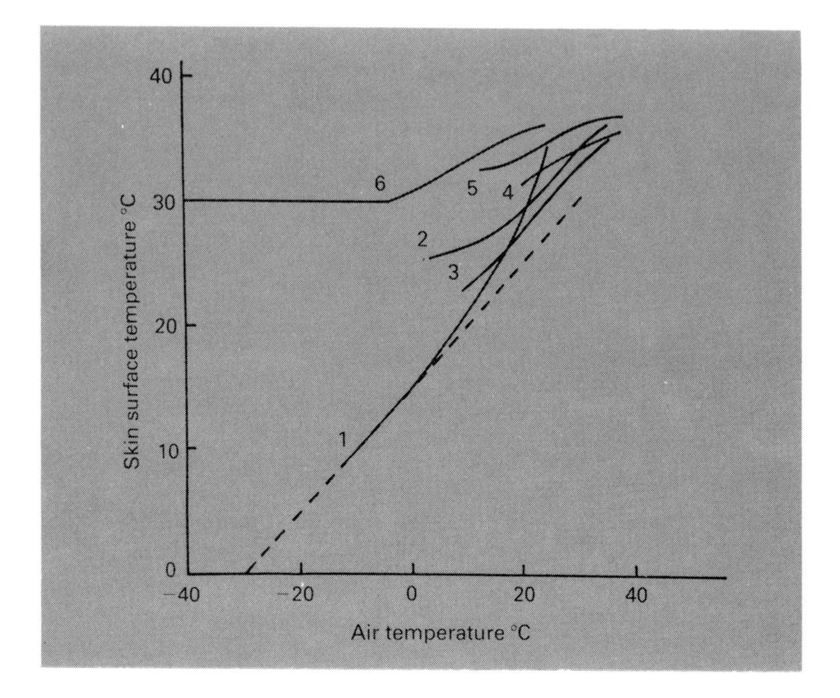

Fig. 4.2 Curves relating skin and air temperature in several mammals. 1, Alaskan pigs; 2, Californian pigs; 3, naked man working; 4, naked man resting; 5, clothed man; 6, thickly furred arctic mammals (Irving, 1956a, by permission of American Physiological Society).

cold although peripheral vasoconstriction occurs.[382] Although the cooled outer layer of tissue acts as an insulator it is not as effective as fat, which has a lower thermal conductivity than other tissues.[85]

Fat is a functional insulator in a naked species like the pig[248] because it works well in the cold, and yet offers little thermal insulation in hot conditions when it is bypassed by blood flowing from deeper to surface tissues through greatly dilated blood vessels (shown for man by Miller and Blyth[363]). The flow of blood in the skin depends partly on whether the arterioles are open to allow blood to pass to the capillaries and partly on whether the arterio-venous anastomoses are open or closed. When the anastomoses are open, they short-circuit the blood flow from artery to vein, so increasing the flow of blood and leading to local warming.[258]

When an animal's poorly insulated appendages, ears and lower parts of the limbs, are exposed to low temperatures, the skin temperatures of the appendages fall to low levels not far above freezing point. Cold-induced vasodilatation then occurs, a phenomenon in which the superficial blood vessels of the part dilate with an accompanying increase in blood flow and rise in skin temperature, in a cyclical fashion (Fig. 4.3). This response has survival value in that frostbite damage to the tissues is prevented, at the cost of an increase in mean heat loss.

Table 4.1 The thermal insulation (°C m^2 W^{-1}) and equivalent resistances (s cm^{-1}) offered by the tissues of various species (from Blaxter, 1977).

	Thermal insulation		Thermal resistance	
	vasoconstriction	vasodilatation	vasoconstriction	vasodilatation
Man	0.10	0.03	1.3	0.4
Steer	0.14	0.04	1.8	0.5
Calf	0.09	0.04	1.2	0.5
Pig (3 months)	0.08	0.05	1.0	0.6
Down sheep	0.08	0.03	1.0	0.4

Tissue insulations for man, cattle, pig and sheep are given in Table 4.1.

Thermal circulation index

In a hot environment, tissue insulation becomes very small due to peripheral vasodilatation, with the result that skin temperature approaches the core temperature. This is the basis for the formulation of the thermal circulation index (TCI):

$$\mathrm{TCI} = \frac{\text{external insulation}}{\text{internal insulation}}$$

$$= \frac{(T_s - T_e)/H_n}{(T_c - T_s)/H}$$

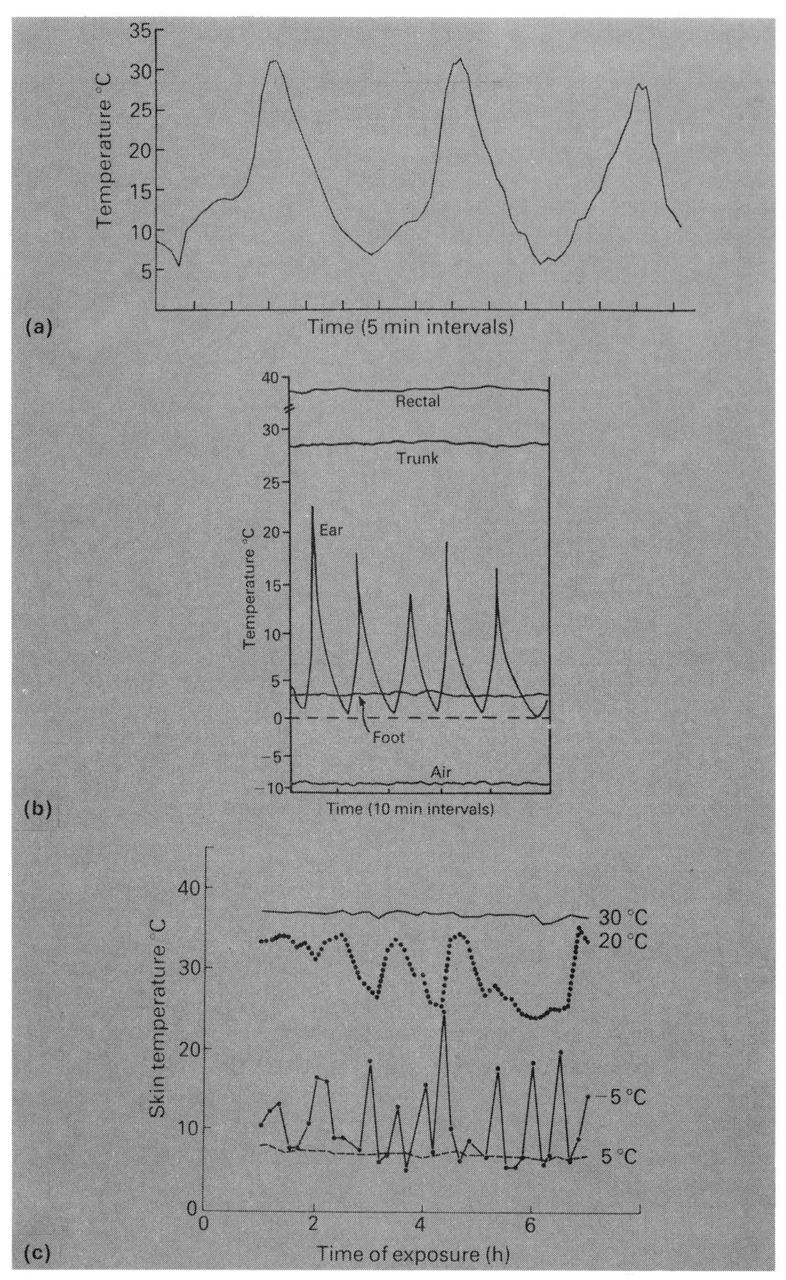

Fig. 4.3 (**a**) The variations in skin temperature on the pinna of the ear of the pig at an environmental temperature of −5°C (Ingram, 1964, by permission of Research in Veterinary Science). (**b**) Rapid fluctuations of skin temperature of the ear in a sheep exposed to cold (Blaxter, 1965b, by permission of the Journal of the University of Newcastle-upon-Tyne Agricultural Society). (**c**) Skin temperature of the ear of the ox at different air temperatures. The air temperatures are shown to the right of the figure (Whittow, 1962, by permisssion of Cambridge University Press).

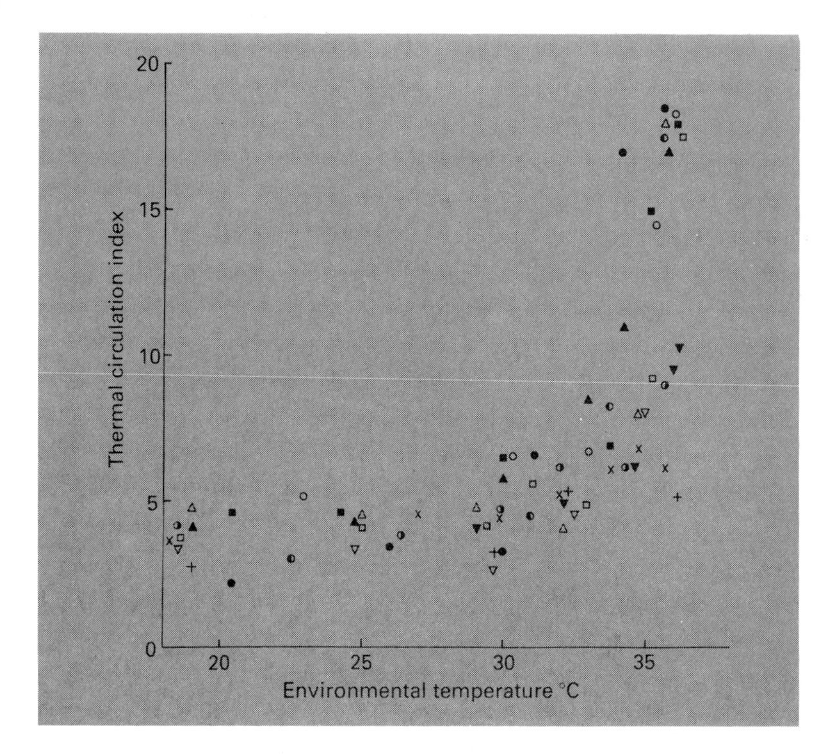

Fig. 4.4 The relation between the thermal circulation index and environmental temperature for twelve baby pigs up to 9 days old; each pig is represented by a different symbol. The skin temperatures from which the index was calculated were mean values of measurements made on the head, both flanks, back and abdomen (Mount, 1964b, by permission of Journal of Physiology).

When H_n is approximated to H, the expression becomes[85, 226]

$$TCI \simeq \frac{T_s - T_e}{T_c - T_s}$$

TCI in the young pig shows a sharp upward turn above an environmental temperature of 33°C (Fig. 4.4), marking the vasodilatation that occurs at the critical temperature. This accords with a critical temperature of about 34°C found from measurements of metabolic rate.[376, 386] TCI does not require the measurement of metabolic rate, but it is valid only under equilibrium conditions and not when temperature is fluctuating.

Counter-current heat exchange

Counter-current heat exchange between arteries and veins acts effectively to impede heat loss. It diminishes heat loss by transfer of heat from arterial to venous blood before the arterial blood reaches the surface of the animal. This depends on the vascular arrangements in the limbs and occurs without mixing of arterial and venous blood. Although many species that are adapted to a cold environment have thick fur coats, their legs and appendages are poorly insulated and, if the animal were to keep these warm, heat loss would be very great (see Fig. 3.2).

Examples of the reduction in heat loss in the cold brought about by counter-current heat exchange, but still with high rates of heat dissipation under warm conditions, are provided by some birds.[265] At low environmental temperatures, down to −10°C, less than 10% of metabolic heat is lost from the legs of herons and gulls, whereas at +35°C almost the entire heat production is dissipated through the legs. The naked legs of these birds are very important heat exchangers (see Chapter 12). In water they lose heat four times faster than in air at the same temperature.[480]

In man, the temperature in the brachial artery can fall by as much as 0.3 $°C\,cm^{-1}$ as a result of heat transfer to venous blood,[37] and a corresponding heat exchange may occur in the legs of pigs and other animals. Counter-current heat exchange contributes to the maintenance of a cooler shell of tissue around the warmer body core, in this way reducing heat loss to a cold environment and so effectively increasing the total thermal insulation (see Fig. 1.1).

Another example of counter-current heat exchange occurs in the scrotum. In most animals, the testes are supported in the scrotum external to the body cavity, and consequently their temperature is lower than that of the body core; spermatogenesis in most mammals is suppressed at the core temperature. The lower temperature in the scrotum is partly due to the cremaster and dartos muscles allowing the testes to drop lower at high environmental temperatures, and partly to counter-current heat exchange between the coiled artery and veins, so that the arterial blood reaching the testes is below body temperature. In the ram, for example, measurements at 21°C environmental temperature have shown that the testes are at a temperature 6°C below that of the core.[518]

The carotid rete may function as a heat exchanger. In its course to the brain, the carotid artery in some species breaks up into a system of small branching and joining blood vessels known as a rete; it was compared by Galen to a fisherman's net (Fig. 4.5). The rete lies inside

T_a = air temperature and D = insulation-wind-decrement:

$$\begin{aligned} H(I_1 + I_2 - D) &= \bar{T}_s - T_a \\ H(I_1 + I_2) &= \bar{T}_s - T_a + HD \\ &= \bar{T}_s - (T_a - HD) \end{aligned}$$

HD, the product of heat flow and the insulation-wind-decrement, is the thermal-wind-decrement, which indicates by how much the air

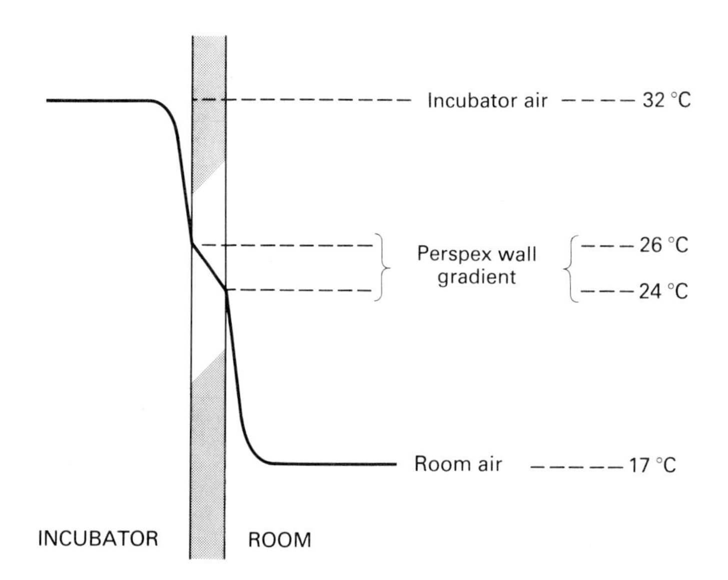

Fig. 4.7 Temperature gradients across the Perspex wall of a heated incubator; the thickness of the boundary layers is not drawn to scale (Hey and Mount, 1966, by permission of The Lancet).

temperature, T_a, would have to be decreased under still-air conditions to induce a heat loss equal to that occurring with the given wind.

The convective part of the air-ambient insulation is associated with the boundary layer of air at the surface of a body.[369] The boundary layer is a layer of air which offers convective insulation because it is stable in a direction normal to the surface.[106] Direct visualization by Schlieren photography (see Fig. 3.11) has shown that the layer is not stationary in a direction parallel to the surface, but that it streams over the surface in consequence of natural convection due to warming by the organism. This does not prevent the layer acting as thermal insulation in a direction normal to the surface. The thickness of the boundary layer depends on the roughness, shape and size of the surface, and on the rate of air movement. It is this boundary layer that confers insulation on a thin dividing partition that has little insulation due to its own material, and it is in the boundary layer that temperature gradients occur (Fig. 4.7).

5

Climatic Zones and Assessment of Thermal Environment

The major industrial developments of the world are in the temperate regions, where climatic conditions do not impose severe demands and where there is a high level of agricultural productivity. The polar regions are cold and barren and the equatorial region includes the hot wet and hot dry areas.

Types of climate are classified in Table 5.1. Some of these will be discussed in more detail in relation to solar radiation, temperature and water.

SOLAR RADIATION

The total radiant power received outside the earth's atmosphere and on a plane at right angles to the sun's rays is termed the solar constant; the best estimate of its value is $1360\ \mathrm{W\,m^{-2}}$ (W = watt). The waveband of visible light is 0.4–0.7 μm; this is concerned with photosynthesis in plants, vision in animals and photoperiodic effects in both plants and animals. The atmosphere scatters the blue light (the shorter wavelengths of the waveband) more than the red (longer wavelengths), so that the sky appears blue, and the setting and rising sun appears red. The potentially harmful effects of ultraviolet radiation are diminished both by its absorption by a layer of ozone in the stratosphere and by scattering in the atmosphere.

The solar radiation that is received at the earth's surface depends both on the elevation of the sun above the horizon and on the

Table 5.1 Types of climate as classified by Köppen (Gates, 1968, in Adaptation of Domestic Animals, ed. E. S. E. Hafez, by permission of Lea and Febiger).

Climate		Temperature	Rainfall and vegetation	
Hot, rainy	Tropical forest	Coldest month above 18°C; annual variation <3°C	Heavy rain; minimum 6 cm/month	Dense forest; heavy undergrowth
Hot, dry	Tropical savanna	Above 18°C, annual variation <12°C	Dry season with <6 cm/month	Open grasslands; scattered trees
Dry	Steppe	Variable winters cold	Precipitation <50 cm/year Pronounced wet and dry season	Short grass
	Desert	High summer temperatures Daily variation high	Very dry Few rain showers	Sparse or non-existent
Medium temperature, humid	Warm and dry winter (monsoon)	Coldest month between +18°C and −3°C; warmest over 10°C	Wet summer	Forested
	Warm and dry summer (Mediterranean)		Dry summer	
	Warm and humid	Coldest month between +18°C and −3°C; warmest over 10°C	Ample rainfall throughout year	Cotton, wheat and corn
Cold, humid	Cold—no dry period	Mean temp. of summer and winter 10°C and −3°C	Precipitation throughout year (Canada)	
	Cold—dry winter		Little rainfall in winter; (only represented in N.E. Asia)	
Polar	Tundra	Mean of warmest month less than 10°C; frozen soil throughout year		Moss and small shrubs
	Ice cap	Mean temperature of warmest month below 0°C		

scattering of sunlight by the atmosphere, including the effects of water droplets and ice particles in clouds. The spectral distribution of direct and indirect solar radiation is illustrated in Fig. 5.1.[180, 369]

The albedo of the earth's surface, that is the proportion of the incident solar radiation that is reflected, varies from place to place and with time of year. Snow cover makes a considerable difference; for

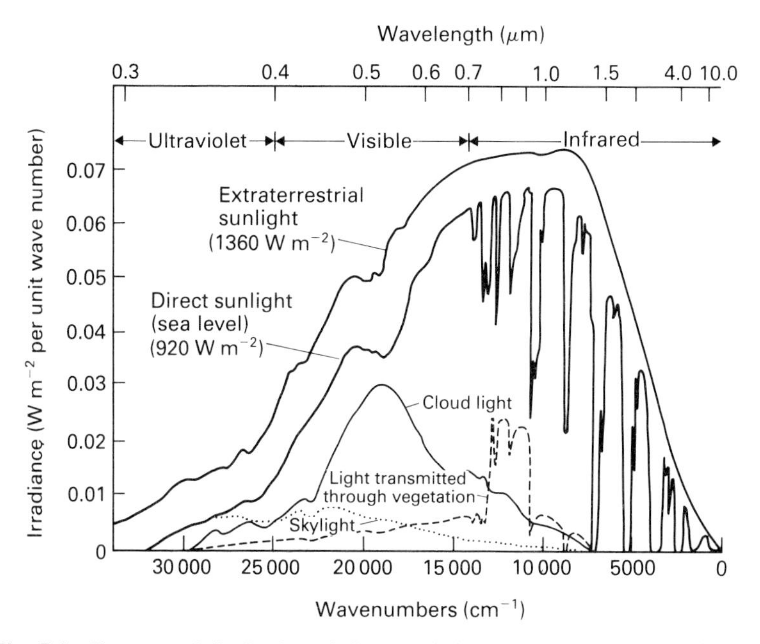

Fig. 5.1 The spectral distribution of direct sunlight, skylight, cloudlight, and light transmitted through vegetation as a function of the frequency of the radiation in wavenumbers. Wavelength is given at the top (Gates, 1968, from Adaptation of Domestic Animals, ed. E. S. E. Hafez, by permission of Lea and Febiger).

example, at Leningrad the value is 0.2 in the summer and 0.7 in the winter.[84] Mean values of albedo for different areas of the land surface in different seasons are given in Table 5.2.

CLIMATIC ZONES

Much of the earth's surface between 23° north and south of the equator tends to be hot and humid and constitutes the tropics. The great deserts lie between 15° and 30°, with the wet tropics on one side and the wet temperate zones on the other. The areas near the equator have two rainy seasons and two relatively dry seasons; between latitudes 5° and 15° there is a monsoonal sequence of hot wet and hot dry seasons, with rain in summer, but food and water are scarce in the dry period. Between 30° and 45° there is predominantly winter rainfall

Table 5.2 Mean values of albedo (Budyko, 1974, by permission of the Copyright Agency of the U.S.S.R.).

Type of surface		Albedo
Stable snow cover in high latitudes (more than 60°)		0.80
Stable snow cover in middle latitudes (less than 60°)		0.70
Forest with stable snow cover		0.45
Unstable snow cover in spring		0.38
Forest with unstable snow cover in spring		0.25
Unstable snow cover in autumn		0.50
Forest with unstable snow cover in autumn		0.30
Steppe and forest during the transition period after snow cover disappears and before the mean daily temperature of the air reaches 10°		0.13
Tundra without snow cover		0.18
Steppe or deciduous forest during the period after the temperature goes above 10° in spring and before snow cover appears in autumn		0.18
Needleleaf forest during the period after the temperature goes above 10° in spring and before snow cover appears in autumn		0.14
Forest dropping leaves during a dry season, savanna and semi-desert	During the dry season	0.24
	During the wet season	0.18
Desert		0.28

with hot dry summers. In the cooler zones there is usually a shortage of food in winter, but water is available throughout the year. Beyond 60° there is a cold desert, with neither food nor water readily available.[347, 350]

Wet tropics

In the wet tropics, with an average temperature between 20 and 30°C, the peak of radiation intensity occurs at noon, but shelter from the sun is provided by vegetation and cloud. Wind velocities are often very low (except for violent storms), averaging 1.5–2.5 $m\,s^{-1}$ in the open and less than 0.5 $m\,s^{-1}$ in dense vegetation. Since air movement influences both convective and evaporative heat transfer, its absence in a hot humid environment leads to decreased heat dissipation and increased discomfort. Temperatures during the day do not usually rise much above about 33°C, but at night do not often drop below 15–20°C; daytime relative humidity is 50–75% and at night the relative humidity rises to 100%.

Deserts

The deserts of the world fall into three classes, cold, warm and hot, all characterized by low rainfall and sparse vegetation. What vegetation does exist is scattered and in the hot desert livestock are affected adversely by high daytime air temperatures and very high solar radiation levels, which can exceed 1000 $W\,m^{-2}$. The arid regions occupy about 20% of the world's land surface, and the semi-arid constitute a further 15%.

Although it is true that the primary characteristic associated with 'desert' is relative lack of water, to consider the climate of the world's deserts only in terms of water shortage is to make an incomplete assessment. The survival of animals and plants in the desert is dependent on their ability to keep cool and avoid desiccation, and is influenced to a large degree by the 24-hourly fluctuation in temperature which can be very marked, reaching 30°C or more in amplitude. This fluctuation is produced by clear skies with high intensities of solar radiation during the day and a high rate of re-radiation from the earth's surface at night.

A hot day in the desert may have an air temperature of 40–45°C with a relative humidity of 10%, although the maximum temperature can rise to 65°C in steeply walled valleys. The nights are cool, with temperatures averaging 15–20°C, although on occasion these drop below freezing point; when the night temperatures fall below the dew-point condensation occurs. The surface temperature ranges from as high as 70°C in some areas during the day down to 0°C at night, although such extremes are not likely to be found in one place, where the average daily range would not exceed about 50°C. Wind velocities are commonly 5–8 $m\,s^{-1}$ and may average 10–15 $m\,s^{-1}$; the dry winds at high air temperatures lead to desiccation. The average desert humidity (Table 5.3) is 15–20 mbar vapour pressure, and does not generally rise above 25–30 mbar; at an air temperature of 45°C this produces a relative humidity of 20–30%. In the driest deserts the vapour pressure may fall below 1 mbar.[258, 316]

Table 5.3 Illustrative temperatures and vapour pressures in deserts (Lee, 1964, by permission of American Physiological Society).

	Yuma, Arizona	Baghdad, Iraq	Aden, Arabia	Teheran, Iran	Karachi, Pakistan	Tabelbala, Algeria(t)† Beni Abbes, Algeria(p)†	Aswan, Egypt	Tombouctou, Sudan(t)† El Oualadj, Sudan(p)†	Fort Lamy, Chad	Cloncurry, Queensland	Laverton, W. Australia	Tenants Ck., N. Australia
Temperatures, °C												
Mean maximum, hottest month	41	43.5	36.5	37	34	44	41.5	43	41.5	38	35.5	37
Mean minimum, hottest month	25	24.5	29	22	28	28	25.5	26.5	23.5	24.5	20.5	24
Mean maximum, coldest month	19.5	15.5	28	7	25	19.5	23.5	30.5	33.5	25	18	24.5
Mean minimum, coldest month	5.5	4	22	−3	13	6.5	10	13	14	10.5	5	10.5
Vapour pressures, mbar												
Mean, ‡hottest month	21	9	33	16	30	11	14	26	16	18	13	16
Mean, ‡coldest month	6	6	24	8	6	6	8	9	8	7	8	8

† Temperature and vapour pressure measured at different stations.
‡ Where only a few measurements are made daily, the morning (7–9 am) values are given.

Polar and sub-polar regions

These regions are characterized by cold and the accompanying cold stress for animals can be severe. Antarctic weather, for example, is reported as typically very cold and windy, with correspondingly high levels of wind chill (see Chapter 7). At one base, the lowest temperature, in August, was −55°C, and the highest temperature, in December, was +3°C.[449] These are extreme conditions, but in sub-polar regions in the northern hemisphere conditions are also highly inimical to terrestrial animal life.[85]

THE ASSESSMENT OF THE THERMAL ENVIRONMENT

There have been many attempts to formulate a physiological temperature scale so that the thermal characteristics of a given environment could be referred to as a single quantity in relation to a living organism. The demand made by the environment on an animal to produce or to dissipate heat, and so to regulate its body temperature, depends on the factors that influence heat exchanges. Is it possible to integrate these components of the environment into a single indicator of environmental thermal demand? The problem is important for man's thermal comfort and for the definition of the partially controlled environments in which many animals are now kept.

However, equivalence between environments with different radiant-convective combinations, for example, would apply only to a particular body owing to the laws that govern heat transfer by radiation and convection. A small sphere would be more sensitive than a large one to a given change in windspeed, although coefficients can be used to relate the results obtained with the smaller sphere to the effect on the larger sphere (see Chapter 3). The alternative is to estimate heat transfer through each of the four channels of radiation, convection, conduction and evaporation separately, and then to combine the results to give the total thermal effect of the environment.[283, 391]

For man, Burton and Edholm[85] describe a method of determining an 'equivalent still-air temperature'. This is derived by adjusting air temperature by subtracting a thermal-wind-decrement that allows for the effect of air movement (see p. 96). In an increment of temperature is now added to allow for radiation from the sun, or elsewhere, the 'equivalent still-shade-temperature' is estimated. These formulations have considerable convenience in assessing man's insulation requirements in differing environments.

An 'effective temperature scale' was developed as a sensory scale of warmth combining air temperature, air movement and humidity into a single index.[559–61] The numerical value of the scale is the temperature of still air, saturated with water vapour, that induces a sensation of warmth or cold like that of the given condition. A 'corrected effective temperature' was also introduced which made allowance for radiant heat.[38] The globe thermometer and the wet bulb are referred to in Chapter 3. Simulated animals have also been constructed and used for assessing thermal environments.[429,523]

OPERATIVE TEMPERATURE

As distinct from the sensory effective temperature scale, the operative temperature scale[172] provides a physical measure of the thermal environment. Operative temperature combines as a single variable the temperature equivalents of the radiant and convective environments. This is achieved by using coefficients to relate radiant and convective heat exchanges to the differences between skin temperature on the one side, and wall and air temperatures on the other. It is possible to estimate radiant exchange with only a small error from a temperature difference, instead of from the difference between the fourth powers of the absolute temperatures (see Chapter 3), when the temperature differences are small:

$$T_o = \frac{K_R \bar{T}_r + K_C T_a}{K_R + K_C} \qquad (5.1)$$

where T_o = operative temperature; K_R = coefficient of heat transfer by radiation; K_C = coefficient of heat transfer by convection, for a given air movement rate; $\bar{T}_r$ = mean radiant temperature; and T_a = air temperature.

This expression does not take account of variations in air movement. The convective heat transfer coefficient is approximately proportional to the square root of the windspeed under conditions of forced convection (see Chapter 3) and the following equation from Gagge's work includes this factor:

$$T_o = \frac{K_R}{K_O}\bar{T}_r + \frac{K_C}{K_O}\left[\sqrt{\left(\frac{V}{V_o}\right)}T_a - \left\{\sqrt{\left(\frac{V}{V_o}\right)} - 1\right\}\bar{T}_s\right] \qquad (5.2)$$

where $K_O = K_R + k_c\sqrt{V_o}$

and $K_C = k_c\sqrt{V_o}$

K_O = standard cooling rate, about 7 W m^{-2} °C^{-1} for man resting in low air movement,[173] and V_o = standard air movement rate.

Equation (5.2) involves the mean skin temperature, $\bar{T}_s$, since the equation describes the equivalent temperature in which a subject would lose the same amount of heat at a standard cooling rate, K_O, as by radiation and convection in the original environment. Operative temperature can therefore be defined in terms of an imaginary environment with uniform air and radiant temperatures with which the subject would exchange the same heat by radiation, convection and conduction as in the actual complex environment containing the subject.

Operative temperature is thus a calorimetrically derived temperature scale, inclusive for the non-evaporative modes of heat transfer but not for evaporative heat loss. Evaporative loss must be assessed separately. A practical disadvantage in using the scale is that the mean skin temperature must be known.

Effective radiant flux

When the mean radiant temperature differs from the air temperature, that part of the radiant heat exchange that is associated with the difference can be determined as the ***effective radiant flux***.[174] This is derived from the heat exchange equation at equilibrium:

$$M_o = K_R(\bar{T}_s - \bar{T}_r) + K_C(\bar{T}_s - T_a) \qquad (5.3)$$

where M_o = metabolic rate less evaporative heat loss, heat storage and external work done (see Chapter 1); K_R = radiation coefficient, W m^{-2} °C^{-1}; K_C = convection coefficient, W m^{-2} °C^{-1}; $\bar{T}_s$ = mean skin temperature, °C; $\bar{T}_r$ = mean radiant temperature, °C; and T_a = air temperature, °C.

Equation (5.3) may be re-written:

$$M_o = K(\bar{T}_s - T_a) - K_R(\bar{T}_r - T_a)$$

where $K = K_R + K_C$ is the combined coefficient for radiant and convective heat transfer. The term $K(\bar{T}_s - T_a)$ describes heat loss to an environment in which air and mean radiant temperatures are equal. The term $K_R(\bar{T}_r - T_a)$ is the effective radiant flux. It constitutes an increased heat loss from the animal when $\bar{T}_r$ is below T_a, and a heat gain by the animal when $\bar{T}_r$ exceeds T_a.

The effective radiant flux and the combined coefficient can be used to relate the air temperature to the operative temperature:[117, 552]

$$T_o = T_a + \frac{K_R(\bar{T}_r - T_a)}{K} \qquad (5.4)$$

where T_o = operative temperature, combining the weighted effects of both mean radiant and air temperatures for a given convection coefficient.

The operative temperature can be calculated if K_R and K are known since $\bar{T}_r$ and T_a are measured. K_R is given by $4\sigma T^3$ (see Chapter 3).

EQUIVALENT EFFECTS OF ENVIRONMENTAL VARIABLES

The relative magnitudes of the heat transfer coefficients for radiation and convection (see Chapter 3) show that for many situations the thermal effect on an animal due to an increase of 1°C in the mean radiant temperature is approximately equal to the effect of an increase of 1°C in the air temperature. This was found to be the case in measurements of radiant and convective heat losses from the young pig.[381]

Lee and Vaughan[317] used the physiological reactions (rectal temperature, pulse rate and sweating) of man under desert conditions to establish the increment of air temperature equivalent to the heat load due to solar radiation. At an air temperature of about 45°C, they found that an increment in air temperature of approximately 8°C was required to produce the same effect as solar radiation. In other experiments, Gagge, Stolwijk and Hardy[176] exposed unclothed human subjects to a variable source of thermal radiation, and measured evaporative loss continuously while the ambient temperature was varied over the range 15–30°C. The subject was allowed to choose a heater setting necessary for comfort and in this way the man was used as a radiometer. As a result, it was possible to determine heat exchange by radiation in a complex radiant environment, and to derive an environmental temperature scale that is useful for comparing physiological responses under widely different physical conditions.

When subjects are allowed to select the radiation intensity in this way, the results can be used to test how uniform an operative temperature is obtained at different air temperatures. Equation (*5.4*) relates the air temperature to the operative temperature, and if the operative temperature were held constant then the plot of the selected radiation intensity against air temperature would give a straight line with a slope of $-K$.[296] Gagge's[173] value for K is $7\,W\,m^{-2}\,°C^{-1}$, and the slope of the line in Fig. 5.2 is $-6.8\,W\,m^{-2}\,°C^{-1}$, close to the expected value. The intercept of about 29°C on the temperature axis is the preferred operative temperature, which is maintained by the

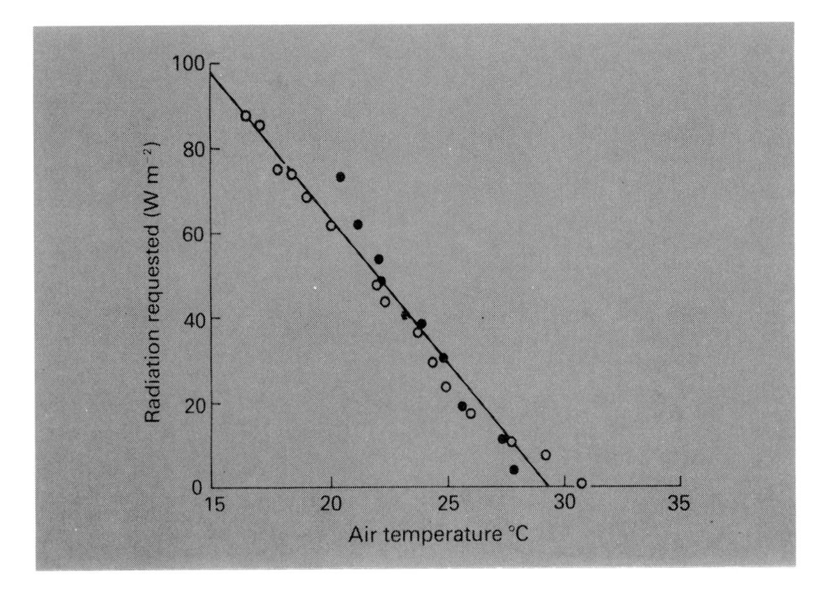

Fig. 5.2 Relation between air temperature and preferred radiation intensity. Subjects were allowed to select the radiation intensity, here expressed as the radiant heat gain, at various air temperatures in an environment of low air movement. Results obtained when the previous air temperature was lower (○), when it was higher (●). If the operative temperature selected by the subject is constant the line should have a slope of $-K\,W\,m^{-2}\,°C^{-1}$, the operative heat transfer coefficient. The slope is $-6.8\,W\,m^{-2}\,°C^{-1}$, close to the expected value. The intercept on the temperature axis is the preferred operative temperature (Kerslake, 1972, by permission of Cambridge University Press).

subject selecting the appropriate radiation intensity at lower air temperatures of 15 to 20°C.

Standardized environmental temperature

One of the concepts that recurs when methods of assessing thermal environments are discussed is that of a standard environment to which real environments can be related. A standard environment can be defined as one where the air and mean radiant temperatures are equal to each other, where there is only natural (free) convection and no forced air movement, where the humidity is held at a constant arbitrary level, for example at a relative humidity of 50%, and where, particularly with livestock, there is an unbroken and thermally insulated floor.

In such an environment the measurement of air temperature by itself gives the standardized environmental temperature. In all other conditions the air temperature differs from the standardized temperature to an extent depending on the other components of the environment that may affect an animal's heat loss, including mean radiant temperature, air velocity at animal level, and the nature of the floor. The extent to which variation in the several components of the thermal environment influences an animal's heat loss is illustrated by measurements that have been made of the critical temperature, defined as the air temperature at the lower end of the zone of thermal neutrality, for cattle exposed to different sets of conditions. A well-fed beef cow in dry calm weather with sunshine has a critical temperature of -21°C, whereas when it is exposed to overcast conditions, rain and a wind of $4.5\,m\,s^{-1}$, the critical temperature is $+2$°C.[524] For sheep with 100 mm fleece depth, Alexander[7] has calculated that the critical temperature is about 0°C in still air, rising to about 23°C in a wind of $7\,m\,s^{-1}$. The critical temperature is affected by the type of floor and bedding; in a group of 40 kg pigs it is 11.5–13°C on straw, 14–15°C on asphalt and 19–20°C on concrete slats.[515] These figures give some measure of the increase in air temperature that is required to compensate for increased heat loss through evaporation, convection and conduction. They emphasize the need to consider the combined effects of heat transfers through the different channels of heat exchange in influencing the total thermal effect of the environment on the animal.

In this connection, Heberden's words in 1826 (quoted by Monteith[370]) are highly appropriate:

> ...'it cannot have escaped the attention of any person moderately conversant with natural philosophy, that the index of a thermometer is a very imperfect measure of what I may call the *sensible cold*. ...For while the thermometer truly marks the temperature of the medium in which it is placed, the sensations of the body depend altogether upon the rapidity with which its own heat is carried off.'

Amongst other things, this passage underlines the need to

differentiate clearly between temperature on the one hand and heat flow on the other. A temperature measured by itself does not necessarily give any indication of heat flow; sensible heat flow is determined by the ratio of a temperature difference to a thermal insulation (see Chapter 4). The realization that temperature by itself is an inadequate description of the thermal environment might be considered a matter of common experience when one compares the subjective effects of a bitterly cold wind with the easily borne conditions when the temperature is the same but the day is calm. 'Wind-chill'[85] (see Chapter 7) is due to the decreased insulation of the air around the organism, and the thermal-wind-decrement can be calculated in the form of a temperature decrement (see Chapter 4).

This means that any attempt to establish a single temperature reading to describe a given thermal environment must take account not only of the several heat transfers but also of the animal to which it relates. For example, some animals are more highly insulated than others, and consequently their convective heat exchange is less affected by variations in air temperature and windspeed and their conductive losses to the floor are less marked.

An example is provided by the pig under farming conditions. The growing pig has relatively little coat and thermal insulation, and consequently tends to respond more markedly than some other animals to changes in temperature and other factors in the thermal environment, although this susceptibility is decreased to some degree by the huddling in the cold and spreading out in the warm that take place when pigs are grouped together (see Chapter 8). Under cold conditions, thermoregulatory heat production is increased, so that energy retention is depressed when food intake is controlled and consequently productivity decreases. To determine whether the environment is too cold for maximum productivity it is necessary to estimate whether the animal has to produce extra thermoregulatory heat in response to environmental demand (see Chapter 2).

This can be done by reducing the animal's actual thermal environment to an equivalent standardized environmental temperature (ESET); this is the temperature of the standardized environment, already referred to, which would exert the same effect on the energy exchange of the animal as the actual environment. ESET can then be used to relate results of calorimetric experiments to particular farming conditions. ESET is obtained by the algebraic summation of radiant and convective variations in the environment with the measured air temperature (Table 5.4), and is then compared with the appropriate effective critical temperature given in Table 8.5, where the values for heat loss, effective critical temperature and the increase in heat loss per

°C below the critical temperature, derived from calorimetric experiments under standardized conditions, are given for pigs of 35 kg mean weight in groups of different sizes.

If ESET assessed from Table 5.4 is below the critical temperature, the extent of the increased heat loss, corresponding to the deficit in energy retention, can be calculated from the values given in Table 8.5.[392]

Table 5.4 Approximate variations to be added to air temperature in °C to give the equivalent standardized environmental temperature (ESET) for a pig pen. $\bar{T}_r$ = mean radiant temperature, °C; T_a = air temperature, °C. The magnitudes of the convective variations are inversely related to the size of the animals (see Chapter 3); these values are for pigs in the body weight range 20–50 kg.

Windspeed m s^{-1}	Variation to be added to give ESET, °C (convective and radiant components)
Still air	$0+0.8\,(\bar{T}_r-T_a)$
0.3	$-4+0.6\,(\bar{T}_r-T_a)$
1.0	$-8+0.5\,(\bar{T}_r-T_a)$

This particular assessment is aimed at determining the effects of thermal environment on pig productivity. It is also possible to reach a generalized specification of the thermal environment, and the following approach is adopted by Monteith.[370]

Monteith takes the heat balance equation (excluding conductive loss):

$$M = H_c + H_e + H_r$$

where M = metabolic rate less external work; H_c = rate of convective heat loss; H_e = rate of evaporative heat loss; and H_r = rate of radiative heat loss; and re-writes each form of heat loss to give the thermal relation between an animal's surface and the environment:

$$H_c = K_C(T_s - T_a)$$

$$H_e = K_E(P_{vs} - P_{va})/\gamma$$

$$H_r = K_R(T_s - T_a) - R_F$$

where K_C, K_E and K_R are the corresponding heat transfer coefficients and where T_s = surface temperature, T_a = air temperature; P_{vs}

= vapour pressure at the surface; P_{va} = ambient vapour pressure; γ = psychrometer constant (mbar $°C^{-1}$, 0.66 at 20°C and 1013 mbar pressure, introduced to give K_E the same dimensions as K_C and K_R); and R_F = effective radiant flux (the radiative energy that the surface would absorb if its temperature equalled the air temperature).
When $K_E = K_C = K_1$

$$\text{then } H_c + H_e = K_1(\theta_s - \theta_a)$$

where θ_s = equivalent surface temperature; and θ_a = equivalent air temperature.
The equivalent temperature, θ, is defined by

$$\theta = T + P_v/\gamma \tag{5.5}$$

where T = dry bulb temperature and P_v is the vapour pressure.

The equivalent temperature, θ, has the same significance in equations for non-evaporative and latent transfers combined as real temperature in equations for non-evaporative heat transfer alone. In Fig. 5.3, which is an extract from Fig. 3.1, X is the state of a sample of air at DB temperature M and WB temperature Y, with the values given in the legend. If the state of the sample of air changes adiabatically (that is, without change in the total heat content of the system) the DB temperature rises and water is condensed until at Z the humidity is zero. The temperature θ given by the point Z is called the equivalent temperature. The slope of the line AZ is $-\gamma$, and the equation of the line is given by equation (5.5). When K_C differs from K_E, an apparent equivalent temperature can be calculated. Finally, R_F can be allowed for by substituting an increment in temperature. Monteith's[370] paper should be studied for further details.

MICRO-ENVIRONMENT

Any attempt to assess an animal's thermal environment must recognize that an animal is reacting less to the general environment and more to those immediate environmental factors that constitute the animal's particular micro-environment. The ***micro-climate*** comprises the conditions immediately around the given animal or object under consideration, and partially defines the animal's ***micro-environment*** as a compound attribute of those environmental quantities that impinge on the animal, including also structural and biological factors.

The reaction of the animal to the environment is generally such as to result in an effective micro-climate most in keeping with its 'comfort', that is a situation in which external demands on the animal are reduced to a minimum. Under hot conditions an animal will lie down, relaxed, on cool ground if this is available; in the cold, the same

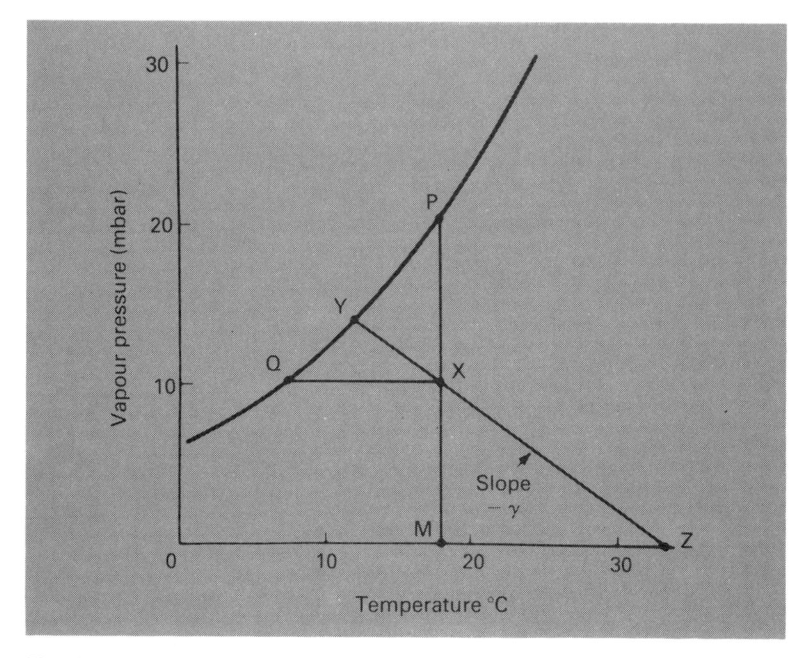

Fig. 5.3 The relation between dry bulb temperature, wet bulb temperature, equivalent temperature, vapour pressure, and dew point. The point X represents air at 18°C and 10 mbar vapour pressure. The line YXZ with a slope of $-\gamma$ gives the wet bulb temperature from Y (12°C) and the equivalent temperature from Z (33.3°C). The line QX gives the dew point temperature from Q (7.1°C). The line MXP gives the dry bulb temperature from M and the saturation vapour pressure from P (20.6 mbar) (Monteith, 1973).

animal will either stand, or rest on the ground with a minimum of contact, thus behaving in a manner that leads to a reduction in conductive heat loss. Moving behind a wind-break and seeking shade from solar radiation reduce convective heat loss and the radiant heat load respectively; keeping dry reduces evaporative heat loss.

It would seem that the most appropriate way to define the thermal micro-environment is in terms of an animal's heat exchanges through the channels of radiation, convection, conduction and evaporation, and in terms of the factors influencing these exchanges (see Chapter 3).

The combination of behavioural responses and apparently minor physical features of the environment can affect very considerably the physiological responses normally necessary for the animal's thermal stability or evan for its survival; even out on the hillside sheep find those areas where the immediate windspeed is lower and where conditions are warmer.[55] In this connection, Cresswell and Thomson[122] suggested from measurements of windspeed at different heights above the ground that even so apparently minor a feature as a tussocky pasture affords shelter to sheep, not only from wind but possibly also from blown rain and snow.

Because the micro-environment is the immediately effective environment of the organism, measurements with meteorological instruments that are placed even short distances from the animal could produce large errors in computing its heat exchanges. Findlay[154] remarks that normal meteorological observations should be capable of interpretation in the field in terms of the limits of heat tolerance of animals. An animal, however, can produce much greater variation than a plant in its exposure to features of its micro-environment by virtue of postural or other behavioural changes, so that it would be difficult to infer an animal's micro-climate solely from the macro-climate and the local physical terrain, as might be done with vegetation. The micro-climate of plants is the result of exposure to external heat sources and sinks; micro-climates of mammals and birds, on the other hand, are affected to a large degree by their own metabolic heat production.

6

Adaptation to the Thermal Environment

Adaptation of an animal to its environment leads to a change in the response to a given stimulus. Adaptation may involve acclimatization of the animal to a changed environment, or genetic adaptation that involves forces of selection operating in successive generations of animals. Adaptation may be localized to certain tissues or it may affect the whole organism.

Usually adaptation implies that an organism's survival is favoured by the changes that occur in its responses. For example, in the cold a mammal exhibits an acute metabolic response and so regulates its body temperature. In addition, vasoconstriction and pilo-erection in the cold are physiological responses that increase insulation and so favour thermoregulation. The animal also shows behavioural responses in changing its posture from extended to compact, huddling with its litter-mates, or burrowing into bedding. These responses increase the effective insulation between the animal and its cold environment and so reduce the rise in metabolism necessary to maintain body temperature; the behavioural responses are in this way adaptive in respect of the metabolic response to cold. If the animal is reared in the cold its body conformation is more compact than that of an animal raised in the warm, constituting a morphological adaptation that allows a smaller metabolic response to cold than would otherwise be necessary.

Prosser[424] has defined adaptation along the following lines:

(1) Adaptation decreases effects that would otherwise be exerted by a changing environment.

(2) In living organisms those alterations that favour survival in a changed environment are said to be adaptive. Adaptive variations that

appear to be similar to each other may be genetically determined or they may be induced by environmental factors. Those that occur in response to environmental change vary depending on the magnitude of the environmental change and on how long it lasts. A small change operating for a long time can be equivalent in its effects to a large change operating for a short time.

Apart from genetic adaptation, there are three types of adaptation that can be recognized as occurring during the individual organism's lifetime. First, acclimatization is the process of adapting to an environment, usually a complex environment, involving a number of interrelated responses. Second, there is conditioning, which involves the transfer of an existing response to a new stimulus. The classical example occurs in Pavlov's dogs: they came to associate a neutral signal, such as the ringing of a bell, with the presentation of food, so that the neutral signal became a conditioned stimulus and the dogs would salivate in response to that stimulus by itself in the absence of food. Operant conditioning is another form of conditioning, in which the animal learns to make responses to produce a given result, such as obtaining food or changing its environment.[242, 434, 472] Rats in the cold learn to press a lever to obtain heat[532] and the method has been used to study the pig's temperature preferences[24] (see Chapter 8).

The third type of adaptation is habituation. This does not involve active adaptation but is instead characterized by the diminution of existing responses to particular environmental stimuli. Habituation occurs when the environmental factor concerned has no significance in relation to any response the animal may make. For example, pigs exposed to the noise of jet engines are disturbed at first, but rapidly become habituated to repetition of the noise and exhibit no differences in their behaviour patterns in feeding, reproduction and activity from animals that have not been exposed to the noise.[71] However, if the noise was always associated with the entry of the stockman with food, it would cease to be neutral and would rapidly become a conditioned stimulus evoking the behaviour associated with the anticipation of food. Learning is also sometimes referred to as a form of adaptation.

The process of acclimatization includes modifications of behavioural, physiological and morphological characteristics that can be considered in turn.

BEHAVIOURAL ADAPTATION

An animal modifies its posture and orientation to wind and sun depending on the thermal demand of the environment. The compact posture seen in the cold effectively reduces the animal's area for heat exchange; contact with a cold floor is also kept to a minimum (Fig. 6.1).

The huddling of animals together effectively modifies the impact of the environment and lessens the need for metabolic response in the cold (Fig. 6.2). The smaller metabolic response to cold in grouped

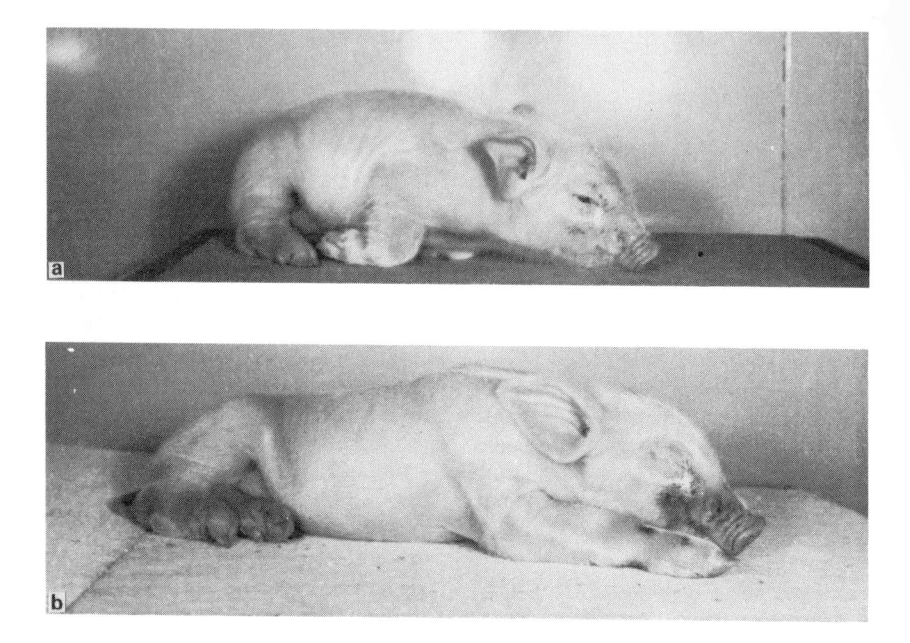

Fig. 6.1 The postures adopted by newborn pigs on cold and warm floors. (**a**) At 15°C on a concrete floor, the pig holds its thorax and abdomen out of contact with the floor. (**b**) In thermal neutrality at 35°C on a sheet of expanded polystyrene, the animal lies in a relaxed position in full contact with the floor (Mount, 1968a).

compared with single animals (see Figs 8.3, 8.4 and 8.5) is due to the increased overall insulation (see Chapter 4) that follows from the behavioural response of huddling. Huddling and nest-building are important behavioural adjustments in the struggle of the homeotherm against cold.[207,554] For example, the reduction in heat production in the cold due to huddling in mice is matched by a reduction in the demand for food.[426]

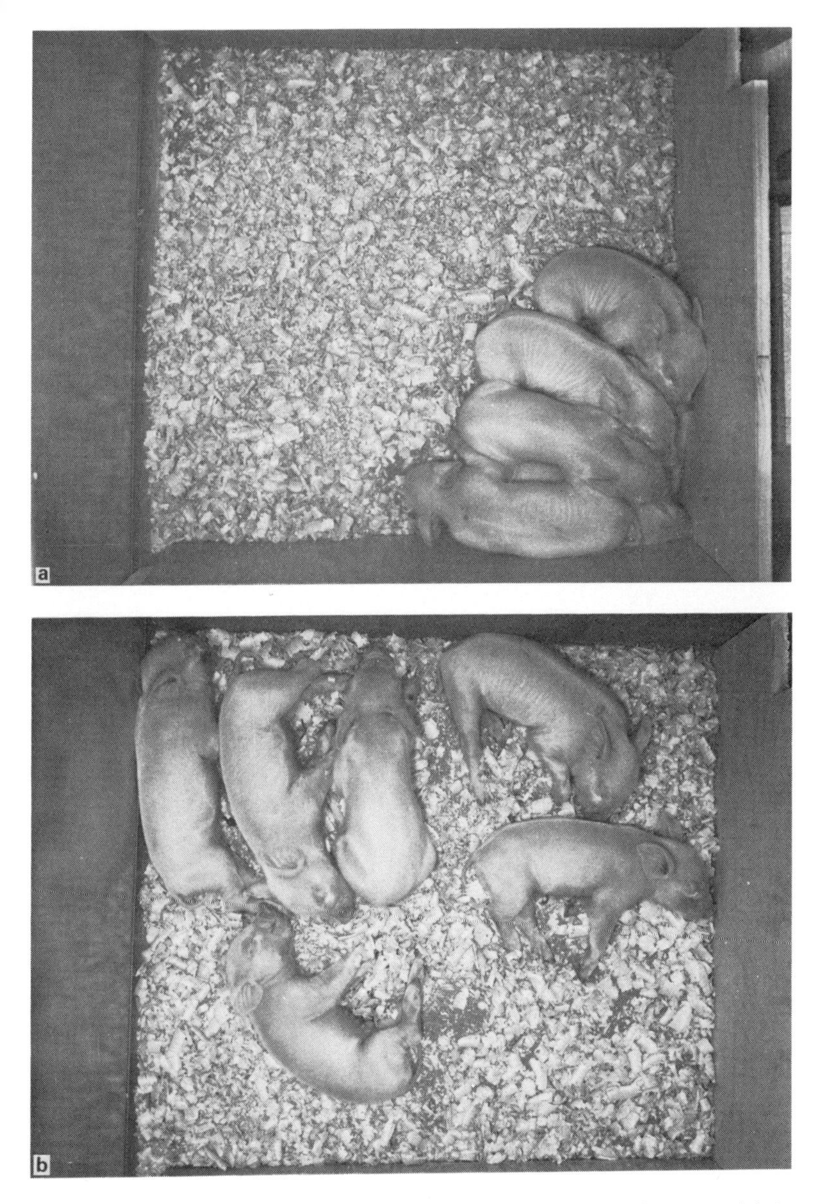

Fig. 6.2 The response of a group of newborn pigs to a cold environment. (**a**) At 15°C, the pigs are huddled together closely. (**b**) In thermal neutrality at 35°C, the animals are spread out in relaxed postures (Mount, 1968a).

Animals use the diversity of their environment for adaptive behaviour. Out-of-doors they do this by making use of sun and shade, wind-breaks, wallows and burrows. For example, in the desert, with a daytime temperature of 44°C, a mouse that can retreat to its burrow which is at 26°C is not exposed to thermoregulatory stress. A deer mouse in its sub-alpine habitat could be exposed in summer to a 24-hourly range of temperature between about 4 and 20°C, whereas its burrow temperature remains close to 10°C both day and night.[211] Indoors, animals seek draught-free areas and use bedding if it is available; some choice is necessary for animals kept in captivity if they are exposed to fluctuations in temperature.

Behavioural adaptation in animals can be investigated by the method of multiple choice, where the animal can select one of a range of pre-existing conditions; as an example, Table 6.1 gives the ambient

Table 6.1 Ambient temperatures preferred by young pigs, both singly and in groups of five; numbers choosing different temperatures (from Mount, 1963a, by permission of Nature).

Age of pigs days	Ambient temperature, °C				Mean±S.E. temp., °C
	23–28	29–31	32–34	35–37	
Single pigs					
<1	1	2	5	2	32.3±0.9
1–7	8	16	4	0	29.3±0.4
8–41	16	33	11	6	29.8±0.6
Groups					
2–7	2	3	3	0	30.1±1.0
8–41	3	9	4	1	30.4±0.7

temperatures selected by young pigs in a thermocline. However, the principal methods currently used to investigate behavioural adaptation are those of classical (Pavlovian) conditioning and operant conditioning. A summary of behavioural responses to high and low temperatures is given in Table 6.2.

PHYSIOLOGICAL ADAPTATION

The modes of adaptation of the animal to its thermal environment through physiological means include metabolic, insulative and heat storage.

Table 6.2 Behavioural responses to high and low temperatures in mammals and birds (modified from Hafez, 1964; Baldwin, 1974).

High ambient temperatures		Low ambient temperatures	
Mammals and birds		**Mammals and birds**	
Reduced food intake		Body flexure, hunched posture	
Increased water intake		Huddling	
Group dispersion, avoidance of body contact		Extra locomotor activity	
Decreased locomotor activity		Nest building	
Seeking micro-climates with lower temperatures, may include burrows in mammals		Seeking micro-climates with higher temperatures	
Avoidance of direct sunlight		Increased food intake if available	
Mammals	**Birds**	**Mammals**	**Birds**
Wallowing	Extension of wings	Hibernation	Torpidity
Rooting up cooler subsoil and lying upon it	Sitting on cool soil		Fluffed feathers
Covering the body surface with saliva	Water splashing		Postures which cover the feet and legs with feathers
Night grazing	Postures which facilitate heat loss from the feet (resting on heels)		Head tucked under wings when resting
Flipper waving (aquztic)			
Lying down in extended postures			
Aestivation			

Metabolic adaptation

The rise in metabolic rate as an acute response to cold can, if repeated, bring about metabolic adaptation in the form of an increased resting metabolism at thermal neutrality; the maximum cold-induced metabolism is also increased, enhancing the animal's survival at low temperatures (Fig. 6.3). When this occurs simply as a metabolic response it is termed 'metabolic acclimation', implying adaptation to only one variable, usually environmental temperature, a situation that is more likely to arise under experimental conditions than in the field. By its nature, metabolic acclimation is a form of adaptation that is dependent on a liberal food supply if the animal is not to waste progressively through using food energy to produce heat in the cold at the expense of body tissue.

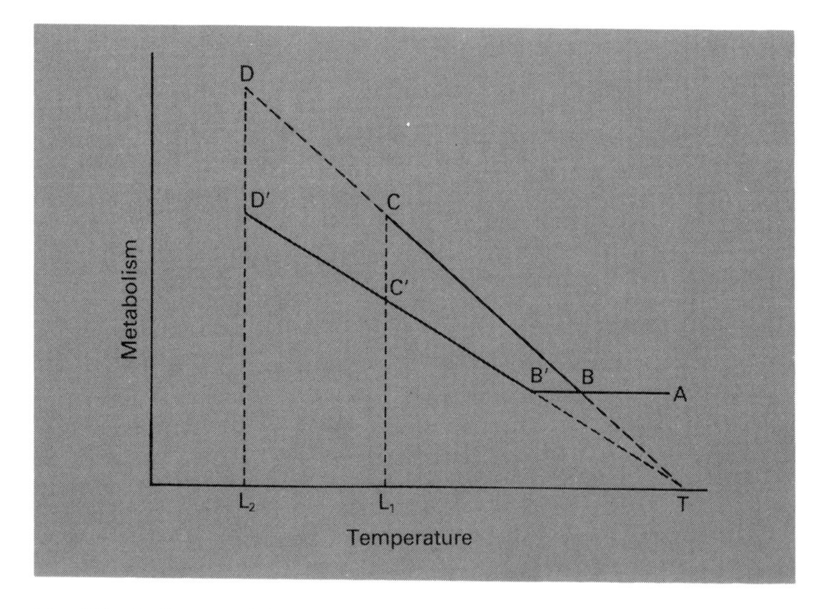

Fig. 6.3 A diagram of the relation between an animal's metabolic rate and the environmental temperature. As the animal's insulation increases, the slope moves from BC to B'C', pivoting approximately (see Chapter 2) on the animal's deep body temperature, T. The critical temperature falls at the same time from that opposite B to B', and the cold limit falls from L_1 to L_2, at the same maximum metabolism; this is insulative adaptation to cold. In metabolic adaptation, the curve BC is extended to D; at this point, the cold limit is extended from L_1 to L_2 by a rise in metabolic rate from C = D' to D. Variation in the level of minimal metabolism, A, would move B relative to the temperature scale and thus change the critical temperature (Hart, 1964, by permission of Society for Experimental Biology; see also Fig. 2.5).

Animals increase their food intakes when they are exposed to cold for long periods.[29, 145] Sometimes there is a delay in the response (Fig. 6.4); when rats are exposed to cold they do not immediately increase their food intake, but do so after about a week.[322]

Under hot conditions, food intake is reduced; because the resting metabolism is influenced by the plane of nutrition, the metabolic rate

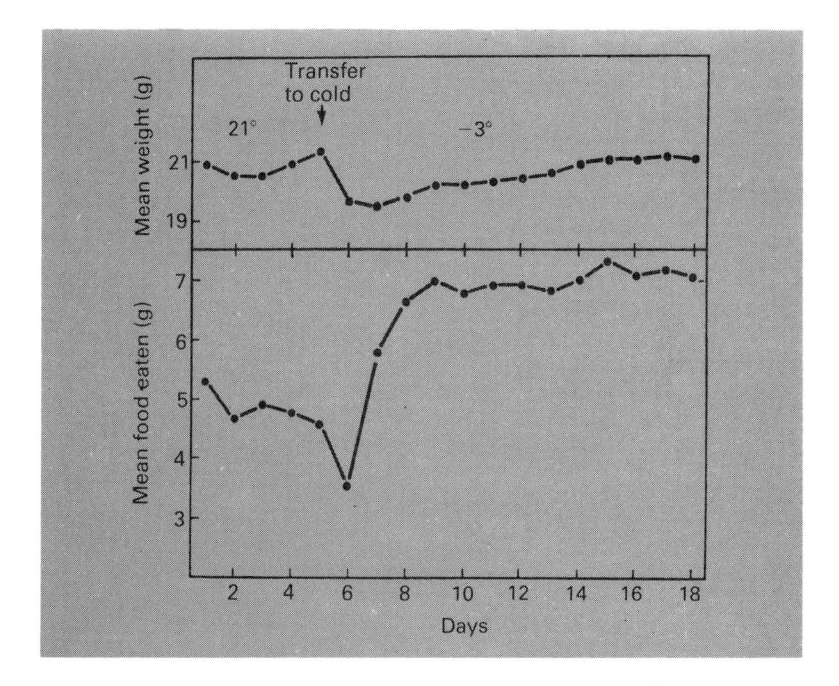

Fig. 6.4 Mean food consumption and body weights of ten mice, aged 5 weeks, transferred from 21° to −3°C. There was a delay in the adjustment of food consumption (Barnett and Mount, 1967, by permission of Academic Press).

is also reduced unless the animal becomes hyperthermic (see Chapter 2). One possibility is that the reduction in metabolism of animals living above their critical temperature is related in part to a diminished rate of secretion of thyroid hormone. A comparison of pigs reared at 25 and 35°C showed that thyroid secretory activity and oxygen consumption were always greater in the animals housed at 25°C.[260] Evidence suggests that thyroid activity may be related not so much directly to metabolic rate but rather to the level of food intake, which tends to be higher in the cold.[255, 350]

Insulative adaptation

When rats are exposed to a lowered environmental temperature with other environmental factors maintained constant, metabolic acclimation occurs. However, if rats are exposed to the mixture of low temperature, rain and wind that normally constitutes their winter environment, their longer term adaptation, while including the potential of increased metabolism, also involves insulative adaptation with the development of more coat insulation coupled with endocrine changes.[221] This combined adaptational response to several factors in a complex environment is what is usually termed acclimatization. In this example the increased insulation that forms part of the adaptation has considerable survival value because the need for extra food is diminished. Mammals raised in the cold have thicker coats than those kept in the warm (see Chapter 9); this can be regarded as a form of morphological adaptation.

Adaptation by heat storage

In the different forms of physiological adaptation to the thermal environment the primary interacting variables are the rate of heat production, thermal insulation, the thermal demand of the environment and the deep body temperature. If in addition to variations in peripheral body temperatures (see Chapter 1) the deep body temperature is allowed to vary to some extent with the environmental temperature, heat production is spared in the cold and the need for heat dissipation under hot conditions is reduced.

This form of thermal adaptation by changes in heat storage has been studied for man in the cold in the Australian Aborigine and the Kalahari Bushman (see Chapter 7). These people allow their extremities to cool at night and the deep body temperature to fall, variations that cannot easily be tolerated by Europeans. Another example of tolerance of lowered deep body temperatures occurs in the ama, Korean diving women.[241]

The tolerance of local cold and decreased tissue temperature has been studied in the hands of Gaspé fishermen in Quebec[314] and in the herring gull, which displays a remarkable degree of temperature adaptation in its legs, with its feet in freezing water and the upper parts of the limbs at deep body temperature.[136, 480]

The best-known example of an animal that allows its body

temperature to rise under hot conditions is the camel, in which the temperature may rise several degrees during the heat of the day and fall during the cold night of the desert. This capacity effect of storing heat reduces the animal's dependence on scarce water for sweating[258] (see Chapter 11).

MORPHOLOGICAL ADAPTATION

Both behavioural and physiological modes of adaptation are often combined with features of morphological adaptation, for example the growth by mice of a thicker coat in the cold and longer tails in hot conditions.[203] One of the clearest examples of morphological changes induced by high and low temperatures is that of litter-mate pigs raised in the two conditions (Fig. 6.5). The cold pig was held at 5°C environmental temperature from age 4–11 weeks, and the warm pig was kept at 35°C for the same period. The cold pig was fed *ad libitum* and the warm pig was fed so that it maintained the same body weight as the cold pig. The cold pig had smaller ears and shorter limb bones than the warm pig; it had 8.9 mg hair per cm^2 body surface compared with the warm pig's 6.8 mg cm^{-2}. The cold pig also had significantly fewer blood-vessels in the skin.[261] Fuller[168] has shown that pigs reared in the cold have smaller ears, shorter limbs and are more hairy than those raised in the warm.

Body type in man influences thermal responses. Thus Cannon and Keatinge[90] found that although fat men increased their metabolic rates in water cooled below 33°C, they did not show maximum tissue insulation until the water temperature had fallen to as low as 12°C. The thinnest men in their experiments, on the other hand, showed both a rise in metabolic rate and developed maximum tissue insulation at much the same temperature, close to 33°C. Thus, in the fat men there were two distinct thresholds, the 'metabolic threshold temperature' and a 'maximal tissue insulation temperature', but in the thin men these two were similar. The effects are illustrated in Figs. 7.9 and 7.10.

GENETIC ADAPTATION

So far the examples discussed have been of changes occurring in the life-time of the individual animal. Genetic adaptation, on the other hand, refers to those heritable characteristics that are transferred from

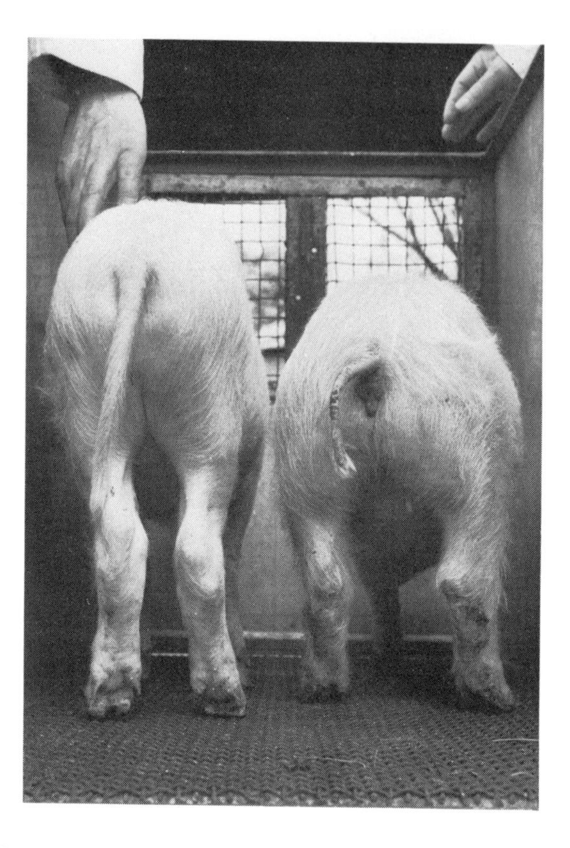

Fig. 6.5 Two littermates, the pig on the left exposed to 35°C for seven weeks and the pig on the right to 5°C; the pig in the warm was fed an amount to keep it at approximately the same body weight as the pig in the cold (Mount, 1968a; Weaver and Ingram, 1969).

generation to generation and that are subject to selection, either natural or practised by man in animal husbandry. Some of the unfavourable environments in which some species live are listed in Table 6.3, and some of the adaptations that make their survival possible are given in Table 6.4.

The characteristics of animals found in different climatic regions can be described partly by some general laws. Bergmann's rule states that members of a taxon tend to be larger in the colder regions of its distribution, and Allen's rule states that the colder the environment is,

Table 6.3 Examples of unfavourable environments in which some domestic and feral mammals can endure (Hafez, 1968b, by permission of Lea and Febiger).

Type of unfavourable environment		Locality	Animal
High temperature	52°—58°C	Lybia, Africa	Camel
		Death Valley, California	Cattle
		West Australia and Queensland, Australia	Cattle, sheep, horses
	No rain, 19 years	A locality in Sudan	Sheep
	50°C during day −18°C at night	Kalahari Desert	
Low temperature	−80°C	Siberia	Sled dog
		South Pole at 3000 m altitude	Eskimo huskies
	−60°C	Alaska	Arctic tundra wolf
High altitudes	3500—5000 meters	Mt. Evans, Colorado, USA	Bighorn sheep Rocky Mountain goat
		Andes, South America	Alpaca, llama, vicuna, huanaco
Arctic tundra	Cold, treeless plain with sparse vegetation	North Alaska	Moose, caribou, Dall sheep, Tundra hare

Table 6.4 Main morphological, anatomical and functional adaptations in domestic animals (Hafez, 1968, by permission of Lea and Febiger).

Environmental stress	Adaptive mechanisms	Animal (breed)
Solar radiation	Long limbs	Camel
	Short reflecting coat	Gazelle
High temperature	Hair shedding in summer	Ungulates
	Increased surface area in skin folds	Cattle (Brahman)
	Small body, long ears	Donkey
	loose, coarse wool	Sheep (Awassi)
	Fine, dense wool	Sheep (Merino)
High humidity	Dark pigmentation, sparsely haired	Buffalo
Low temperature	Long hair intermixed with fine hair	Cattle (Scottish Highland)
	Minimum exposed extremities	Yak (Tibetan)
	Good grazing behaviour	Musk-ox (Arctic)
	Thick subcutaneous fat	Arctic species
	Abundant brown fat	Neonate of several mammals
	Thickset, heavy coat	Horse (Shire)
Seasonality in available feed	Ruminant stomach	Ungulates
	Adipose tissue reserves—hump(s)	Camel
	fat-tail	Sheep (Awassi, Kudri, Masai, Karakul)
	fat rump	Sheep (Somali, Sudan, blackhead)
	in rumen	Antelope
Deserts		
(thorny vegetation)	Thick skin, hard tissue around mouth; Thick mouth, lined with long papillae	
(water scarcity)	Increased drinking capacity, Hump (for pseudo-water storage) conservation of metabolic water Ability to survive dehydration	Camel
High altitude	Increased O_2 carrying power in blood through increased concentration of red blood cells Ability to transfer O_2 from capillary blood to tissue cells at a lower partial pressure High efficiency in extracting nutrients from feeds	Llama, alpaca

then the shorter the appendages and the more compact the body. These general statements have been critically discussed.[460, 530]

ADAPTATION OF THE NEWBORN ANIMAL TO ITS THERMAL ENVIRONMENT

When it leaves the thermal stability of the uterine environment, the newborn mammal is precipitated into the thermally complex external environment. This is perhaps the most marked thermal change that the organism is exposed to in the course of its life, not so much in terms of the absolute change in environmental temperature but certainly in terms of exposure to the fluctuating external environment in place of the constancy of the uterus to which it has been accustomed. The thermal responses of the newborn are a part of the many changes that occur at birth, including those in respiration, circulation and muscular activity. The process of thermal adaptation involves particularly metabolic and behavioural adjustments.

The newborn animal appears to be weak and helpless, but coupled with the apparent weakness is a degree of physiological tolerance that is not seen in the adult.[331] For example, the body temperature of the newborn in a cold environment can fall to levels of hypothermia from which an adult could not be expected to recover, but from which the newborn animal can be re-warmed with a high chance of subsequent normal development. Greater degrees of tolerance are found in those newborn animals that are the less mature at birth, rodents for example.

There is less stability of the internal environment in the newborn than in the adult, with reduced costs of maintaining stability. This is evident in thermoregulation, where the high metabolic cost of maintaining a stable temperature in a small animal is avoided. However, the newborn is incapable of self-help once it is hypothermic, when it must be rescued from its cold state by the mother providing re-warming and food.

There is a considerable range of body weight amongst individual newborn mammals of different species (Table 6.5). However, the double logarithmic plot of total weight of the newborn young against maternal weight for a wide range of mammals is linear, with a slope of 0.83 indicating that total newborn weight increases progressively less rapidly than maternal weight (Fig. 6.6); relatively, the larger mammals carry a smaller weight of young (Fig. 6.7). The lactation demand is also relatively less for the larger mammal: for the range of body size from the mouse to the cow, milk yield (kg per day) is given by

$0.1\,W^{0.79}$, and milk energy' (kJ per day) by $527\,W^{0.73}$, where W = body weight, kg, of the mother.[325] The composition and energy values of milks from a number of species are given in Table 6.6.

Table 6.5 Approximate body weights of some adult and newborn mammals and litter size (Hull, 1973, copyright Academic Press).

Mammal	Body weight of newborn (g)	Litter size	Bodyweight of adult (g)
Mouse	1	6–10	35
Rat	5	6–10	300
Rabbit	55	4–9	3 000
Guinea pig	90	2–5	750
Cat	110	3–5	3 000
Coypu	200	3–10	6 000
Monkey	450	1	6 000
Pig	1200	6–12	200 000
Human	3500	1	60 000
Sheep	4000	1–2	60 000

The development of thermoregulation is influenced by the environment to which the newborn animal is exposed. Newborn rats exposed to cold once a day showed a more rapid development of effective thermoregulation than similar animals kept in uniformly warm surroundings.[191] Exposure to cold may increase energy loss, but it is also the required stimulus for eliciting thermoregulatory responses. When applied to farm animals a balance must be struck, because a newborn pig, for example, may die if exposed to cold for too long, although some variation in environment might be expected to produce a more hardy animal. An extreme case of exposure of the mammalian newborn to cold occurs in caribou,[321] where the newborn is delivered into the snow.

Following birth, metabolic rate increases: this is referred to particularly in Chapter 8 on the pig. The rise in metabolic rate from the level found in the period following birth appears to be a general phenomenon in the homeotherm.[495] It occurs not only in pig, sheep and man, but also in the monkey,[131] and in the puppy.[181, 337]

The development of heat conservation mechanisms lags behind the development of metabolic rate in the puppy, and this is true for the pig, since although the pig shows peripheral vasoconstriction in the cold from birth, it is only after some days that its subcutaneous fat reaches a level that allows the insulative value of vasoconstriction to become effective.[386] The calf and the lamb have by virtue of their coats an insulative advantage in the development of homeothermy over the less well insulated newborn of other species.

All newborn mammals show some rise in metabolic rate when the environmental temperature is lowered, but in those species where the newborn are small and relatively immature the cold limit for the maximum metabolic response is high. This led to the earlier

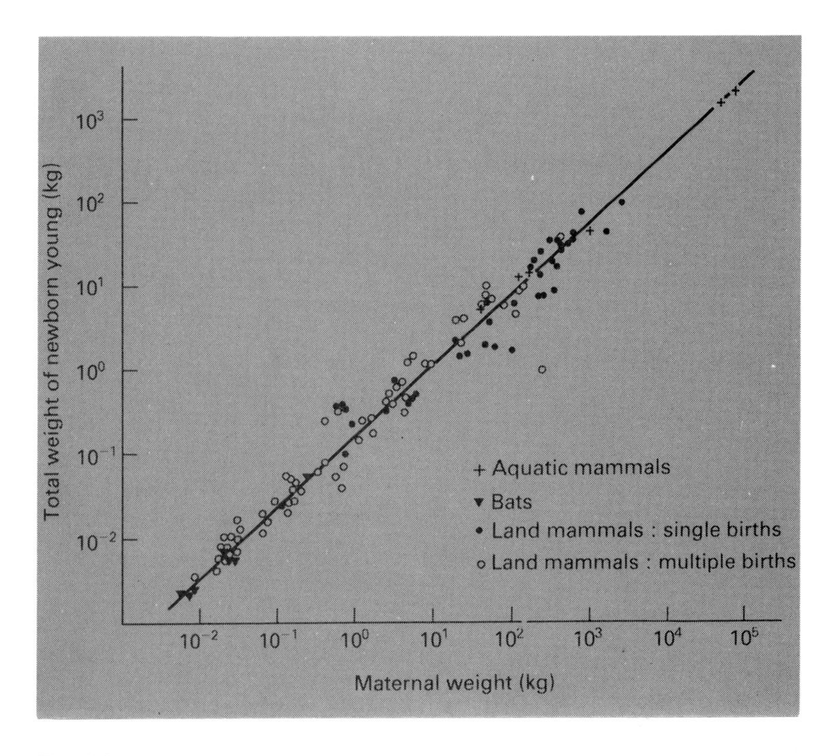

Fig. 6.6 The logarithm of total weight of the newborn plotted against the logarithm of the weight of the mother; both land and aquatic mammals are represented (Leitch, Hytten and Billewicz, 1960, by permission of the Zoological Society of London).

conclusion that the newborn rat and mouse are poikilothermic; this false conclusion was shown to have arisen because at environmental temperatures below about 30°C the animals' limits for heat production were reached so that they became hypothermic.[479, 495] The newborn guinea-pig has a lower cold limit, at about 10°C, so that it is easily recognized as a homeotherm. The lamb's cold limit is even

Table 6.6 Energy and protein in milk (from Kleiber, 1975, table compiled by Dr J. Luick; by permission of John Wiley and Sons Inc.).

Source of milk	Composition			Energy in components per kg of milk			Energy per kg of milk (MJ)
	Fat (%)	Lactose (%)	Protein (%)	Fat % × 0.385 (MJ)	Lactose % × 0.165 (MJ)	Protein % × 0.245 (MJ)	
Rhinoceros	0.3	7.2	3.2	0.12	1.19	0.79	2.09
Mare	1.7	6.6	2.2	0.65	1.09	0.54	2.28
Bison	1.8	4.6	4.0	0.69	0.76	0.98	2.44
Woman	3.8	7.0	1.2	1.46	1.15	0.29	2.91
Cow	3.7	4.8	3.3	1.42	0.79	0.81	3.03
Goat	4.1	4.7	3.3	1.58	0.78	0.81	3.16
Sheep	6.2	4.3	5.4	2.38	0.71	1.32	4.42
Sow	7.0	4.0	6.0	2.69	0.66	1.47	4.83
Musk ox	11.0	3.6	5.3	4.23	0.59	1.30	6.13
Rat	9.3	3.7	8.7	3.58	0.61	2.13	6.33
Water buffalo	12.0	4.0	6.0	4.62	0.66	1.47	6.75
Deer	10.5	4.5	9.0	4.04	0.74	2.20	6.99
Elephant (African)	20.5	7.3	3.2	7.89	1.20	0.79	9.88
Reindeer	22.5	2.5	10.3	8.66	0.41	2.53	11.60
Fin whale	32.0	0.3	13.0	12.31	0.05	3.19	15.56
Blue whale	42.0	1.0	12.0	16.12	0.17	2.94	19.28
Porpoise	49.0	1.3	11.0	18.86	0.21	2.70	21.77

Milk constituents: lactose, 16.5 $kJ\,g^{-1}$, casein, 24.5 $kJ\,g^{-1}$, butter fat, 38.5 $kJ\,g^{-1}$

lower[8] and this is shown together with the metabolic responses to cold and the cold limits for the newborn of several species in Fig. 6.8.

The influence of thermal insulation on the cold limit is clearly demonstrated in the human infant. The naked baby can thermoregulate over the environmental temperature range 36 down to 26°C, but when thinly clad and wrapped in a blanket this temperature range

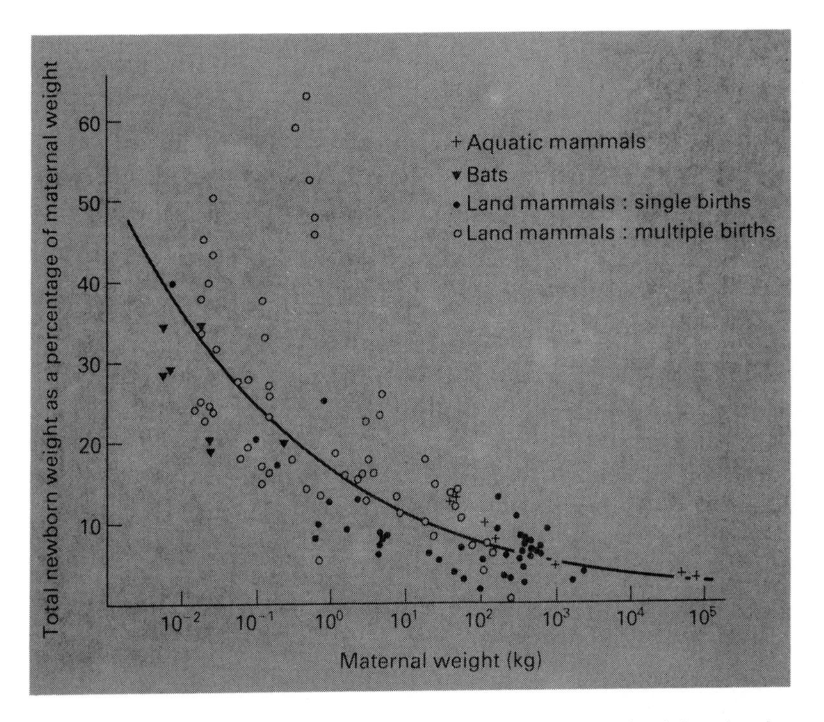

Fig. 6.7 The total weight of newborn, as a percentage of maternal weight, related to maternal weight; both land and aquatic mammals are represented (Leitch, Hytten and Billewicz, 1960, by permission of the Zoological Society of London).

becomes 30 down to 10°C[230] (see Chapter 7). Mice are well known for their ability to raise their young in cold stores, living in nests that provide adequate thermal insulation and warmed by the older animals.[28] The young of species that bear litters huddle together effectively providing insulation by diminishing the total surface for heat exchange with the environment[377] (see Fig. 6.2).

Most newborn mammals that have been investigated can increase

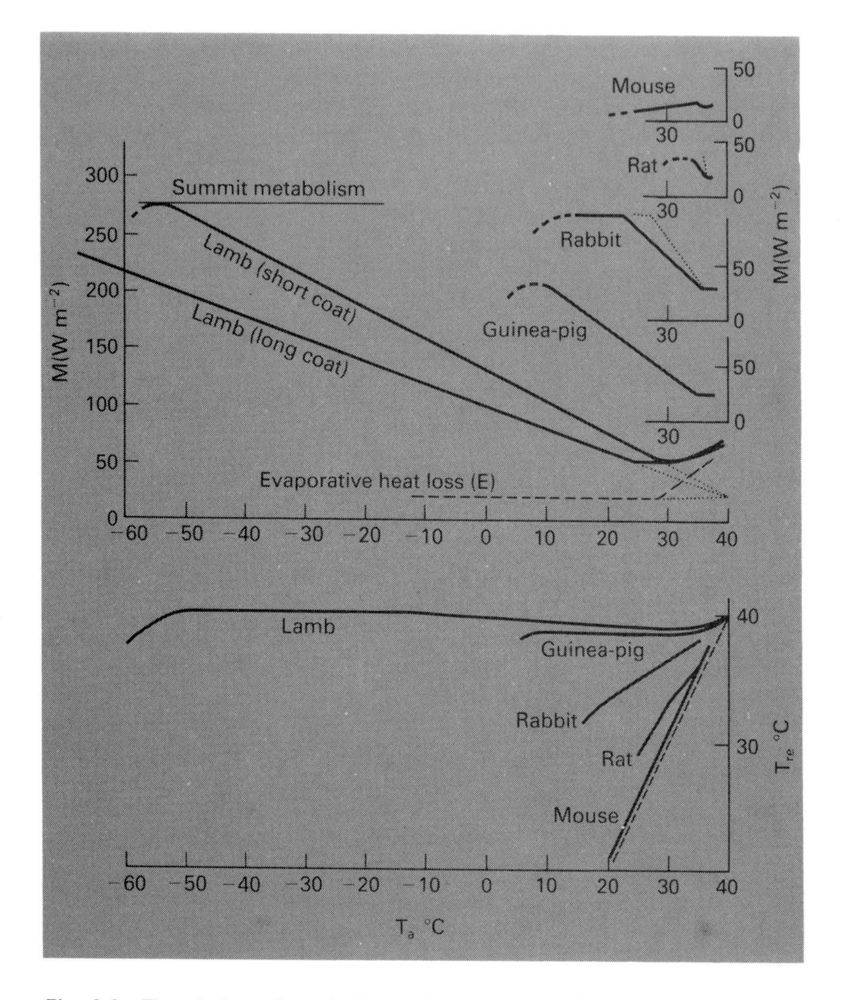

Fig. 6.8 The relations of metabolic rate (M), watts per m^2 body surface, and deep body temperature (T_{re}) to environmental temperature (T_a), for the newborn mouse, rat, rabbit and guinea-pig, and newborn lambs with either short or long coats (Alexander, 1975, by permission of British Medical Bulletin).

heat production without apparent shivering, that is by non-shivering thermogenesis, which appears to occur to a large degree in brown fat (see Chapter 2).

ADAPTATION TO LOW TEMPERATURES

Adaptation to cold can include the various forms of behavioural, physiological and morphological adaptation.[29] Acclimatization to

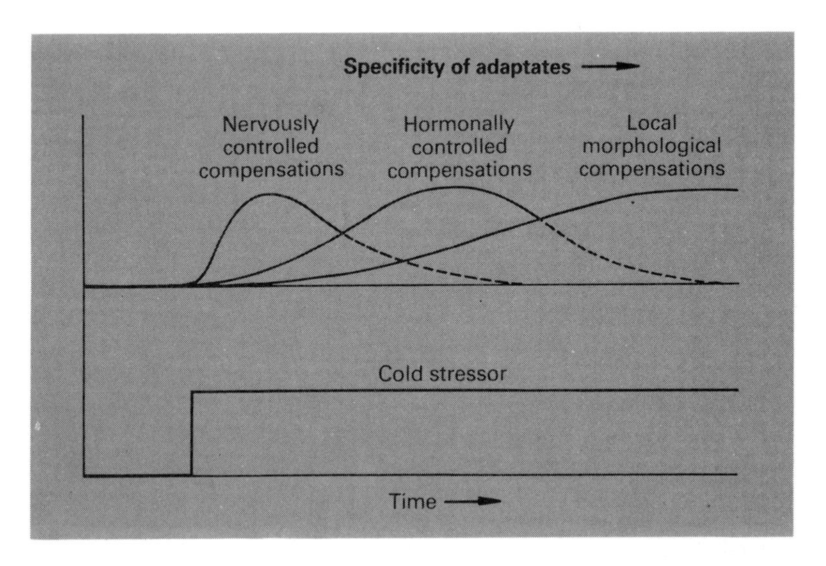

Fig. 6.9 The temporal succession of specific nervous, hormonal and morphological adaptations that occurs in response to chronic exposure to cold (Hensel, 1968, adapted from Hildebrandt, 1967, by permission of Lea and Febiger).

cold may be acute or chronic; the acute responses tend to be metabolic and behavioural, whereas chronic responses are more likely to involve morphological change (Fig. 6.9). Acute exposure to cold results in shivering, with increased heat production, and an increased thermal insulation due to pilo-erection and peripheral vasoconstriction. Chronic exposure to cold results in morphological changes (see Fig. 6.6) and in increased thermal insulation due to increased growth of coat and increased subcutaneous fat; non-shivering thermogenesis has been described in Chapter 2. Different species can tolerate low temperatures to varying degrees depending on their body size and thermal insulation. A rat, for example, can tolerate an environmental temperature of −25°C for at least one hour without becoming hypothermic, whereas for naked man the corresponding temperature is −1°C and for the arctic white fox it is −80°C.[220]

The chemical composition of body fat changes during adaptation to cold or heat. Pigs kept at low temperatures have more unsaturated fatty acids in their body fat than animals under warm conditions; the fat of the cold-exposed pigs has a lower solidifying temperature (see Chapter 8). A corresponding difference has been found in membranes in the brains of goldfish kept at 5°C or 25°C. At 5°C the membranes contain more unsaturated fatty acids and are less viscous than the membranes from goldfish at 25°C.[118] In both the homeothermic and poikilothermic examples, the higher content of unsaturated fatty acids confers an adaptational advantage.

Endocrines

The urinary excretion of catecholamines increases in the cold to a degree that is proportional to the intensity of the stimulus. Catecholamines stimulate lipolysis and glycogenolysis in the liver.

Thyroxine secretion rate relative to its plasma concentration is relatively slow (Table 6.7), with a biological half-life of released thyroxine of 30–40 hours. High thyroxine concentrations are not necessarily related to thyroxine secretion rate. When thyroxine turnover remains constant, acclimatization to cold can still occur.[222] There is synergism between the thyroid and sympatho-adrenal system.[284–5]

The secretion of cortisol increases considerably in the cold and the secretion rates of insulin and growth hormone are probably also increased by cold. The glucocorticoids, cortisol and corticosterone, are the steroid hormones most concerned in the metabolic response to cold; they stimulate the breakdown of protein and enhance gluconeogenesis in the liver.[525] Acclimation to cold is accompanied by endocrine changes and by the transfer of heat production from shivering to non-shivering thermogenesis.[205]

Aquatic mammals

Except for the hippopotamus, a feature of amphibious mammals of freshwater is the dense waterproof fur that keeps the skin both dry and warm; the fur rapidly dries after immersion. The marine fur seal also has dense water-repellent fur, but the sea lion has a thin wettable coat and the walrus has no coat. The polar bear's coarse hair is wetted

Table 6.7 Plasma concentrations and secretion rates of hormones regulating the metabolism of ruminants (Webster, 1976a, by permission of Swets and Zeitlinger).

Hormone	Species	Environment	Plasma concentration μg $litre^{-1}$	Secretion rate μg kg^{-1} body weight per day
Thyroxine	Cattle	18°C	64	3.2
		1°C	64	4.4
	Sheep	Thermoneutral	54	2.9
		Moderate cold	72	4.5
Cortisol	Sheep	Thermoneutral	9	326
		Very cold	31	1500
	Reindeer	Summer	22	276
		Winter	34	452
Insulin	Sheep	Thermoneutral, fed	1.1†	16
		Thermoneutral, fasted	0.3†	3
Growth hormone	Cattle	18°C	15 (variable)	43

† Assuming 24 mU per μg.

in water and the thermal insulation drops considerably, unlike the seal,[463] but the bear's coat dries out quickly on landing (see p. 86).

High resting metabolic rates are characteristic of the marine mammals; in seals and porpoises, for example, the rates are double those expected from the surveys by Benedict,[41] Brody[77] and Kleiber[302] (see Chapter 2). Amongst freshwater mammals, the metabolic rate of the hippopotamus corresponds to that of land mammals. Subcutaneous fat provides thermal insulation in whales and porpoises, and counter-current heat exchange regulates the vascular transport of heat to the surface of aquatic mammals.[266]

These adaptations for a water environment are disadvantageous on a hot shore, where some of the species breed, particularly since their evaporative cooling mechanisms are relatively ineffective with the result that hyperthermia can easily occur.[539]

A thermoregulatory appendage that acts as a cooling device occurs in the beaver. When totally surrounded by air, the beaver can maintain a normal rectal temperature of 37°C if the environmental temperature does not rise above 20°C. In warmer conditions the animal becomes hyperthermic. When the beaver is allowed to keep its tail in cool water, however, with its body in air, it maintains a normal rectal temperature at an ambient temperature of 25°C, losing 20% of its total heat production through its tail.[481]

ADAPTATION TO HIGH TEMPERATURES

In a hot wet climate, lack of water is not a problem, but high humidity at high temperatures is not conducive to human activity and presents some disadvantages for animal production (see Chapter 11). In a hot dry climate water economy is of great importance, and the consideration of adaptation to desert conditions must take into account adaptation to drought and dehydration as well as adaptation to heat and hyperthermia.

Dehydration can occur as the result of water deprivation combined with a high rate of sweating or panting. There are various types of adaptation evident amongst animals by which dehydration is avoided. Some animals adopt a nocturnal habit and shelter in their burrows during the heat of the day. In the camel, heat dissipation is reduced by the body temperature rising during the day and falling at night. In both cases evaporative loss is lessened. Man in the desert wears loose clothing that provides shelter from radiant heat during the day and protection against cold at night.

Adaptation of man, ungulates and birds in hot conditions is

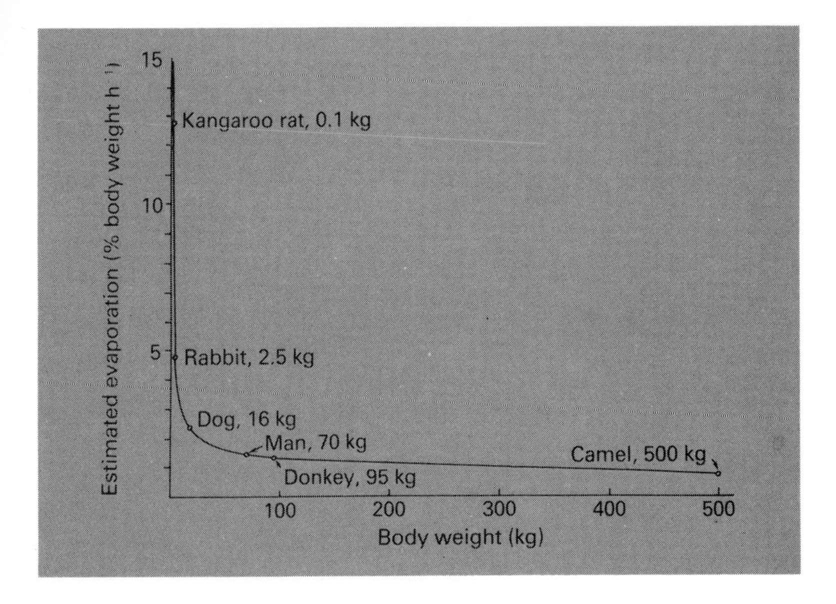

Fig. 6.10 Relation between body size and the evaporation estimated to be necessary for the maintenance of a constant body temperature in a hot desert climate, showing the rapid increase in the theoretical cost of heat regulation in small mammals (Schmidt–Nielsen, 1964, by permission of Oxford University Press).

discussed in succeeding chapters. Here it is appropriate to consider adaptations made by some other animals as indications of the various mechanisms that have developed in different species.

Rodents

Desert rodents are small animals that lack sweat glands and on the whole avoid exposure to heat. They cannot afford to use water for thermoregulation on account of their large surface areas relative to their mass; calculation of the evaporative rate that would be required for thermoregulation shows that this becomes increasingly prohibitive as a percentage of body weight as the animal becomes smaller (Fig. 6.10). Most burrows are at a depth of 0.5 to 1.0 m where there is little temperature fluctuation: at 0.5 m depth the daily fluctuation is about 1°C and at 1 m depth it is zero. At the surface, on the other hand, fluctuation may be of the order of 50°C, with the surface temperature rising as high as 75°C. The maximum temperatures in the burrows at the height of the summer are in the region of 29 to 31°C.

Most desert rodents are nocturnal in habit and in this way the daytime heat load is avoided. Those that are not primarily nocturnal adopt an intermittent type of activity pattern; the antelope ground squirrel, for example, after spending a period above ground, when its body temperature rises, descends into its burrow and lies in close contact with the ground, with the result that its body temperature falls, so producing a see-saw type of temperature control. If the animal is exposed continuously to 40°C it succumbs within a few hours.[258]

Desert rodents are not particularly tolerant of hyperthermia. The lethal limit of body temperature is 42–43°C and at temperatures approaching this some rodents spread saliva over the body, with resulting evaporative cooling. However, this can be only an emergency measure since repetition would produce fatal dehydration.

Desert rodents fall into two groups: the water conservors, including the jerboa, gerbil and kangaroo rat, and those that live on water-containing food, including the pack rat, sand rat and grasshopper mouse. The only source of water available to the water conservors is the metabolic water derived from the food, and Schmidt–Nielsen[455] has shown how this is sufficient to keep the animal in water balance; even when water is present the animal seldom drinks it. Water loss is diminished by reduction in respiratory evaporative loss by counter-current heat exchange in the upper respiratory tract (see Chapter 3), by producing a very concentrated urine and by excreting a small amount of very dry faeces. In addition the kangaroo rat practices coprophagy (eating its own faeces) as do many other rodents, but in this instance a saving of water is achieved. The rodents that live on water-containing food do not drink water but rely for food on cactus plants that have a high water content.

The maximum concentrations of salt and urea that can be achieved in the urine of the kangaroo rat can be compared with those of man and the white rat:

Animal	Salt	Urea
Man	2.2%	6%
White rat	3.5%	15%
Kangaroo rat	7.0%	23%

The kidney of the jerboa can produce a 25% urea solution. The ratio of the urine/serum osmolar concentrations is as high as 11 in the kangaroo rat, 10 in the antelope ground squirrel, and 8 in the white rat.[156] The highest concentration of urine recorded is 9.37 osmole per litre in *Notomys alexis*, a seed-eating mouse of the Australian desert,[344] compared with 1.16 osmole per litre as a maximum for man.[533]

Kangaroos

The thermal exchanges of two kangaroos, the Red and the Euro, have been discussed by Dawson[132-3]. They live in the arid interior of Australia, but whereas the Red kangaroo lives on the open plain, the Euro lives in rocky hill country and uses caves as refuges from heat. Both animals have the low resting metabolic rates characteristic of marsupials and their body temperatures at thermal neutrality are approximately 35.5°C, clearly lower than the level usually found in eutherian mammals. The Red kangaroo is exposed to solar radiation; its fur is nearly twice as reflective as that of the Euro and in addition is also more resistant to penetration by solar radiation. Licking or wiping of saliva on to the forelimbs occurs in both animals during heat stress; it is more noticeable in the Red kangaroo and this animal has a higher evaporative loss than the Euro, although the Euro has the higher panting rate. Sweating in the kangaroo begins some time after exercise begins and then stops when the exercise stops, even when body temperature is high and the animal is panting and licking its forelimbs.

TORPIDITY

Some mammals and birds enter into periods of torpidity at certain times, when they become quiescent, with low metabolic rates and low body temperatures. Amongst the mammals there are hibernators, which become quiescent during the winter, and aestivators, which become quiescent during the summer. Hibernators are sometimes also aestivators.

Hibernation is a method of avoiding the need for the considerable quantities of food that are required to maintain summer levels of metabolism during the winter. Aestivation avoids the difficulty of thermoregulation with high levels of metabolism during the hot summer. Some desert rodents aestivate, that is they become dormant during the hottest and driest part of the summer, with a fall in body temperature, a decreased metabolism and as a result a sparing of evaporative loss.

Other animals, including some bats and, amongst the birds, the humming-bird, become torpid during their resting periods and so avoid high metabolic demands during that part of the 24 hours when they are not collecting food. The animals in which this phenomenon occurs are usually of small body size and they have high metabolic rates per unit of body weight. However, a nocturnal decline in body temperature also occurs in the hypothermic adaptation seen in the Australian Aborigine, without an increase in metabolic rate (see Chapter 7).

Hibernation

There is a difference between tolerance of environmental cold and tolerance of hypothermia. Tolerance of cold occurs as the result of the metabolic production of heat and the use of thermal insulation, so that the deep body temperature is maintained. By contrast, in hibernating animals active thermoregulation is replaced by tolerance of hypothermia when the environmental temperature falls below a certain value. Above that value, hibernators behave as typical homeotherms.[235, 291, 330]

The metabolic rate during hibernation is about 1–3% of the resting value whilst awake and it varies with the body weight instead of with a metabolic body size that is more closely proportional to surface area (see Chapter 2). During hibernation, arousal (Fig. 6.11) occurs if the environmental temperature falls below about −2°C or rises above about 30°C.[29]

Deep hibernation is characterized by a deep body temperature that is only a few degrees above freezing, maintained for several days or weeks before spontaneous arousal. Successive periods of arousal and dormancy occur during the winter. There appears to be an upper size limit to hibernation, although amongst large animals a lethargic state may occur in the winter. Thus the bear, *Ursus americanus*, has a winter deep body temperature of 30°C, with a correspondingly low metabolic rate and a reduction in food requirement, but it can be aroused readily if necessary for self-defence. There is a considerable range of tolerance of hypothermia, extending from the mild hypothermia of the bear to the deep hypothermia of a true hibernator such as the chipmunk, hamster and ground squirrel.[243] Brown fat is important in the hibernator as the tissue that produces the heat for the animal's arousal (see Chapter 2).

POIKILOTHERMS

Poikilotherms also develop adaptation to the thermal environment.[258] An example of genetic adaptation occurs in the desert snail,

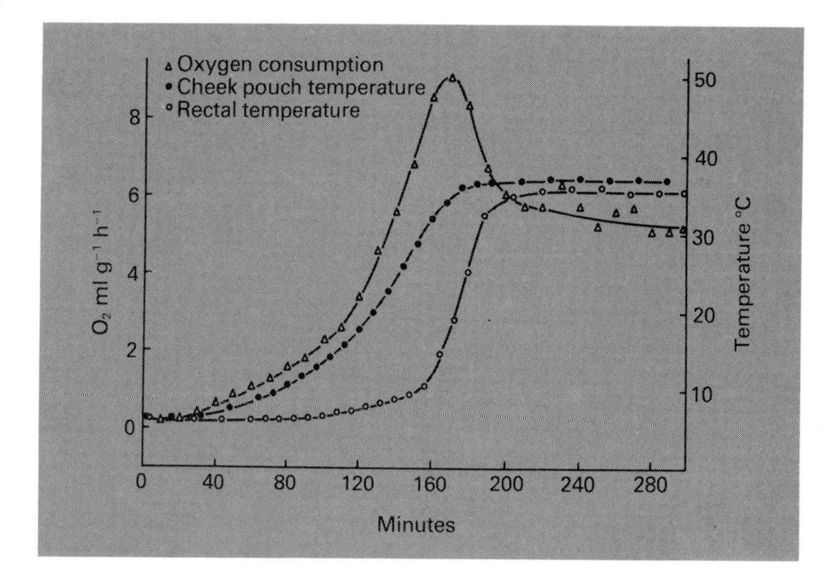

Fig. 6.11 Oxygen consumption and body temperature of the hamster *Mesocricetus auratus* during arousal from hibernation. Note overshoot of oxygen consumption (△) and lag in rectal temperature (○). (Lyman, 1948, by permission of Journal of Experimental Zoology).

which can remain alive in the desert with a sealed operculum for up to two years in the absence of water, although it may be exposed to high levels of solar radiation. The snail's shell is pale and reflecting, and it is likely that this leads to reflection of most of the incident radiation.[459] Metabolic acclimation occurs in the goldfish; the goldfish adapted to 10°C has a higher metabolic rate measured at 25°C than a goldfish adapted to 30°C;[167] lizards respond similarly. Activity provides a form of thermal adaptation in some flying insects: a period of intense muscular activity leads to a rise of the animal's body temperature with correspondingly greater flying effectiveness.[115]

Behavioural adaptation occurs in poikilotherms. For example, the desert iguana moves in and out of the sun, thereby maintaining a relatively stable body temperature and the desert wood-louse seeks burrows and crevices that constitute friendly micro-environments quite distinct from the hostile desert outside. Burrowing is the most important form of behavioural adaptation in invertebrates. Desert invertebrates characteristically have high lethal temperatures, as in the scorpion where the lethal temperature is 47°C.

Adaptation to drought may present a greater challenge than adaptation to heat. Diapause, a dormant resting phase, occurs in desert arthropods, and it is primarily an adaptation to drought rather than to high temperature.[115] Louw[326] has described the dependence on fog water of certain water-storing beetles and lizards in the Namib desert on the south western coast of Africa. The fog, which is produced by a cold ocean current, condenses on the animal to form water droplets that the animal drinks. The fog also sustains succulent plants that support other animals such as the ostrich and Namib gerbil.

Reptiles have no sweat glands, but they may lose appreciable quantities of water through the skin and the respiratory tract. The oxygen consumption of the reptile is much less than that of a mammal of the same size (see p. 6), but the heliothermic reptile can maintain thermoregulation by using sun and shade.[496] As the day becomes warmer so less time is spent in the sun, until midday all activity is in the shade. The animal may partially bury itself in sand so reaching cooler sand below the surface, and when too hot the lizard retreats underground.[258]

7

Man

The consideration of man's responses to climate in general and to the thermal environment in particular is beset by certain peculiarities. One of these, shared with a relatively small number of other mammalian species, is that man is bare-skinned; often he has only little subcutaneous fat and the combination leads to susceptibility to cold. A second peculiarity is that man can sweat at a higher rate than any other mammal, and so through evaporative cooling can maintain a normal body temperature in hot environments. A third peculiarity is that in many instances where adaptation to environment is required, man achieves adaptation by clothing, shelter and other engineering arrangements, instead of by adapting physiologically. It is true that animals use the diversity of their environments to offset influences that would otherwise threaten survival, including sheltering behind wind-breaks or in burrows, using whatever bedding may be available, and huddling. However, man does much more than this in response to environment; he can live on the sea, under the sea, in the air, in interplanetary space, as well as anywhere on the Earth's surface, not by virtue of physiological adaptation but through the application of environmental engineering, developed through common experience earlier in his history and by conscious scientific effort in more recent times.

If it were altogether the case that man so arranged his micro-environment that the environment at large never impinged on him, the study of his physiological responses to environmental variation would be academic. That this is not the case is obvious when one compares living in the tropics, even in an air-conditioned house, with living in polar regions, even in heated insulated buildings, and when one compares the climate of a Mediterranean country with that of a northern city. It is possible that changes and contingencies might lead to the need for man to confine himself more and more in enclosed environments, but in that event the study of his responses to such environments, already well advanced, would be essential for his health and survival. This is the state of affairs that now exists for an increasing part of the world's farm livestock; they are kept under confined conditions for purposes of management and in an attempt to achieve high levels of productivity, and increasing attention has in consequence been paid to their responses to environment.

METABOLIC RATE

The effects of environmental temperature on the energy exchanges of unclothed resting male subjects are shown in Fig. 7.1. The minimal metabolic rate is about 45 W m^{-2}, which for the average surface area of 1.8 m^2 gives a rate of 81 W per man. The increase in skin conductance occurs above 28°C, body cooling occurs below 28°C, and the increase in evaporative loss takes place between 26 and 28°C. These measurements therefore indicate a critical temperature close to 28°C, although as the temperature falls the metabolic rate begins to rise only below 25°C, sometimes with delay.

The rectal temperature is at a minimum of close to 37°C at an environmental temperature in the region of 28°C, rising in environments that are either warmer or colder (the so-called 'paradoxical' rise in the cold). Above an environmental temperature of 28°C, the skin temperature over the body surface is fairly uniform, whereas at lower temperatures, for example at 18–20°C, the feet may be 6–12°C cooler than the forehead.[201]

The effects of activity and exercise on heat loss measured in a calorimeter are shown in Fig. 7.2. The female subject had a lying resting rate of heat loss of 55 W following a night's sleep in the calorimeter. Body weight was 49 kg and the body surface area 1.5 m^2, giving a rate of 37 W m^{-2}, rather less than the rate found for the resting male subjects of Fig. 7.1. The activity associated with movement and breakfast raised the heat loss to a peak of 128 W, succeeded by a rate of 85 W when the subject sat quietly for a period following the meal. Cycling on a bicycle ergometer in the calorimeter raised the rate to 164 W, about three times the initial lying resting metabolism.

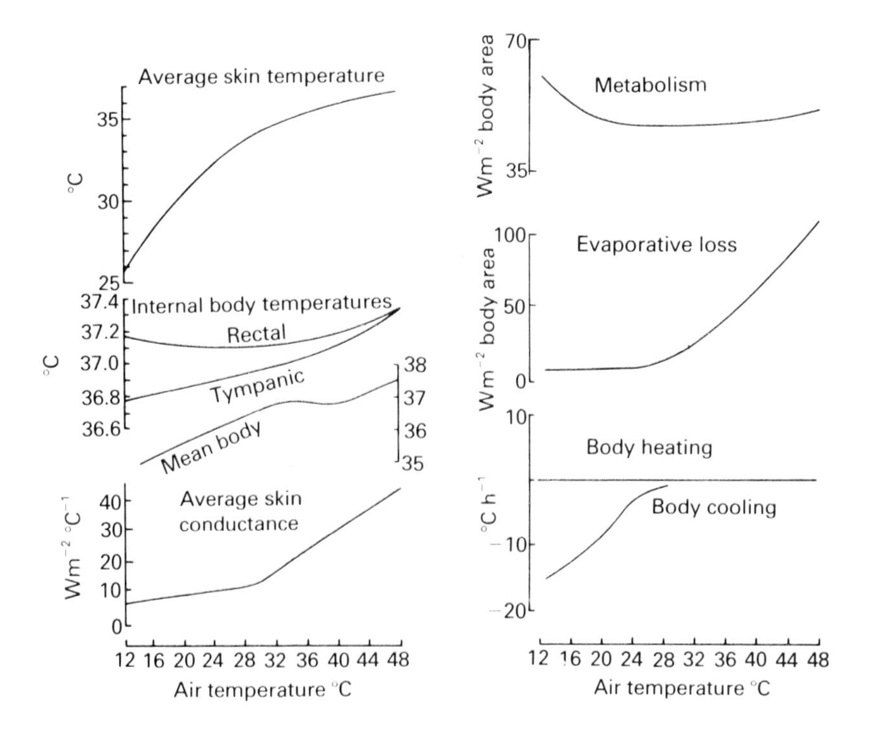

Fig. 7.1 Thermoregulatory responses of unclothed, quietly resting male subjects to air temperature 12–48°C, in still air, and at low relative humidity (Hardy, Stolwijk and Gagge, 1971, by permission of Academic Press).

The metabolic rates reached by athletes during maximum effort can be 20 times the resting rate, equal to 1600 W. This exceeds probable environmental heat loads, but clearly such rates are not maintained for long. During exercise the rectal temperature is related to the metabolic rate and not to the environmental temperature, even under very cold conditions; the regulation fails in very hot conditions. A resting man with a rectal temperature of 37.1°C and a metabolic rate of 80 W might have a rectal temperature of 38.5°C when engaged on heavy work.[296, 319, 409]

Man's metabolic and thermoregulatory responses will be considered in two parts, in relation first to cold and then to hot environments. There is wide individual variation in both energy expenditure and food intake, with considerable variation from day to day in any one individual. However, a considerable degree of energy balance is established over a period of about a week.[147]

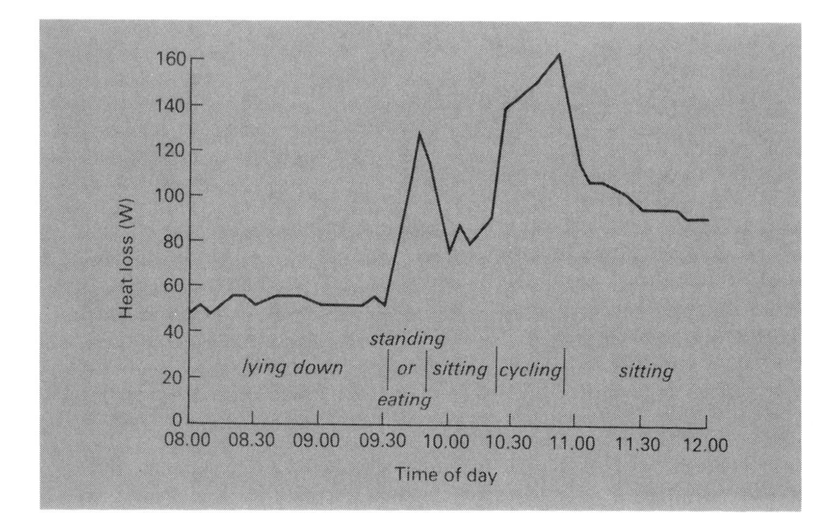

Fig. 7.2 Rate of heat loss recorded from a young woman in a calorimeter that was equipped for 24-hourly occupation by human subjects. The non-evaporative (sensible) heat loss was measured with a heat-sink system, and the evaporative heat loss was estimated from the water vapour added to the ventilating air. Body weight, 49 kg; body surface area, 1.5 m^2 (by courtesy of Dr M. J. Dauncey).

MAN IN A COLD ENVIRONMENT

In this connection, mention must be made of a book with this title[85] that has been valuable not only to those working on man but also to those whose primary concern is with animals. In addition to basic considerations, the book contains many references to practical investigations that were carried out during the Second World War, and the urgency and importance attached to these studies produced a rapid development of ideas.

Effects of wind at low temperatures

Man as a bare-skinned mammal is particularly susceptible to the effects of air movement. Exposure to wind at low temperatures greatly enhances the cold effect; when the face is exposed to wind at low temperatures it shows marked falls in skin temperature that vary with the different parts of the face. Values for skin temperatures on the nose, cheek and forehead are given in Figs 7.3, 7.4 and 7.5. Sensations

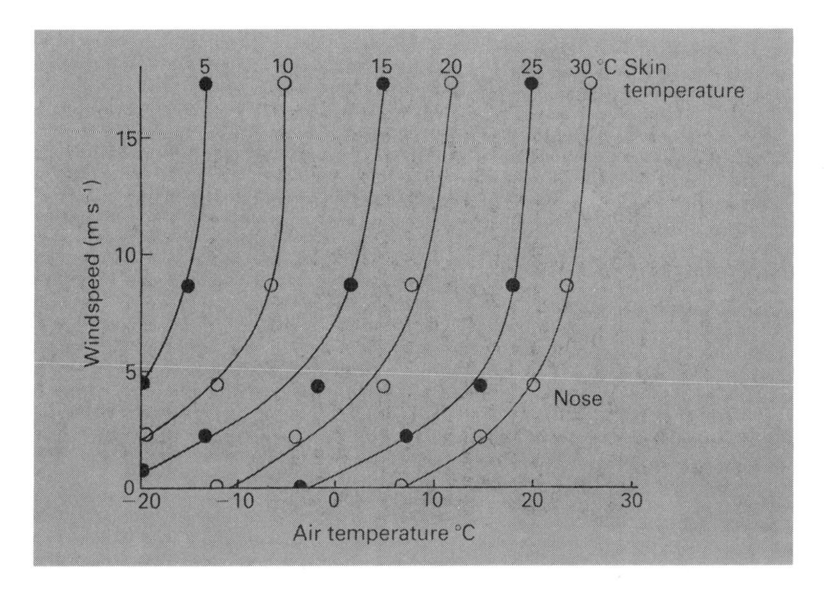

Fig. 7.3 Diagram showing the skin temperature of the nose for various combinations of wind and air temperature (from LeBlanc, J., Man in the cold, 1975. Courtesy of Charles C. Thomas, Publisher, Springfield, Illinois).

of cold appear to be perceived more acutely on the forehead than on the nose or cheek.

Assessment of the effects of wind on the exposed parts has given rise to the 'wind-chill index' that is used empirically to estimate weather conditions (Fig. 7.6). The chilling effect of wind in fact depends on the insulation of the clothing: wind has relatively little effect on a well-clothed man or thickly coated animal (see Chapter 4). However, man's face and hands are sometimes unavoidably exposed, and it is here that the effect of chill is experienced. When the temperature of the

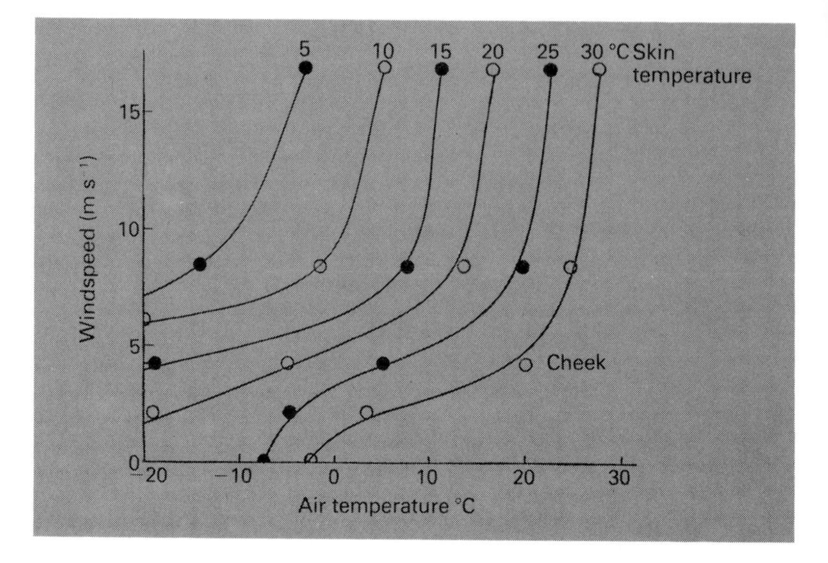

Fig. 7.4 Diagram showing the skin temperature of the cheek for various combinations of wind and air temperature (from LeBlanc, J., Man in the cold, 1975. Courtesy of Charles C. Thomas, Publisher, Springfield, Illinois).

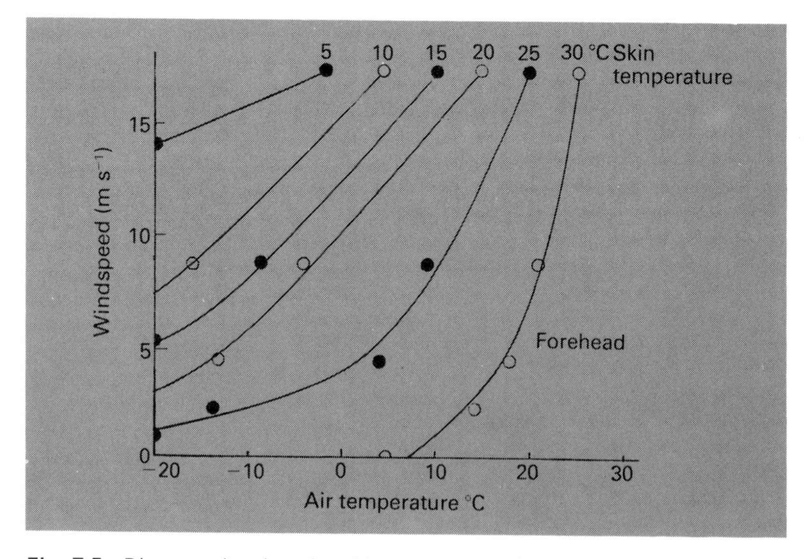

Fig. 7.5 Diagram showing the skin temperature for the forehead for various combinations of wind and air temperature (from LeBlanc, J., Man in the cold, 1975. Courtesy of Charles C. Thomas, Publisher, Springfield, Illinois).

forehead, normally one of the warmest areas of the body surface, falls to 15°C (compare the average skin temperature in Fig. 7.1) the subjective evaluation is 'bitterly cold'; this corresponds to a wind of $9\,m\,s^{-1}$ at -8°C ambient temperature on the wind-chill scale. Measurements of skin temperature can therefore be used as well as the

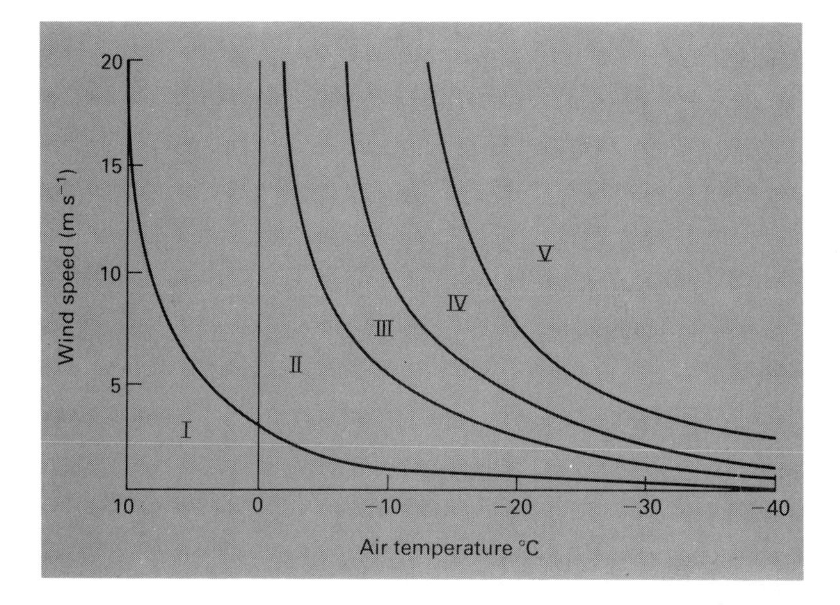

Fig. 7.6 The combinations of wind and temperature are divided into five zones of increasing environmental severity. I, comfort with normal precautions; II, very cold; III, bitterly cold; IV, freezing of human flesh begins; V, survival efforts are required; exposed flesh freezes in less than one minute. The chart can be used to make decisions about activity and clothing in very cold weather (Folk, 1974, by permisssion of Lea and Febiger).

wind-chill system to indicate the cooling effect of wind and temperature.[314]

The very high air velocities of up to $33\,m\,s^{-1}$ that can be experienced by men below hovering helicopters can lead to convective cooling coefficients as large as $80\,W\,m^{-2}\,°C^{-1}$ (see Chapter 3). Under these conditions, the warm unprotected area over the face is a region of major heat loss, accounting for up to 40% of the total body heat loss.[107]

Polar human biology

Clothing, shelter and heating are essential for man's survival in the polar environment. The efficiency of the clothing depends on its ability to trap still air, to provide an outer windproof layer and to be permeable to water vapour. Several layers are preferable to one thick layer and for the polar traveller would typically include a string vest,

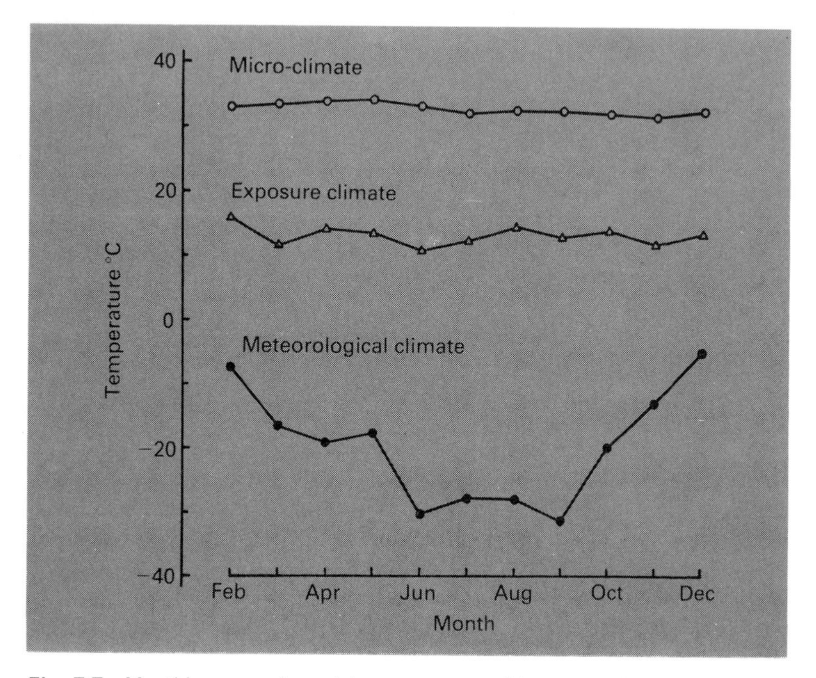

Fig. 7.7 Monthly mean values of the temperature of the meteorological climate, the exposure climate (within the station) and the micro-climate (under the clothing) to which four subjects were exposed during one year at an Antarctic station (Norman, 1965, reproduced by courtesy of the Editor of the *British Antarctic Survey Bulletin*).

long underpants, a flannel shirt, two pairs of trousers, a zip jacket, a windproof outer garment (parka), three or four layers of gloves or mittens, two or three pairs of socks and leather boots or canvas foot-coverings. The need for the clothing to be permeable to water vapour is emphasized; condensation and freezing of water vapour in clothing and sleeping bags leads to loss of the high thermal insulation that would otherwise be due to trapped air (see Chapter 4). During muscular work increasing the ventilation of the clothing leads to reduced condensation by allowing the water vapour to escape more easily.

The rigours of the polar climate can be considerably reduced by the high level of direct solar radiation, which during calm weather in Antarctica can lead to solar heat gains two to four times greater than the heat gained by a nude man in the desert. This is due to the low solar altitude in polar regions and the large surface area of an upright man exposed to the sun. However, information from polar bases shows that individuals spend only between 6 and 15% of the time out of doors, depending on the season. The total energy expenditure per man per day at base is about 15 MJ, and during sledging is up to about 23 MJ. There are corresponding requirements for food intake and under sledging conditions a high fat diet is well tolerated; walking on snow or ice in the Antarctic results in a rate of energy expenditure that is about double the rate during walking in a temperate climate.[147] Dehydration can occur from lack of fuel to melt snow and ice for water, and from the high rates of sweating that accompany sledging.

The evidence for general adaptation to cold amongst men at polar bases is inconclusive.[183] The micro-climate to which they are exposed is in any case not the external climate (Fig. 7.7), so the adequate stimulus may not be present. However, local adaptation to cold probably does occur in the hands.[149]

There has been considerable research on human biology in the polar regions, particularly in the Antarctic where there are a number of bases established by several countries. Adaptation to cold, isolation, sleep patterns and 24-hourly variations have been studied, as well as more specific physiological and medical problems.[148]

Cold injury

Local cold injuries include chilblains (the mildest form), immersion foot and frostbite.[85] Immersion foot is produced by wet cold at temperatures above freezing. There is initially cold vasodilatation, followed by ischaemia, then hyperthermia, pain and loss of sensation. It is produced by exposure to cold in the sea, or on land (trench-foot); chilblains resemble mild immersion foot.

Frostbite is the most severe form of local cold injury. At very low temperatures, circulation through the cooled part decreases, and if ice is formed there is complete vascular occlusion and subsequent death of tissue. Modern polar clothing is very efficient, so that frostbite is not now as great a hazard as might be expected. It has occurred extensively in the past particularly in military campaigns where large numbers of men have been placed in conditions of extreme cold to which they were unaccustomed.

ACCLIMATIZATION TO COLD

The critical temperature of the modern European or American, non-fasting, is about 27–29°C for the naked subject (see p. 146). This means that even the warmest climates on earth (tropical rain forest and tropical savanna) have night-time environmental temperatures that might be 5–10°C below this critical temperature. In the colder regions, some groups of primitive man have survived without the usual forms of protection from the cold and Hammel[197] has discussed the forms of physiological adaptation that they have employed.

The Aborigines of Australia were investigated in their nomadic state before their life style was changed by modern civilization.[150, 231, 462] These people lived naked in Central and North Australia. In Central Australia, the night-day air temperature range is 22–37°C in January and 4–21°C in July. The Aborigines slept on the ground behind a wind-break, with fires nearby; during the night the mean body temperature would fall about 2°C and the mean skin temperature about 4°C. Body cooling without metabolic compensation occurred in the Aborigines in summer to the same degree as in winter. They slept comfortably, in contrast to Europeans who exposed themselves to similar conditions and experienced shivering and discomfort, with smaller falls in both deep body and skin temperatures and a marked metabolic response. Europeans adapted to cold could sleep, with a raised metabolic rate and maintained foot temperatures. Australian Aborigines could sleep although their foot temperatures fell considerably; they showed no rise in metabolic rate. The form of thermal adaptation of the Europeans in this situation can be described as metabolic acclimation. The metabolic response of Aborigines living on the tropical north coast of Australia was intermediate between that of the central Aborigines and that of the Europeans.

Hammel[197] has classified the type of acclimatization to cold shown by the Aborigines as 'insulative-hypothermic'; Kalahari Bushmen in Africa exhibit a similar form of acclimatization. This form of acclimatization involves the insulation that is provided by a cooled body shell (see Fig. 1.1), a fall in body temperature and a lack of metabolic response to exposure to cold. It is in contrast to 'metabolic acclimatization', where there is a sustained rise in metabolic rate as a result of chronic exposure to cold, combined with maintenance of body temperatures. Some degree of metabolic acclimatization was found in a group of Norwegian youths exposed to cold on a mountain plateau. This form of acclimatization was also found in the Alacaluf Indians living in Tierra del Fuego in South America, in a cold wet climate; they showed a metabolic response to moderate cold and a high foot temperature. Metabolic acclimatization also occurs, but to a lesser extent, in the Arctic Indians, and in the Eskimos, who are clothed and therefore have no chronic exposure to marked cold.[156]

When compared to Kalahari Bushmen, it is found that in a standard cold test Eskimos shiver more, and maintain higher skin and rectal temperatures. The Eskimo lives in clothing of such high insulation that he is comfortable at environmental temperatures down to −50°C. He virtually has a tropical micro-climate inside his clothing; it is consistent with this that he begins to shiver at the same skin temperature as people from temperate zones and shows little difference in thermoregulation, with similar tissue conductances.[433] Hand blood flows are however higher in the Eskimo, associated with reduced cold sensation and ability to work easily with the hands under cold conditions. When the hand was placed in a 10°C water bath, the Eskimo experienced cold but no pain, whereas white subjects developed severe cold sensation followed by a deep aching pain and more marked changes in blood pressure (Fig. 7.8). Gaspé fishermen in Quebec, who immerse their bare hands in water at 9°C for many hours while pulling nets, maintain higher skin temperatures than control subjects when the hand is immersed in cold water, without pain and with a smaller rise in blood pressure. This is a specific local adaptation in the extremities, as in Eskimos, because when exposed naked the fishermen gave no evidence of cold adaptation. Adaptations are highly specific; for example, adaptation to cold immersion of the left hand is specific, with no development of adaptation to subsequent immersion of the right hand.[314]

Other European subjects exposed to cold, either intermittently or continuously, were found to react by lowering the mean body temperature levels at which increased metabolic rate and shivering occurred. This hypothermic type of adaptation to cold was accompanied by a reduction in thermal discomfort and cold sensation. Separate observations on long-distance runners also showed a reduction in the threshold body temperature for shivering and for cold sensation, as well as a reduction in the sweating threshold that was comparable to the reduction found in heat adaptation. The runners behaved as if the 'set-point' of their thermoregulatory system had been reset at a lower level. There is thus cross-adaptation between cold and physical exercise, producing a hypothermic adaptation that is quite distinct from metabolic adaptation.[36, 83] Mean skin temperatures during rest and activity have been measured on athletes by infra-red colour thermography.[108] The method offers a means of obtaining a comprehensive picture of skin temperature and its variations that

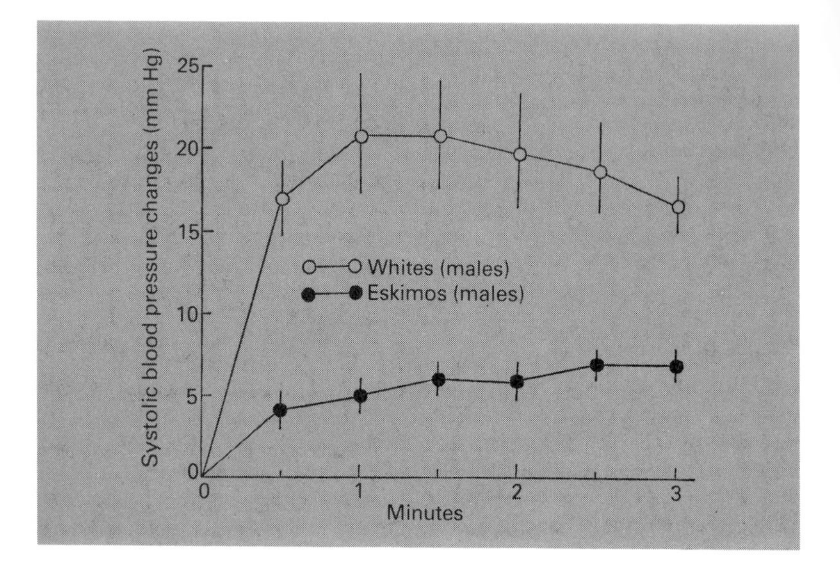

Fig. 7.8 Changes in systolic blood pressure in white subjects and Eskimos during immersion of the hand in water at 4°C for 3 minutes (from LeBlanc, J., Man in the cold, 1975. Courtesy of Charles C. Thomas, Publisher, Springfield, Illinois. From LeBlanc, 1973, by permission of Wm. Heinemann Medical Books Ltd).

could be more useful than information derived from thermocouples in studies on metabolic adaptation (see Fig. 3.9).

Another form of adaptation is seen in successful long-distance swimmers. These persons typically have considerable subcutaneous fat, of mean thickness about double that of the general population. They can maintain high levels of metabolism and this together with their considerable thermal insulation allows the maintenance of rectal temperature to be more readily attained than in lean subjects[266] (Fig. 7.9). The larger falls in rectal temperature that occur in leaner people exposed to cold water are related to thickness of subcutaneous fat in Fig. 7.10.[293] In this connection, Jéquier, Pittet and Gygax[275] have demonstrated calorimetrically that obese people have a higher thermal insulation and lower metabolic rate per unit area than lean people.

The elderly subject

The elderly person is particularly susceptible to hypothermia, the causes being physiological, social and economic. Physiologically, the older person has a lower metabolic rate;[302] socio-economically, the elderly individual may live alone, may not pay sufficient attention to an adequate diet, and may not be able to afford sufficient food and fuel.[160]

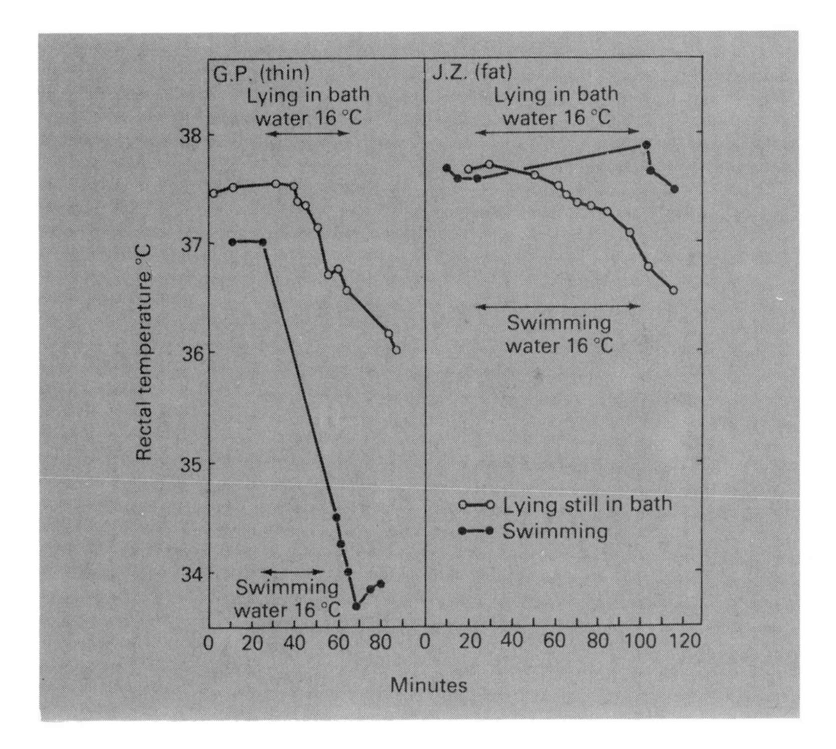

Fig. 7.9 Rectal temperatures of subjects G. P. (thin) and J.Z. (fat) in water at 16°C, with and without swimming (Pugh and Edholm, 1955, by permission of The Lancet).

MAN IN A HOT ENVIRONMENT

In contrast to man's susceptibility to cold, man's tolerance of hot conditions is considerable, largely on account of his ability to sweat at a high rate. Sweat production per unit area of body surface can reach $1000\,g\,m^{-2}\,hr^{-1}$ or more, and if this were all evaporated at the skin surface the rate of evaporative heat loss would be close to $700\,W\,m^{-2}$, which is about 14 times the resting metabolic rate. The following account of man's responses to a hot environment is largely taken from '*Man and Animals in Hot Environments*'.[258]

There are broadly two situations in which man is exposed to hot conditions. The first is one in which the individual is accustomed to living in a hot environment; he is in equilibrium with it and adapted to it. He may live in the desert or in a tropical country, or he may live in a temperate climate and yet spend a part of each day at high

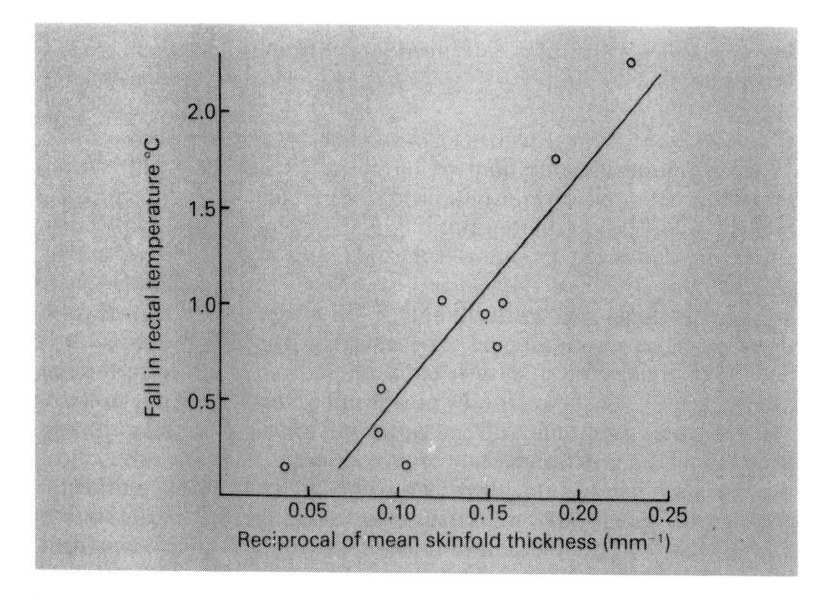

Fig. 7.10 The relation between subcutaneous fat thickness and fall in body temperature during 30 minute immersions of men in stirred water at 15°C. Skinfold thickness is the mean of readings at 4 standard sites (Keatinge, 1960, by permission of Journal of Physiology).

temperatures, as in the case of a foundry worker. The second situation is where the temperature rises from the equable cool level to which a man is accustomed to a height that imposes such demands on thermoregulatory mechanisms that adaptation becomes necessary. The adaptation that takes place includes initial acute responses followed by more persistent chronic changes that constitute acclimatization to the changed conditions. The second situation, that is where environmental temperature rises, indicates the effects of high temperatures more usefully than the first equilibrium situation because it involves change and therefore contrasts physiological responses occurring before, during and after acclimatization. These effects are experienced by people moving between temperate and hot climates.[332]

ACCLIMATIZATION TO HEAT

It is a matter of common observation that when the weather suddenly becomes hot the individual may be overtaken by lassitude and a relative inability to perform effectively tasks that were carried out at a high level of efficiency under cooler conditions. Clothing is discarded, sweating is noticeable and the individual becomes irritable; physical work that was undertaken previously without error now becomes a burden, and may be accompanied by nausea and dizziness. After a few days discomfort decreases as acclimatization takes place, with both psychological and physiological adjustments. The principal changes are: (1) fall in body temperature, (2) fall in heart rate and (3) increase in the rate of sweating, with a more rapid onset of sweating. Corresponding adaptation occurs in men who live in hot climates.

Bass[34] has discussed acclimatization to heat in a man undertaking physical work. The effects produced by heat during work are much more pronounced than those in the sedentary individual. Physical work imposes an additional heat load and so enhances the rise in body temperature and rate of sweating. Bass describes an unacclimatized individual asked to walk at 5.6 km h^{-1} for one hour at 49°C. The subject first experiences severe discomfort, then dizziness, nausea and even collapse, accompanied by high deep body and skin temperatures, very rapid pulse and an inadequate secretion of concentrated sweat. If the walk is repeated every day, limited as necessary to what is within the capability of the individual, dramatic improvement in the performance of the task takes place within a few days. Subjective discomfort tends to disappear as the fall in body temperature and increased sweating take place.

Acclimatization to heat begins with the first exposure, and then progresses rapidly. It can be induced by short daily bouts of work in hot conditions. Acclimatization is more rapid in subjects in good physical condition, whose maximal work load can be reached quickly by daily increases; it is well developed in 4–7 days, and then even without further exposure it is retained for about two weeks before it is progressively lost. The length of daily exposure required to bring about acclimatization has been the subject of investigation: 100 minutes in a single exposure is recommended as the most economical, with no improvement in acclimatization as the result of two such exposures per day.[319]

Acclimatization is often said to be specific for the conditions of exposure, both for the particular energy expenditure and the particular environment. Sometimes acclimatization to one level of energy expenditure and one hot environment does not provide physiological adaptation either for a higher level of expenditure or for a hotter environment. For some occupations, such as working in hot mines, acclimatization can be achieved in climatic rooms before exposure to actual working conditions, so allowing close control of heat exposure and work rate.[558] Although heat acclimatization tends to be linked to a given rate of working, acclimatization does occur with the technique of controlled hyperthermia, where body temperature is raised passively by exposure to heat without work being undertaken by the subject.[159]

Sweating

The essential thermoregulatory response in the heat is sweating, coupled with cardiovascular changes.

Evaporative heat transfer provides the only effective channel through which heat can be dissipated to the surroundings when the temperature of the surroundings is above that of the body. Under these conditions non-evaporative heat transfer is in the form of an added heat load on the individual instead of providing channels for heat dissipation.

In a cool environment, evaporative heat loss accounts for about 25% of the heat produced by the resting metabolism in man; as the environmental temperature rises so an increasing proportion of the heat lost is by evaporation until under hot conditions this mode accounts for the total heat loss. As a result of his highly developed sweating function and the resulting high cooling rate, man can withstand very high temperatures in a dry atmosphere. Blagden[53–4], (quoted by Ingram and Mount[258]) tolerated a dry atmosphere of 130°C for 15 minutes during which a beefsteak that was also exposed to that temperature was cooked.

Human sweat is more than 99% water, with sodium chloride as the principal solute but in highly variable amounts. The variations are related to the rate of sweating: in an unacclimatized man, as the rate increases, the concentration also increases, and the two quantities subside together. After frequent exposure to heat, however, although the rate of sweating increases the chloride concentration tends to decrease; in the summer the concentration is lower while the sweating rate is higher, and in the winter the reverse is true.[304, 563] In a hot desert climate, the chloride concentration in the sweat in normal subjects has been found to lie mainly between 0.2 and 0.35%, with extremes of 0.1–0.6% (Fig. 7.11). The actual salt loss in a hot environment may reach 10–30 g per day and be even higher under extreme conditions.

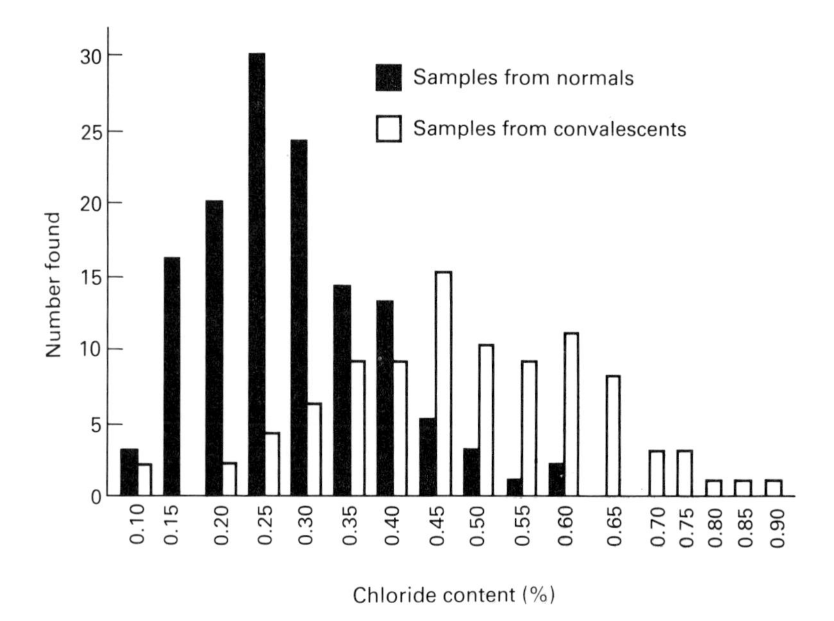

Fig. 7.11 The distribution of chloride concentration in the sweat among British Army personnel in Iraq in 1943. The sweat from subjects convalescent from heat exhaustion was in general found to contain more chloride than sweat from controls (Ladell, Waterlow and Hudson, 1944, by permission of The Lancet).

The maximum sweating rate that can occur at rest has been investigated by Kerslake and Brebner.[297] They heated subjects in a water bath until the mouth temperature was 38°C and then measured the sweat loss by weighing. Sweat rate was independent of mouth temperature provided this exceeded 38°C, but it varied with skin temperature, increasing by about 11% per degree Centigrade rise. The maximum sweat rates were about 40 g min^{-1}, or rather more than 2 kg per hour, for each of two subjects weighing 69 and 76 kg; the maximum rates were lower in the afternoon than in the morning. Maximal rates of sweating of 1.3 kg m^{-2} hr^{-1} have been observed on men walking on a treadmill.[447] In other investigations it has been found that the maximal ability of man to sweat increases with acclimatization that is brought about by daily exposure to severe heat stress.

The rate of sweating often declines during exposure to heat, a phenomenon termed hidromeiosis. The effect is more pronounced in humid environments where unevaporated sweat drips from the body; it is associated with swelling of the epidermis due to wetting, and can be induced in an area of skin by soaking in water. Hidromeiosis may be due to blockage of the duct of the sweat gland by particles from the gland. Recovery is rapid.[296]

Hot dry conditions: dehydration

Men working in the desert can lose 10–12 kg water per day in sweating and although some saving might be made by the reduction of urine output it is obvious that extra water must be taken in if dehydration is to be avoided. The rate of evaporation from men and women sitting in the sun, at about 40°C in the desert, has been found to be half the rate when they were walking.[476] Table 7.1 gives a series of rates of sweating for men clothed and unclothed, active and inactive, at high temperatures in the desert; the rate of evaporation falls below the rate of sweating when sweat runs off the body. 1.47% of man's body weight is given as the quantity of sweat to be evaporated per hour for a constant body temperature to be maintained under desert conditions at 40°C, compared with 8.5% of the body weight for the rat and 0.77% for the camel; the percentage required is inversely related to body weight (see Fig. 6.11).

Dill[141] remarks that thirst does not always cause a man to keep his water intake up to the rate of loss. This is true for the unacclimatized man, although the heat-acclimatized man produces a dilute sweat, loses relatively more water than salt and is much more thirsty for a given water deficit and more nearly maintains his water balance by voluntary drinking. Under conditions of very high sweat rates man simply cannot drink water fast enough to keep in balance; efforts that he makes to do so may lead to vomiting. Men acclimatized to desert conditions, working at a moderately severe level, have a chloride concentration in the sweat of between 1–1.5 g $litre^{-1}$, so that loss of sodium chloride in the sweat is likely to be 10–15 g per day, which is the upper part of the usual dietary intake for Europeans and Americans. Table 7.2 gives the suggested salt additions to the conventional diet for a range of conditions.

Studies on man subjected to a dehydration of 10% of body weight in the desert have shown that in spite of the severe thirst and mental derangement that accompany the maximum water deficit, recovery is complete within an hour of drinking. The signs of thirst are very strong when as little as 2% of body weight has been lost, but do not get progressively worse as dehydration proceeds; after 4% weight loss

Table 7.1 Sweating rates in summer at Yuma, Arizona, vapour pressure 7–20 mbar; subjects nude or in light weight clothing; marching 5.6 km h^{-1} on level surface; air velocity from fans 5 m s^{-1} (Lee, 1964, by permission of American Physiological Society).

Conditions	Rate of sweat loss, g hr^{-1} 35°C	40°C	45°C	Rate of sweat evaporation, g hr^{-1} 35°C	40°C	45°C
Nude, marching, sun	965	1210	1450	935	1165	1375
Clothed, marching, sun				640	910	1120
Clothed, marching, shade				490	730	960
Nude, sitting, sun	455	685	915	385	615	800
Clothed, sitting, sun	280	500	730	280	460	610
Nude, sitting, shade	220	380	540	220	360	475
Clothed, sitting, shade	245	435	620	245	305	375
Marching, shade						
Shorts, fans	690	930	1120	675	910	1100
Shorts, no fans	640	850	1015			
Clothed, fans	570	805	990	500	735	925
Clothed, no fans	575	775	940	580	740	860

the mouth is very dry and at 8% the tongue is swollen and speech difficult. Observations made on men who have been lost in the desert suggest that at 12% loss of body weight recovery is possible only with some assistance and it may be necessary to give water by injection or *per rectum*. The actual lethal limit of dehydration, however, is probably as much as 15–25% of body weight.[2]

There is very little tendency for sweat rate to be diminished during dehydration. A comparison of a group of men who received water *ad libitum* and a similar group without water in the desert showed that although the men without water used about 10% less for cooling they became more dehydrated than the other group. Schmidt-Nielsen[455] has pointed out that the idea that man might train himself to use less water during journeys in the desert is attractive, but is based on wishful thinking rather than sound physiological knowledge.

One tendency that would lead to economy in water would be the reduction of urine flow due to the excretion of a concentrated fluid. In man, urine volume is reduced to about half the usual quantity in a hot climate even when water is freely available; if he is doing hard work there is a further reduction, but the amount does not fall below the dangerously low level of 250 ml per day unless dehydration is very

Table 7.2 Suggested daily salt addition to normal diet (Lee, 1964, by permission of American Physiological Society).

Conditions	Acclimatization	Work	Addition, g
Warm humid	Acclimatized	Moderate to heavy	0
Warm humid	Unacclimatized	Moderate to heavy	7
Hot dry	Acclimatized	Sedentary	0
Hot dry	Acclimatized	Moderate to heavy	7
Hot dry	Unacclimatized	Moderate	7
Hot dry	Unacclimatized	Heavy	14
Very hot	Acclimatized	Moderate to heavy	14
Very hot	Unacclimatized	Moderate to heavy	21

severe. The extent to which urine flow can be reduced depends on the ability of the kidney to concentrate the urine and in man the ability to produce a concentrated urine is rather limited.

The limitation of concentrating power explains why the drinking of sea-water is disadvantageous under conditions of dehydration. Sea-water has a salt concentration that is greater than the maximum salt concentration that can be achieved in the urine by man, and the elimination of this additional salt requires more water than is taken in with the sea water. Since sweat production and consequently salt loss is high in a hot climate, however, some salt is lost through the sweat glands and under these conditions a slightly saline drinking water is an advantage in the maintenance of salt balance. The important factor is the balance between the concentration of salt in the drinking water and the elimination of salt in the sweat. The drinking of concentrated urine is completely useless since the solutes are returned to the body and have to be excreted again in the same amount of water.[258]

In preparation for a journey under desert conditions perhaps the most important measure man can take is to dress in clothes that reduce the radiant heat load and so reduce the quantity of heat that needs to be lost by evaporation. In comparison with the unclothed state, light clothes with long trousers and a long sleeved shirt reduce radiant heat load by half and water loss by two-thirds.[455] Drinking as much water as possible before starting out on a journey in the desert is likely to lead to a diuresis, but if conditions are already hot and sweat rate high some additional water can be drunk without affecting urine flow.

In a comparison between man and dog in the desert[142] it was found that, in contrast to man, the dog drinks, at the first opportunity, almost the exact amount of water it has lost during a day in the desert. After exposure to extreme heat the course of dehydration in the dog follows a similar pattern to that in man and at a water loss of 14% of body weight there is an explosive rise in temperature that would be lethal if dehydration were continued. Although when man becomes dehydrated the amount of water lost in sweat decreases very little, body temperature nevertheless tends to rise. Resting men exposed to desert conditions have rectal temperatures that increase progressively as dehydration proceeds (Fig. 7.12).

According to Schmidt–Nielsen[455] the reduced capacity to withstand dehydration in a hot climate is related to the reduced blood volume which is no longer capable of transferring heat to the surface. The dog that is continuously exposed to a high temperature would thus die from circulatory failure rather than from dehydration. The cat can withstand a slightly greater degree of dehydration than the dog and the kidney can concentrate the urine to a greater extent, so that the animal is better able to withstand hot dry conditions.

Hot wet conditions

In a tropical climate, under hot-wet conditions, a man is not nearly so likely to be exposed to a high radiation gain as he is in the desert. In a hot-wet climate, resting man can maintain thermal equilibrium for most of the day without sweating but with peripheral vasodilatation. If he becomes active he will sweat, although the high ambient humidity limits the possible evaporative cooling. Convective cooling is almost impossible for much of the day.[305]

Circulation

In addition to sweating, the other thermoregulatory response that is of great importance in the heat is the redistribution of blood in the circulation. The transfer of metabolic heat from the deep tissues to the skin, where it is lost by evaporation, depends on skin blood flow, which under hot conditions is increased relative to other organ blood flows. The early stages of acclimatization to heat are marked by an inadequate cardiovascular response; there is a considerably increased vascular volume relative to blood volume due to widespread peripheral vasodilatation. As acclimatization proceeds the blood volume increases, the pulse rate decreases and a new cardiovascular equilibrium is established.

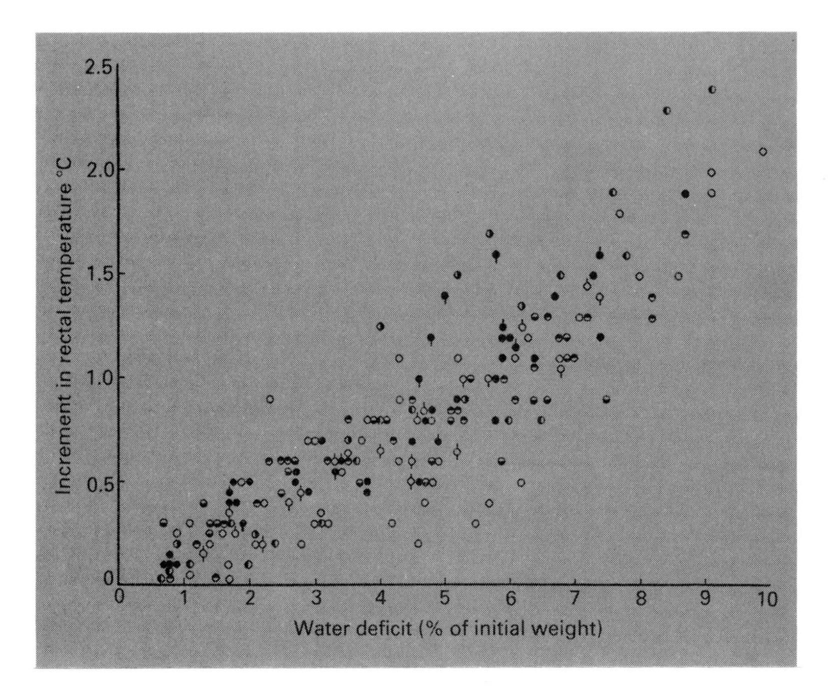

Fig. 7.12 The increments in rectal temperature that occurred at various water deficits in 27 laboratory dehydrations of 11 men undergoing exercise. Control subjects who worked at the same rate but drank water to maintain their hydration showed no progressive changes in rectal temperature after the first 30 min of exposure. A water deficit equal to 10% of initial weight was associated with an increment of 2°C in rectal temperature. Each symbol represents one subject (Rothstein and Towbin, 1947, by permission of Professor E. F. Adolph).

The cardiovascular inadequacy of unacclimatized man results from the high skin blood flow leading to impaired circulation in the brain, working muscles and other organs. The overall blood flow to the skin ranges from 0.16 litres $m^{-2} min^{-1}$ in the nude resting man at 28°C to 2.6 litres $m^{-2} min^{-1}$ in a very hot environment. An acclimatized man working for 6 hr at a metabolic rate of 220 W m^{-2} at 50°C with 18% relative humidity can maintain thermal equilibrium with a total skin blood flow of 1.2 litres m^{-2} min^{-1}, which is about 20% of his total cardiac output of 11 litres min^{-1}.[448]

Blood flow to the hands and feet is increased under hot conditions by the release of vasoconstrictor tone. Robinson[448] states that vasodilatation in the hand has its onset at an environmental temperature below the critical level; the vessels of the arms dilate and the legs and feet show vasodilatation at temperatures up to 30°C with maximum vasodilatation about 31–32°C. Compensatory vasoconstriction in the viscera is illustrated in resting man at 50°C by the reduction that occurs in renal blood flow to 60–75% of the flow found in a cool environment. Exercise in the heat reduces renal blood flow to 40–50% of control values and dehydration brings about a further reduction.

The rate of blood flow in the skin is the factor that has the chief influence on the internal conductance of the body; the higher the blood flow and the conductance then the greater is the rate of transfer of metabolic heat from the tissues to the skin for a given temperature difference. An acclimatized man working under conditions of extreme heat stress can conduct metabolic heat to the skin at a maximal rate of 90 W m^{-2} per °C temperature difference between rectal and skin temperatures. This coefficient is more than twice as large as that in a man at rest and about three times as great as those determined on the same men in corresponding metabolic states in a cool environment in which cutaneous flow is reduced by vasoconstriction.[447]

Exercise

There is an interaction between the blood-flow demands of muscle for activity and of skin for heat dissipation when exercise is performed in the heat.[465] In a comfortable environment the cardiac output is determined largely by metabolic changes in the active muscle, resulting in vasodilatation in the active tissues and vasoconstriction elsewhere. In the heat, skin blood flow must also be increased. However, with identical levels of working in hot and comfortable environments the cardiac output is not larger in the heat, and renal and hepatic blood flow show greater reductions during exercise in the heat than under comfortable conditions.

The maximal capacity for work in the heat is reduced because the maximal heart rate and maximal decrement in visceral blood flow are reached at lower levels of work than in a cooler environment. A criterion for the limitation of tolerance of heat lies in the heart rate; a reasonable limit for the heart rate, which might apply for prolonged periods, is 100–110 beats min^{-1} while the subject is sitting, and

120–130 while working, although higher rates might be borne without undue disturbance for shorter periods.[39]

Women appear to be affected more adversely than men by short-term exposure to heat or to work in the heat; this may be partly because women sweat less than men in severe heat. However, body temperature and circulatory reactions after acclimatization are similar in both men and women. This suggests that unacclimatized women are further from acclimatization than unacclimatized men, possibly on account of generally lower levels of activity.[101]

Blood volume

There is an increase in blood volume early in acclimatization to heat.[319] The range of increase during heat exposure extends from 5–40%, with acclimatization producing greater increases in blood volume than those found in short exposures and returning to control levels within 2–4 days of exposure to a cool environment. An important consequence of dehydration is a reduction in blood volume with an increased viscosity of the blood, leading to increased work by the heart in maintaining the circulation. Pulse rate increases and stroke volume decreases, the cardiac output remaining about the same. This increased load on the heart occurs at a time when there is a large skin blood flow associated with heat-induced peripheral vasodilatation; because blood volume is reduced a point can be reached at which the system fails and body temperature increases above the lethal limit. Dehydration leads to a proportionately greater reduction in plasma water than in total body water and if the subject is not well hydrated under hot conditions the consequent dehydration prevents the adaptive increase in blood volume, leading to cardiovascular insufficiency and lack of tolerance to heat.

HEAT STRESS

Several indices of heat have been developed to describe the stress imposed by a hot environment. One of the earliest was the wet-bulb temperature, which was used by J. S. Haldane[193] to assess the effects of high air temperature on man. His conclusions were that the upper limits of wet-bulb temperature that could prevail without a rise in body temperature were 31°C with the subject clothed and at rest in still air, 34°C at an air movement of $0.8\ m\,s^{-1}$, and 25°C during the performance of moderate work in still air; these values are still generally acceptable. It has been found in South African mines that heat stroke becomes a serious problem above a wet-bulb temperature of 31°C, the incidence rising sharply above 32°C and becoming very high at 34°C.[159] Indices of heat stress, other than wet-bulb temperature, have involved the deep body temperature, heart rate and other factors such as skin temperature and sweating; these have been discussed by Kerslake.[296]

Scales that depend on the effective temperature have become the best known measures of heat stress and these, together with an account of other systems, have been discussed recently by Belding.[39] The basis on which the scales were made was to assign equivalent effective temperatures to combinations of conditions, each combination producing an equivalent thermal sensation, with particular attention to comfort limits but also extending to hot and cold zones. Several modifications of these scales were introduced later to take account of factors which were under- or over-represented in the original scales; modifications for radiant heat load included the 'corrected effective temperature' and the 'wet bulb: globe temperature index' (WBGT).

The WBGT requires three readings: standard black globe thermometer (T_g), shaded dry bulb (T_a), and wet bulb without artificial ventilation (T'_{wb}).

The index is then given by:

$$\text{WBGT} = 0.7T'_{wb} + 0.2T_g + 0.1T_a$$

If the normal wet bulb temperature (T_{wb}) is used, from a forcibly ventilated wet bulb not exposed to radiation:

$$\text{WBGT} = 0.7T_{wb} + 0.3T_g$$

The WBGT has been valuable in reducing heat casualties during army training; its particular value, coupled with its simplicity, is that it indicates environmental conditions that may cause casualties through heat illness. Its application is therefore particularly appropriate in conditions where heat stress reaches a critical level.

The 'wet, dry index' (WD) is given by:

$$0.85T_{wb} + 0.15T_a = \text{WD}$$

The WD gives index numbers that correspond to those of effective temperature, and the WBGT corresponds to corrected effective temperature.

Kerslake[296] has given an extensive treatment of the subject of heat stress and the indices that are used to indicate the intensity of heat stress. Because in practice the precision of a heat stress index should be matched to the situation, Kerslake considers that something like the WBGT might be the most suitable index for many purposes.

Heat tolerance

There is considerable variation between individuals in respect of their tolerance of hot conditions. Men with low maximum oxygen intakes develop higher body temperatures in the heat and consequently show less heat tolerance. Other factors associated with intolerance are previous work in cool conditions, and overweight. The prediction of intolerance to heat is improved if all three factors are used in classification.[557]

How well a man can perform tasks at high temperatures depends on his degree of acclimatization. Men acclimatized to the heat and stripped to the waist become less efficient in a variety of tasks when the effective temperature rises above about 30°C, whereas the performances of unacclimatized and clothed man deteriorate at lower temperatures.[420] The severity of hot conditions for prolonged and intermittent exposures to heat has been subdivided into three grades.[324] Intolerable situations are those in which only short exposure leads to fainting or other effects, and in which thermal man-environment equilibrium cannot be established, as in fire-fighting emergencies. The just tolerable situation is that in which equilibrium is established at the limits of thermoregulatory capacity, in certain industrial or military situations, in which only intermittent exposure can be permitted. Dehydration is the principal danger in the just tolerable situation, because sweat losses are so high that drinking cannot be expected to make good the water deficit; it is therefore undesirable to prolong such exposures beyond about 4 hr. Easily tolerable situations are those in which continuous exposure is accompanied by thermal equilibrium without undue physiological effort, as in some everyday industrial work.

Heat disorders

Illness or disorders associated with heat are produced by metabolic and environmental heat that together constitute an excessive load on the body. The clinical effects arise not only from the failure of thermoregulation, but also from its success, as in the loss of water and electrolytes in sweat. Syncope may result from cutaneous vasodilatation in response to heat in the unacclimatized man, or it may be caused by the circulatory insufficiency that occurs in salt depletion. Heat exhaustion may or may not be accompanied by a rise in body temperature.

Leithead and Lind[319] refer to these characteristics and describe the classification and incidence of heat disorders.

1) Disorders complicating thermoregulation. These include:
   (a) Heat syncope, occuring without observable depletion of water or salt, resulting in giddiness, acute fatigue, or loss of consciousness.
   (b) Heat exhaustion, related to depletion of water or salt. Water depletion is associated with thirst and pyrexia, leading eventually to delirium and death, whereas salt depletion is associated with nausea, vomiting, giddiness, muscle cramps (the well known miner's cramp), and eventually circulatory failure.
   (c) Prickly heat, a skin condition associated with prolonged wetting of the skin by sweat and characterized by a prickling sensation when the patient is sweating.
2) Disorders resulting from failure of thermoregulation. These are heat stroke and heat hyperpyrexia. Heat stroke comes on suddenly following exposure to a hot environment, with a high body temperature, absence of sweating and disturbances of the nervous system. It is frequently fatal. In heat hyperpyrexia, the patient is conscious and may be sweating; the rectal temperature, although high, is usually lower than in heat stroke.
3) Psychological effects of heat. These are apathy, fatigue, and poorer performance of skilled tasks; willingness to work is reduced rather than the capacity to work.

THERMAL COMFORT

A thermal balance involving man's heat production and his exchanges of heat with the surroundings can be attained over a wide range of environmental conditions. The achievement of such a balance, however, does not necessarily imply that the individual is in thermal comfort. Fairly narrow intervals of skin temperature and rates of sweat secretion correspond to thermal comfort. At skin temperatures between 32 and 35°C there is little sensation of either warmth or cold, with a maximum degree of comfort at about 33°C. Behavioural responses to environmental conditions in man may depend only on skin temperature, so that he adjusts his clothing and position accordingly. Subjective feelings about skin temperature can be affected by core temperature: the hyperthermic subject finds water at 10°C pleasant, but when he becomes hypothermic water at 10°C is unpleasant, and water at 40°C, which is unpleasant to the hyperthermic man, is pleasant when he is hypothermic.[86]

Man's reactions to environment are also influenced by 24-hourly rhythms in metabolic rate and body temperature. As a diurnal (or perhaps crepuscular) mammal, man has a higher metabolism and body temperature during the day than during the night and the 24-hour rhythm in heat balance does not depend on alternating states of wakefulness and sleep.[16] The time of maximum body temperature is usually in the late afternoon, with the minimum at about 4 am, the difference being of the order of 0.7°C. Work at polar stations has contributed to knowledge of the effects of variations in daylength.[148]

Subjective comfort can be related to level of activity, clothing, air temperature, air velocity, mean radiant temperature and humidity, either by equation or graphically.[152] In practice it is simpler to use diagrams instead of the 'comfort equation', and a set of such diagrams for persons with medium clothing is reproduced in Fig. 7.13. The lines on the graphs are 'comfort lines'. The comfort lines cross each other (a and b) when the air temperature is equal to the mean temperature of the outer surface of the clothed body, since at this point the convective heat transfer is zero independently of the air velocity. To the left of the crossing-point, surface temperature is higher than air temperature, and a rise in air velocity requires a rise in air temperature for thermal comfort. To the right of the crossing-point, the air temperature is higher than the surface temperature, so that heat is being transferred to the body and an increase in air velocity needs a fall in air temperature for the same degree of comfort.

The zones of dry-bulb and wet-bulb temperatures that are comfortable for individuals who are acclimatized either to a temperate climate or a tropical climate are summarized in Fig. 7.14. Information like that given in Figs 7.13 and 7.14 allows the selection of desirable indoor temperatures and environments as combinations of air temperature, air velocity, mean radiant temperature and humidity, for people at different levels of physical activity.

CLOTHING

The world has been divided into seven 'clothing zones' by Siple,[471] from whom the following account is largely taken. The zones are:

1) Minimum clothing: humid tropical and jungle;
2) Hot dry clothing: desert;
3) One layer clothing: subtropical;
4) Two layer clothing: temperate cool winter;
5) Three layer clothing: temperate cold winter;
6) Four layer maximum clothing: subarctic winter;
7) Extreme cold: arctic winter.

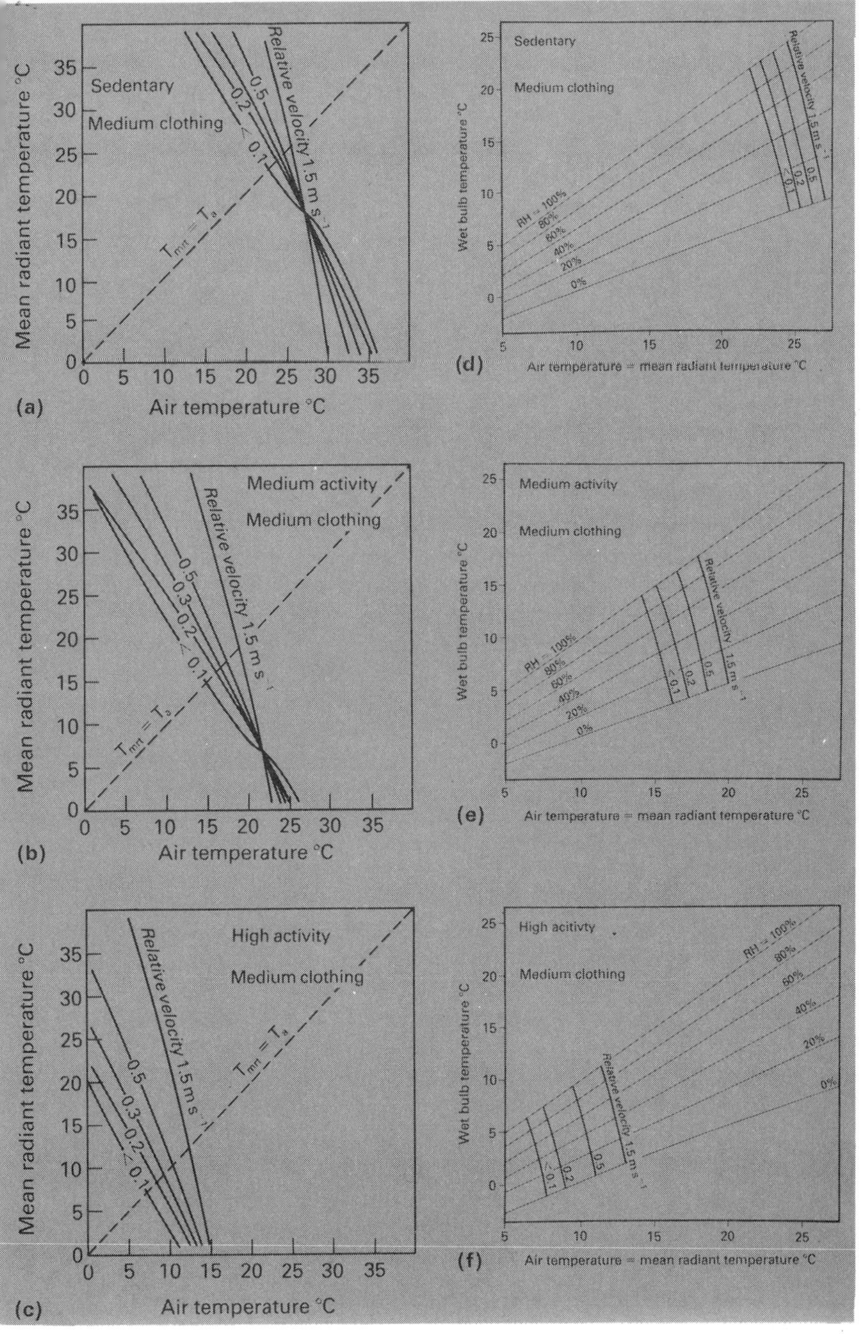

Fig. 7.13 (**a**)–(**c**) Comfort lines (air temperature versus mean radiant temperature with relative air velocity as parameter) for persons with medium clothing at three different activity levels: (**a**) sedentary, (**b**) medium activity, and (**c**) high activity. T_{mrt} = mean radiant temperature, T_a = air temperature. (**d**)–(**f**) Comfort lines (environmental temperature versus wet bulb temperature with relative air velocity as parameter) for persons with medium clothing at three different activity levels: (**d**) sedentary, (**e**) medium activity, (**f**) high activity. RH = relative humidity (from Fanger, P.O., Physiological and Behavioral Temperature Regulation, 1970. Courtesy of Charles C. Thomas, Publisher, Springfield, Illinois).

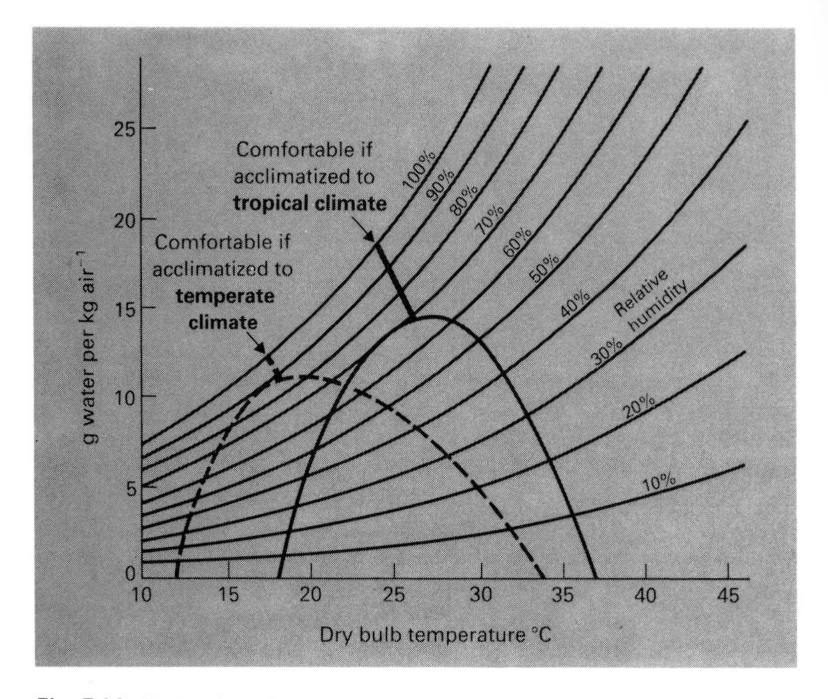

Fig. 7.14 Zones of comfort for lightly clothed individuals in terms of both dry bulb temperature and humidity (Folk, 1974, after Lemaire, 1960, by permission of Lea and Febiger).

Zones 1 and 2 are those in which man is exposed to a hot environment. In zone 3 thermoregulation is maintained with minimum physiological control, and in zones 4, 5, 6 and 7 protection against cold is required to varying degrees in order to conserve heat (see Chapter 5 for details of climatic variables and p. 152 for clothing in polar regions).

The principles of thermal insulation, in terms of non-evaporative insulation and evaporative impedance, are discussed in Chapter 4. The plane of nutrition together with the level of physical activity determine the amount of clothing insulation that is required both for comfort and for the avoidance of injury like frostbite.

Burton and Edholm[85] remark that the thermal insulation of clothing is proportional to the thickness of dead air that is enclosed, and they comment on how the design of clothing for the cold has been improved following the recognition of this fact. Both external wind and internal air currents due to movement can greatly lower the insulation of clothing. This is not always a disadvantage if the decreased insulation accompanies activity and its associated increased heat production, and this factor contributes to the success of Eskimo winter clothing. Where sweating occurs, it is essential that the clothing is permeable to water vapour, so that moisture does not accumulate, when it may lessen the insulation of the clothing or even freeze. The histories of polar exploration provide graphic accounts of the hazards due to failure of clothing in respect either of thermal insulation to non-evaporative heat flow or of vapour permeability.

Providing that clothing, food and shelter are available, heat conservation for man is often less of a problem than the requirement for heat dissipation in regions of the earth where air conditioning is not available and where environmental heat loads may be considerable. The remaining part of this discussion is therefore devoted to clothing in hot climates.

Regions in which minimum clothing is used (zone 1) are the humid, wet tropical areas. The need for clothing under these conditions, excluding cultural considerations, is only to provide protection from strong sunlight and from insects and other damaging material; apart from these limitations the naked state is preferable.

Under hot desert conditions, clothing must insulate against radiant, convective and conductive heat transfer from the environment, but it must also allow the vaporization of sweat while controlling air movement, since when the air temperature is above body temperature increasing air movement at the body surface leads to an increased heat load. Adequate ventilation is, however, necessary if evaporation is to be maintained. The voluminous, loosely worn thick wool or mohair robe of native desert people is admirably suited to the purpose. The thickness of the robe acts as insulation against heat gain during the day and as insulation against heat loss during the night. During the day its loose folds allow ventilation at the body surface with control of the strong hot winds impinging on the outside of the robe. Evaporative transfer is helped further by the removal of water vapour from the air next to the skin by the turbulent air movements produced by walking and other activity.

The surface of the robe is dark, with the result that both short-wave and long-wave radiation are absorbed. A light-coloured robe would allow reflection of some of the incident short-wave radiation, producing a rather lower temperature at the surface of the robe and so leaving a smaller quantity of heat to be lost by re-radiation or convection. One reason why a dark robe is used may be that it is possible to make the material into an opaque densely woven sheet, whereas a material lighter in colour may allow some transulcence with consequent radiant effects directly on the skin surface. The fully absorbent dark, but opaque, material may then be preferable to a partially reflective, but partially translucent, material. This corresponds to the relative merits of dark and light coats in animals exposed to the sun[258] (see Chapter 3).

Under hot desert conditions clothing reduces the rate of heat gain from the environment, and there is a correspondingly lower rate of sweating in clothed men than in nude men when resting or working at low rates. During exercise, however, clothing may hinder the evaporation of sweat, so that evaporative cooling is incomplete. This is more likely to be the case when the ambient humidity is rather higher than the low levels usually found in the desert, although under dry indoor experimental conditions higher sweating rates have been found in men in poplin uniforms than in men walking in shorts. Robinson[447] found that clothing reduced heat absorption of resting men in the summer sun, but the advantage of clothing disappeared at a metabolic rate of 280 $W\,m^{-2}$ and at higher rates of work clothing increased the evaporative requirements. Figure 7.15 illustrates this;

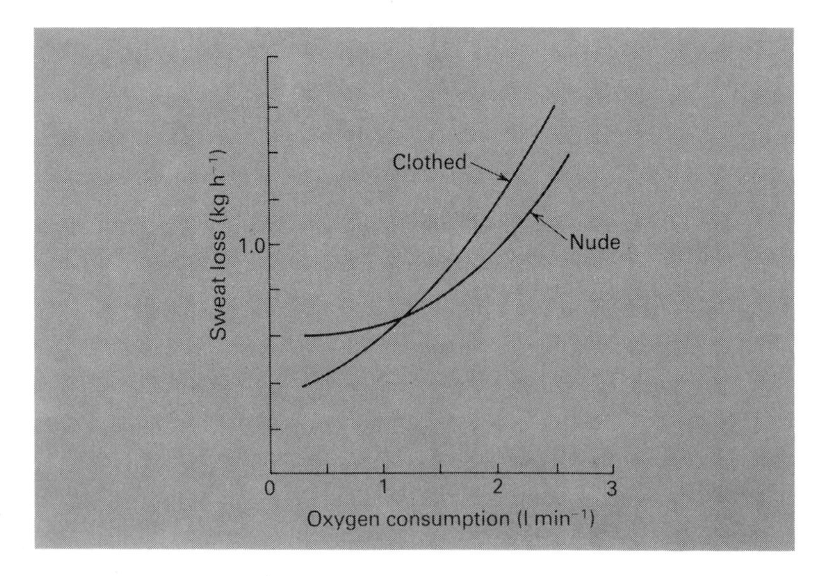

Fig. 7.15 The effect of clothing on the sweat loss of subjects in the desert. At low work rates, the clothing is beneficial because it acts as a radiation screen. At higher work rates this effect is overridden by the restriction of evaporation (Adolph, 1949, by permission of W. B. Saunders Company).

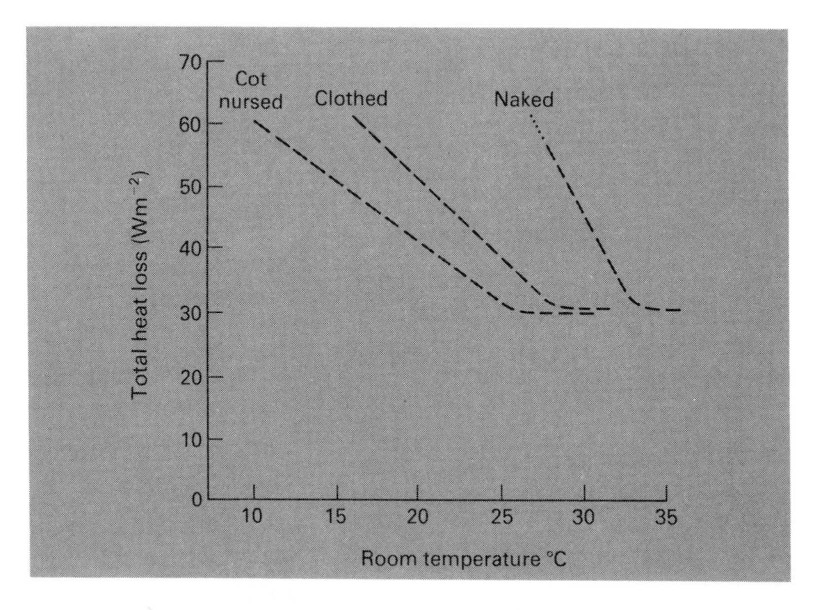

Fig. 7.16 The relation between heat loss and room temperature in a typical baby weighing between 2 and 3 kg when more than 2 days old. Clothes and bedding increase the resistance to heat loss and decrease the temperature necessary to provide thermoneutral conditions (Hey and O'Connell, 1970, by permission of the Editor of Archives of Disease in Childhood).

clothing was beneficial at low metabolic rates, but became an embarrassment at higher rates when evaporation was impeded enough to outweigh the screen to radiation provided by the clothing. Although clothing is preferable both for comfort and for reduction of heat load when the sun shines on the resting individual,[3] it might be a relative disadvantage under some conditions of exercise.

The feet require protection from the hot ground. This can be achieved by sandals, provided that the upper surfaces of the feet are protected from sunburn. Head protection in hot deserts usually takes the form of a turban or cloth which offers some insulation and some reflection of incident radiation, and which is not as likely to be blown off in a strong wind as a helmet with a brim.

The principles of ventilation, wind penetration and radiation characteristics of clothing have been discussed in relation to evaporative heat loss by Kerslake.[296] There is also a great deal of information on clothing, shelter and environmental responses in the book edited by Newburgh.[407]

THE NEWBORN INFANT

The baby is normally clothed and nursed in a cot, although in some situations, particularly with premature babies, the infant is naked in an incubator maintained at the required temperature. The temperatures required for babies nursed naked or in a cot are indicated in

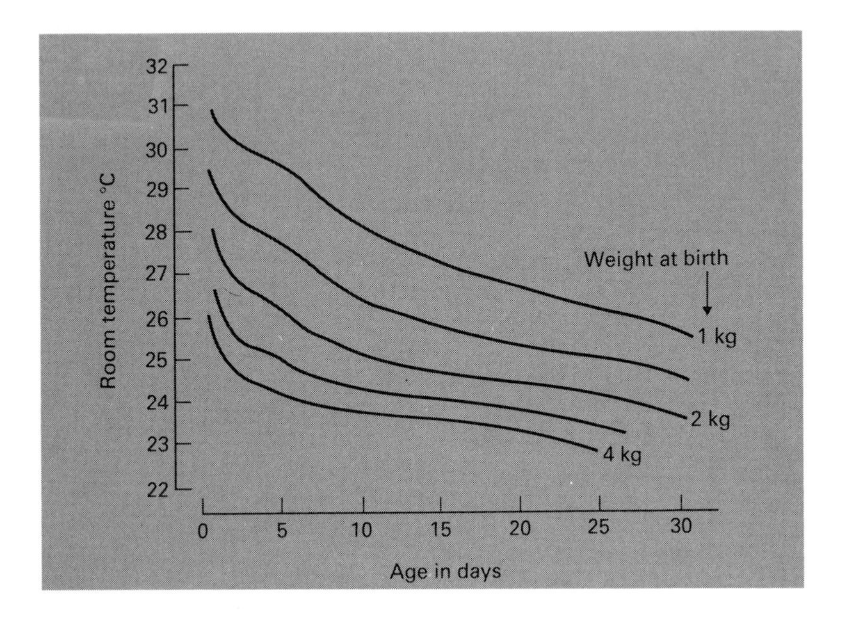

Fig. 7.17 Estimates of the room temperatures to provide a cot-nursed baby with conditions of optimum warmth (thermal neutrality) during the first month of life. The baby is assumed to be clothed and wrapped in blankets under draught-free conditions (Hey and O'Connell, 1970, by permission of the Editor of Archives of Disease in Childhood).

Figs 7.16 and 7.17. For a naked baby weighing 3 kg at birth the temperature to provide reasonable warmth in draught-free surroundings is about 34°C on the day of birth, falling to 32 to 33°C within 2 days, and then changing very slowly.[226] For very small babies a temperature about 1°C higher than this is necessary.[224]

Important factors that affect measurements of the baby's metabolic rate[82] are the level of physical activity[362] and whether the baby is sleeping, either with or without rapid eye movements.[486] These factors

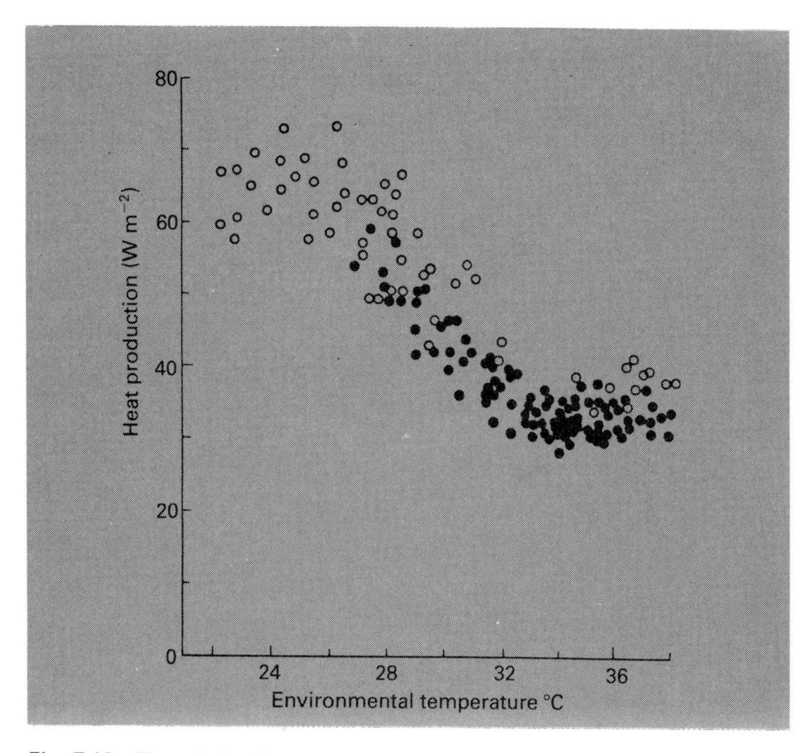

Fig. 7.18 The relation between operative environmental temperature and heat production in 6 babies who weighed approximately 2.5 kg when 7–11 days old. (○) relate to data obtained during periods of physical activity (Hey, 1974).

are important in influencing the rate of heat production in all mammals, and it is necessary in drawing conclusions to allow for the state at the time of measurement. The newborn baby shows increased activity when subjected to thermal stress, either cold or hot, but no-one has ever discerned visible shivering in babies. However, the presence of brown fat allows non-shivering thermogenesis (see Chapter 2). Metabolic rate per kg remains almost constant during the first year after birth, indicating a metabolic rate proportional to body weight (W), later falling to proportionality to about $W^{0.6}$.[234]

The baby has much less thermal insulation than those newborn of a similar size range in some other species, although there is control over vasomotor tone even on the first day after birth.[81] The combination of low insulation with the lower level of maximum metabolic rate (Fig.

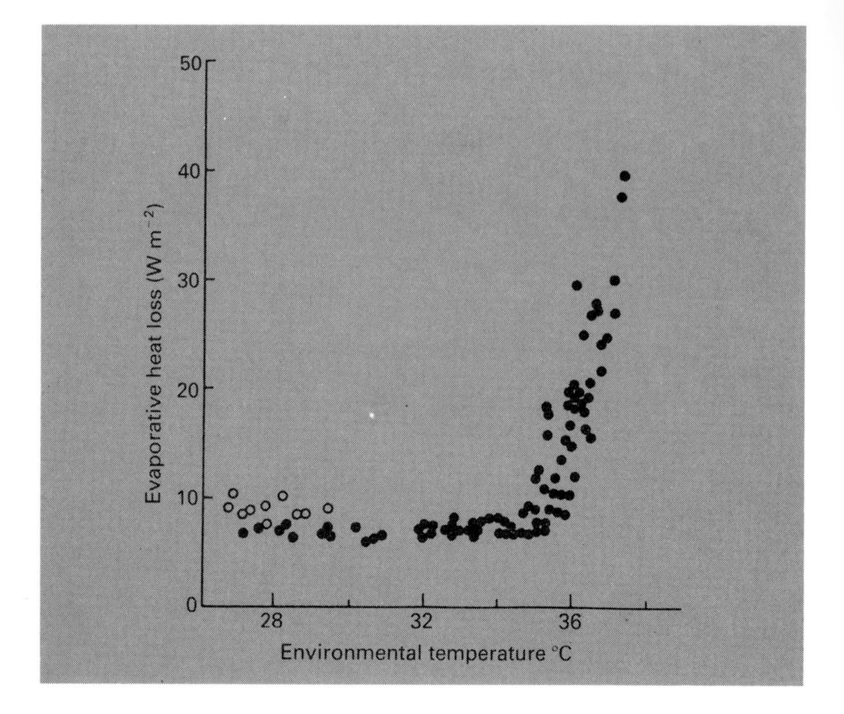

Fig. 7.19 The relation between operative environmental temperature and evaporative water loss when ambient water vapour pressure is about 20 mbar. Subjects and symbols as for Fig. 7.18 (Hey, 1974).

7.18) produces a relative inability on the part of the baby to withstand cold exposure. The temperature difference that can be maintained between the body core and the environment is given approximately by the product of the thermal insulation and the maximum cold-elicited metabolic rate (see Chapter 4). Representative values are given in Table 2.2 for lamb, piglet and baby, and the calculated corresponding cold limits are close to −80°C, 0°C and 25°C, figures that have been estimated or demonstrated for these three species (lamb,[7] piglet,[386, 394–5] baby[224]).

Whereas adult man can increase heat production nearly three-fold in response to severe acute cold stress, during the first two weeks after birth babies cannot often increase their heat production by more than 150%. For the average full-term cot-nursed infant this limit is reached when the room temperature falls to about 10°C, so that in bedrooms that are colder than this babies can easily become hypothermic. For smaller babies, this happens at a higher temperature, about 15°C. When swaddled babies are under severe cold stress they may not cry, so that attention may not be drawn to their falling body temperature.[230]

The appearance of the hypothermic baby can be misleading: the facial colour may be good and the infant is quiet, but if the lowered body temperature is not detected and if rewarming is not undertaken then neonatal cold injury may result.[356]

Hey and Katz[225] found active sweating in babies 0–10 days old when the environmental temperature exceeded 34–35°C and when rectal temperature rose above 37.2°C (Fig. 7.19). These babies were born within 3 weeks of the full term of the gestation period. The threshold rectal temperature for sweating fell during the 10 days after birth. Foster, Hey and Katz[158] could not detect sweating in infants of less than 210 days postconceptual age, even when rectal temperature rose as high as 37.8°C. In full-term babies, 414 active sweat glands per square centimeter were found on the thigh, which was 6.5 times the number found in adults. The mean maximum sweating rate on chemical stimulation, however, was only 2.4 nl per gland per minute, about a third of the maximum rate found in the adult.

8

Pig

The pig's responses to its thermal environment are influenced by a number of special characteristics.[386–7, 394] One characteristic is that under hot conditions there is only a very low rate of sweating with inadequate thermoregulatory compensation by respiratory evaporative heat loss. This particular disability is overcome in the pig's natural habitat because the animal is native to swampy woodland and tropical forest and can wallow in mud or water as necessary; also, wild pigs shelter during the heat of the day and become active at night. Exposure of domestic pigs to hot dry conditions can lead to death in hyperthermia, so that sprays or wallows must be provided.[214a, 252–3] The other characteristics are related to the relatively low thermal insulation of the young pig either as subcutaneous fat or coat. The newborn pig is particularly susceptible to cold, whereas the mature animal is susceptible to heat and hyperthermia. This is true for many species to some degree, but in the newborn pig the cold susceptibility can lead to death in hypothermia. As a member of a litter the newborn pig can huddle with its littermates and so reduce the effects of a cold environment. This behavioural feature is retained throughout the animal's life, although as the pig becomes larger and acquires a layer of subcutaneous fat it does not have to rely so much on social thermoregulation. Furthermore, the pig is adept at burrowing into any bedding or undergrowth that is available; for example, in the natural state, the pregnant sow makes a hollow in which she rears her litter. This nest-building activity provides shelter from environmental extremes.

GENERAL FEATURES

The pig belongs to the order of mammals known as the Artiodactyla (even-toed ungulates) which contains three sub-orders: Suiformes (pigs, peccaries, and hippopotamuses), Tylopoda (camels), and the Ruminantia.[371] Within the Suiformes there is the infra-order of Suina, containing the super-family Suoidae, of which the the family Suidae constitutes the Old World Pigs (fossil pigs have been found in Europe and India, but not in America). This family has as its members:

Bush pig	*Potamochoerus porcus*
Wild boar (and domestic pig)	*Sus scrofa*
Pygmy hog	*Sus salvanius*
Javan pig	*Sus verrucosus*
Bornean pig	*Sus barbatus*
Wart hog	*Phacochoerus aethiopicus*
Giant forest hog	*Hylochoerus meinertzhageni*
Babirusa	*Babyrousa babyrussa*

Of these animals, *Sus scrofa* is the one with which this chapter is primarily concerned. Some classifications have previously subdivided this species into a number of sub-species. Thus the Indian pig was termed *Sus cristatus*, the Chinese pig *Sus indicus*, and the European pig *Sus scrofa*, while the domestic pig became the sub-species *Sus scrofa domesticus*. It is doubtful, however, whether such sub-divisions are either useful or valid on zoological grounds. The true origins of the domestic pig are probably lost in obscurity.

Sus scrofa is distributed over the woodlands of Europe, and in North Africa and Asia, while island races are to be found in Ceylon, Indonesia, Japan, and Taiwan. The wild boar was once very plentiful in Europe, where it has been hunted for centuries, but it is now restricted to a few areas of woodland. It has long been extinct in Britain. Crandall[119] points out that although pigs show wide variation in size, colour, and shape of tusks, they all have in common the flat rounded snout on a flexible muzzle. The upper canine teeth turn upwards to form tusks, and are sometimes very large, particularly in males. The animal has four toes on each foot, with only the third and fourth functional. The hair is coarse and sparse. The pig can be described as omnivorous, although vegetable material forms the bulk of its food; from studies of the rates of water turnover, Yang[562] concluded that the pig is functionally a wet forest type of animal. The gestation period of *Sus scrofa* (wild pig) is quoted as ranging between

101 and 130 days, an average of 116 days, as compared with 112–115 days for domestic pigs of various breeds.

Historical

Intensive selection of pigs for domestic purposes began in the eighteenth and nineteenth centuries with the result that there are now many distinctive breeds that are used for meat production and they vary according to local preference and market requirements. The Large White, known in many parts of the world as the Yorkshire pig, has played a major role in the development of the domestic pig. Its origin is commonly said to derive from the selection efforts of one Joseph Tuley, a Yorkshire weaver. The Duroc pig in the United States has reached the greatest numbers in total production. General descriptions and photographs of the most common breeds are to be found in Davidson's[129] book on the production and marketing of pigs, and in the work on pig diseases by Anthony and Lewis.[13]

The greatest part of the world pig population now consists of the domestic pig. The estimated world population of pigs is in excess of 650 millions, but this is only rather more than half the total number of either sheep or cattle.[157] The modern practice is to confine pigs in groups in houses of various types, with the emphasis on selection of animals that do not become excessively fat on controlled levels of food intake. The gain of body fat overtakes the gain of protein by the time the pig weighs about 30 kg, on usual levels of feeding in farm practice.[506] Whereas the wild pig is vigorous, active, and aggressive, the domestic pig is a very different creature, content largely to feed and sleep, inclined to be obstreperous only if disturbed.

RESPONSES TO THERMAL ENVIRONMENT

Phases of development

During the course of its development from the newly born to the mature animal, the pig shows at least three phases in its responses to environmental stimuli. The newborn phase, which is taken here to cover the first week of extra-uterine life, is the first of these. The second phase is that of the growing pig; it includes weaning (which in commercial husbandry may take place at 5–8 weeks, although sometimes much earlier) and lasts until about seven months of age. This second phase is marked by a rapid increase in size which, by itself, without any other change being necessary, diminishes the animal's cold-sensitivity progressively. The third phase of the pig's development is marked by sexual maturity and reproductive activity. The animal is then relatively insusceptible to cold, in consequence of its large size and increased tissue insulation, but is highly susceptible to heat, environmental temperatures in excess of 30°C proving progressively more embarrassing, and requiring wallows, mud baths, or water-sprays to provide evaporative heat loss from the animal's surface if hyperthermia is to be prevented.

Pigs characteristic of each of these three phases of development differ from each other particularly in respect of their overall thermal insulation. In the first phase, the newborn, insulation is minimal, and cold is resisted to a large degree by the behavioural adaptation of huddling with litter-mates and with the sow, and in farm practice by the provision of supplementary heating and bedding. In the third phase, by contrast, thermal insulation is high, and pigs have even lived in the Arctic with apparently little discomfort.[265] The newborn pig brings out in sharp relief, and quantitatively to a greater degree than the older pig, those features that characterize the relation between the animal and its thermal environment.

NEWBORN PIG

The newborn pig is particularly sensitive to cold. The pig commonly weighs between 0.7 and 1.5 kg at birth, and it is a member of a litter that is usually six to twelve in number. Postnatal growth is rapid, so that after one week the animal has approximately doubled its birth weight.

The pig feeds from the sow shortly after birth and thereafter at one to $1\frac{1}{2}$-hourly intervals, and grows rapidly. Associated with this rapid growth is the availability of milk from the sow; milk can be expressed from the sow's udder usually one to two days before farrowing. The fat content of the pig is very low at birth, only about 1–2%[126, 542] and the resulting lack of subcutaneous fat, together with lack of an effective coat, leads to the animal's characteristically low thermal insulation. However, after one week the body fat content approaches 10%, and by that time the animal's cold resistance has increased appreciably.[386] The sow's milk production reaches a maximum at three to four weeks following farrowing.[4]

The pig can shiver from birth, and the change in the thermal circulation index (see Chapter 4) at the critical temperature suggests that peripheral vasoconstriction occurs under cool conditions. These features indicate that a considerable degree of thermoregulation is already established at birth, and the capacity for thermoregulation develops rapidly in the first two days after birth.[123,125,386–7,402,485] When the pig is born, its rectal temperature is about 39°C, close to that of the mother, but falls rapidly to 37 to 38°C, or even lower under cold conditions, and it is not until the second day after birth that the value of 39°C is regained.[408] This parallels the animal's metabolic rate measured at thermal neutrality: the rate increases during the first day or two after birth. Fasting the pig from birth prevents the increase in body temperature.[388]

Carbohydrate provides the main source of energy for the fasted newborn pig; the respiratory quotient immediately following birth is 0.95, falling after the first day to about 0.8, indicating increased fat metabolism.[124, 388] Fasting can lead to death in hypoglycaemia, particularly in the cold when metabolism is increased, but the tendency to hypoglycaemia diminishes as the pig grows (Fig. 8.1). The newborn calf and foal, unlike the newborn pig, do not readily become hypoglycaemic during starvation, but the newborn lamb does go into hypoglycaemic coma and convulsions when it is fasted.

The relation between metabolic rate and standardized environmen-

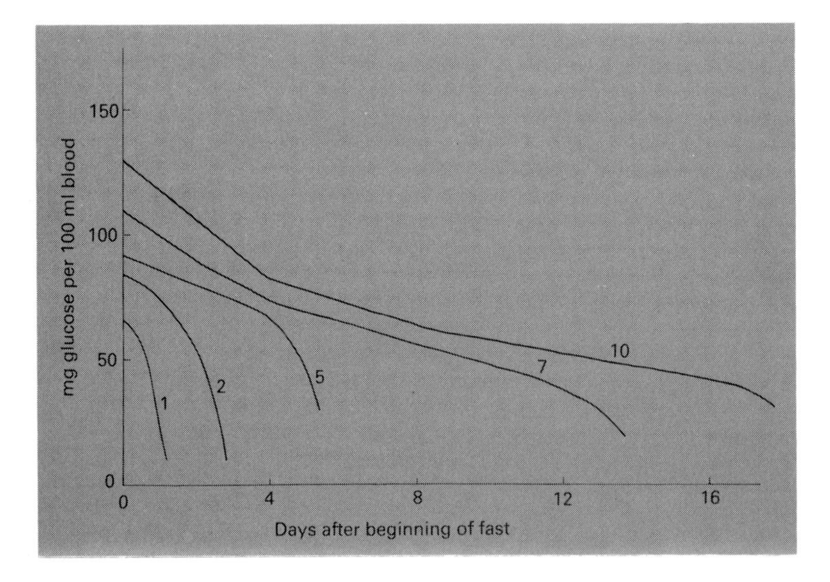

Fig. 8.1 The effect of fasting on the blood glucose level of the newborn pig. The figures on the curves give the age in days at the beginning of the fast; the older the pig at the onset of fasting, the more difficult it is to produce hypoglycaemia (Mount, 1968a, constructed from results of Hanawalt and Sampson, 1947).

tal temperature (see Chapter 5) in the newborn pig at 2 days old is shown in Fig. 8.2. At the critical temperature of 34–35°C heat production is approximately 60 W m^{-2}, a rate that is doubled by lowering the temperature to 20°C.[376] The maximum cold-induced metabolic rate is approximately three times the minimum that occurs in the narrow range of thermal neutrality above the critical temperature. The animal's lack of thermal insulation leads to a steep increase in metabolic rate when the environmental temperature falls, and there is a relatively high cold limit of about −5°C.

As the body weight increases, subcutaneous fat and thermal insulation increase, the critical temperature falls, and oxygen consumption per unit body weight falls. These effects in the young animal are illustrated in Fig. 8.3, which also gives results obtained from groups of four or six 1–2 kg pigs housed together in the metabolic chamber. At the higher temperatures the animals are spread

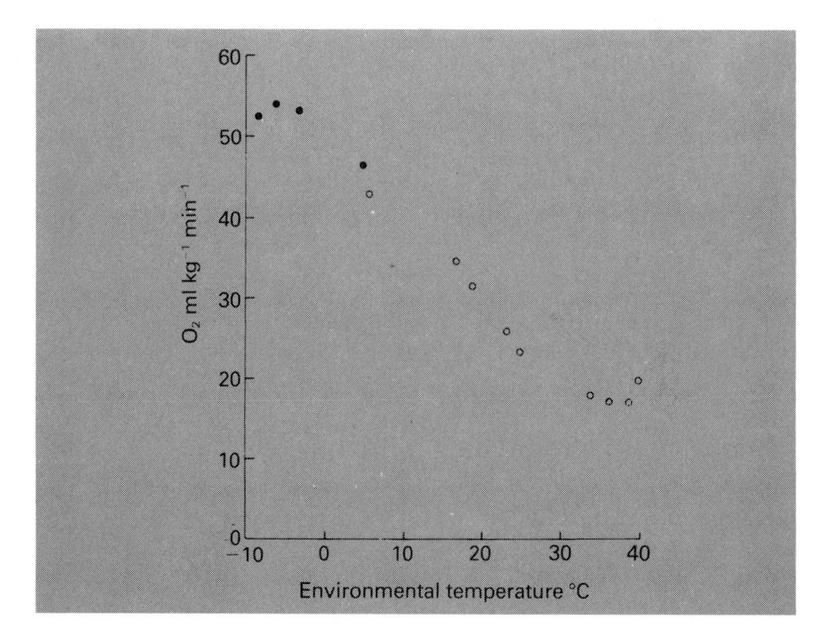

Fig. 8.2 Maximum and minimum metabolic rates in the newborn pig. One pig, 52 hr of age and weighing 1.70 kg (●), was exposed to a decreasing environmental temperature for the detection of the maximum rate. Another pig, 60 hr of age and weighing 1.65 kg (○), was exposed to a rising environmental temperature for the measurement of minimum metabolic rate (Mount and Stephens, 1970, by permission of Journal of Physiology).

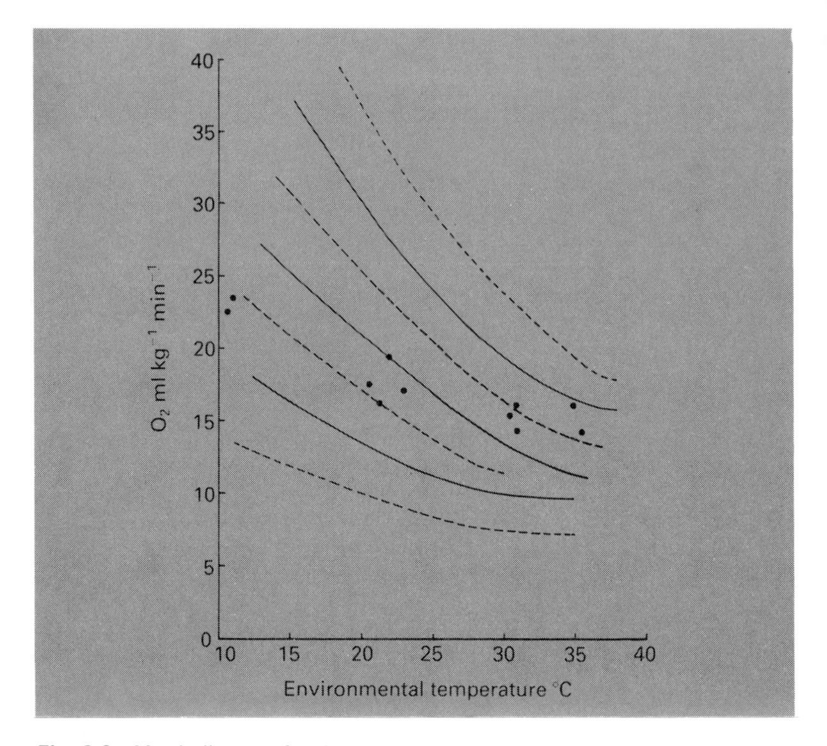

Fig. 8.3 Metabolic rates for three weight groups of young pigs over a range of environmental temperature. Curves refer to results on single pigs, with interrupted lines enclosing approximately 85% of results in each weight-class, 1, 3 and 5 kg from above downwards; points refer to results on groups consisting of either four or six animals, each weighing 1–2 kg, together in the metabolic chamber (Mount, 1963b, reprinted from Federation Proceedings *22*, 818–823).

out in the chamber and the mean metabolic rate per unit weight is similar to that for an individual 1–2 kg pig. As the temperature falls so the animals huddle together increasingly, and the metabolic rate per unit weight falls towards that found for a single larger pig. At low temperatures the group of pigs behaves like a single larger pig, with consequent energy saving, whereas in the warm the group breaks up into individual animals and heat loss is increased.

The degree to which the thermal demand of the environment is countered by the group as opposed to the individual is illustrated in Fig. 8.4. Measurements were made at 20°C, a temperature that is cool for the baby pig, and which therefore elicits cold-resisting behaviour

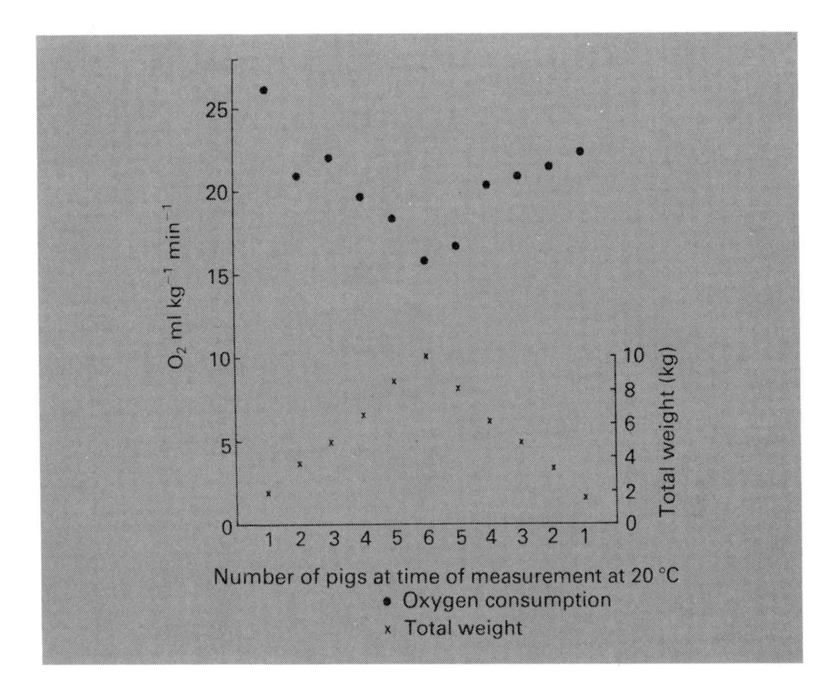

Fig. 8.4 The relation between metabolic rate per unit of body weight and the number of newborn pigs together in a chamber at an environmental temperature of 20°C (Mount, 1960, by permission of Cambridge University Press).

and physiological responses. The number of pigs together in the chamber was first progressively increased and then decreased. Metabolic rate per unit weight fell as group size increased, indicating the saving in energy expenditure conferred by the group at a given environmental temperature.

The importance of the group in the newborn pig's responses to cold is made evident by comparison with a single bacon pig in Fig. 8.5, in which metabolic rate is expressed per unit of surface area so that the rates of animals of different sizes can be compared with each other (see Chapter 2). At an environmental temperature of 20°C the huddled group has a metabolic rate much closer to that of the larger bacon pig than to that of the single newborn pig. As the temperature falls, the single newborn animal reaches its limit of heat production at a point at which there is still reserve metabolic capacity for the members of the group, indicating their ability to tolerate and survive lower environmental temperatures more easily than the single pig.

GROWING PIG

As body size increases, oxygen consumption per unit body weight falls progressively to reach about 5 ml kg^{-1} min^{-1} at a body weight of 100 kg (Table 8.1). For the pig of three months old the critical temperature might be in the region of 20°C with the cold limit at approximately −40°C, although only short exposures to such a low temperature could be tolerated by unadapted animals. These figures are only rough indications, since they depend on the level of feeding and thermal adaptation history, but they indicate the decreased tendency to cold-susceptibility and the potentially increased heat-susceptibility that accompany growth. Pigs adapted to the cold can

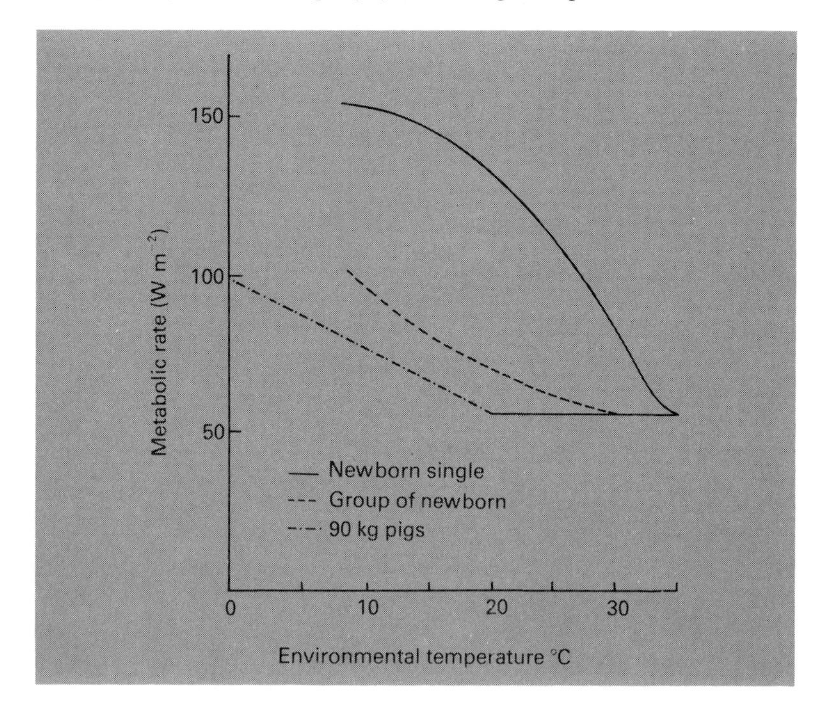

Fig. 8.5 The relation between metabolic rate and environmental temperature for single newborn pigs, a group of newborn pigs, and single 90 kg pigs (Mount, 1965, by permission of European Association for Animal Production).

live out of doors with only the protection of a roof and dry bedding; at an air temperature of -30°C they have skin temperatures that may fall below 5°C.[265] Pigs raised at high temperatures have higher critical temperatures than animals raised in cooler environments.[114,254]

The huddling together of larger pigs in the cold reduces the thermal demand of the environment for heat production, as with newborn

Table 8.1 Resting rates of oxygen consumption at thermal neutrality for pigs fed at usual levels from birth to 100 kg body weight. Values in column A from Brody (1945), B from Thorbek (1969), and C from Mount (1968a) (Mount, 1976c, by permission of Pitman Medical Publishing Company Ltd).

Body weight (kg)	Oxygen consumption ($ml\ kg^{-1}\ min^{-1}$) A	B	C
1.2			15.0
5	11.1		11.3
10	10.1		
15	9.5		
20	9.4		8.7
25	9.1		
35		10.7	
50	8.5	9.2	
60			6.1
75	5.7	7.7	
100	5.1		

pigs, although the temperatures at which the effects occur are lower than those for the newborn.[239] In groups of four pigs at a time, there are effective critical temperatures of 11°C for a feeding level of 52 g food per kg per day and 17°C for a feeding level of 45 g per kg per day.[514] Fig. 8.6 shows the increased rates of heat loss when environmental temperature falls below these values, with rates of heat loss at different levels of feeding converging together, indicating that at temperatures below the critical level it is the thermal demand of the environment, rather than plane of nutrition, that has the major influence on heat loss (see Chapter 2). In addition to body weight, group size also influences the increased rate of heat loss below the critical temperature (Table 8.2). The smaller the pig, the greater is the proportionate effect of a fall in temperature on the animal's heat loss.

Fasting heat loss in $W\ kg^{-0.75}$ has been found to be 5.3 at 20°C environmental temperature and 4.4 at 30°C for 25–40 kg pigs,[112]

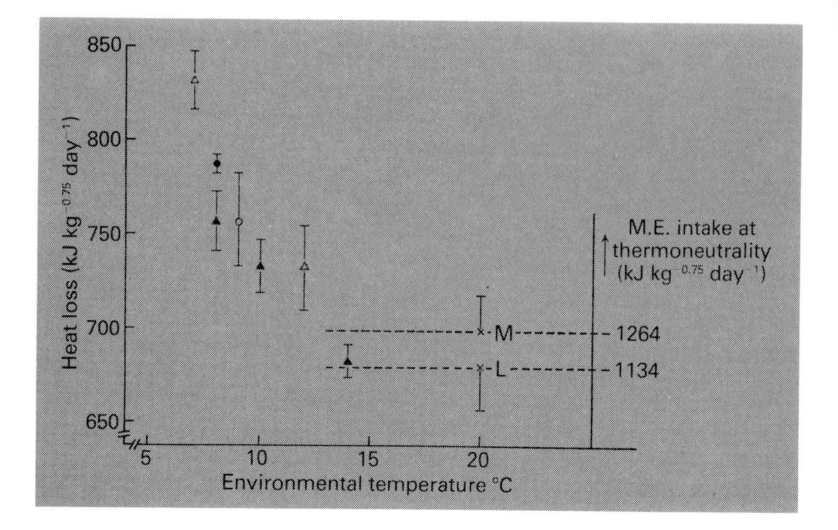

Fig. 8.6 Rates of heat loss per day from groups of pigs at several feeding levels and environmental temperatures, with standard errors, taken from several authors. M.E., metabolizable energy, M, medium level intake, L, low level intake (Verstegen, Close, Start and Mount, 1973, by permission of British Journal of Nutrition).

and 5.5 at 18°C and 4.4 at 26°C for 26–60 kg pigs.[505] These values suggest that the critical temperature for such pigs is about 25°C, much higher than the critical temperatures for the fed pigs in Fig. 8.6.

The principal factors that influence metabolic rate and consequently heat loss are body size, grouping, environmental temperature and level of feeding[112] (see Chapter 2). The 24-hourly variation in metabolic rate must also be taken into account (see Fig. 2.9). At an

Table 8.2 Approximate increases in heat loss for each 1°C below the effective critical temperature for different body weights and groupings of pigs (from Mount, 1976b, by permission of Swets and Zeitlinger).

Individual body weight (kg)	Single or group	W kg$^{-0.75}$ per °C below critical temperature
1	single	0.46
1	group (5 pigs)	0.23
5	single	p.35
35	single	0.20
35	group (4 pigs)	0.14
35	group (9 pigs)	0.10

environmental temperature of 20°C the amplitude of the variation in heat loss from a group of pigs is approximately 30% of the mean 24-hourly value, that is ±15% of the mean. Activity also increases the metabolic rate.

Evaporative heat loss

The pig is unable to sweat to a degree that would be effective for thermoregulation, although there are functional sweat glands associated with the hair follicles.[249,378] Table 3.2 shows the very small increase in cutaneous water loss that occurs from the pig when it is exposed to heat compared with water loss from other species. The pig compensates for its inability to sweat by wallowing in any available water, including wet surfaces, mud and urine. The subsequent evaporation from the animal's surface produces as much cooling as that found in a sweating man, and thermoregulation is maintained (Fig. 8.7); wet mud, by holding water, produces a prolonged cooling.[250] Mud or water wallows are very effective for body cooling, and food intake increases relative to that of the animal exposed to heat without this provision.[253]

Keeping pigs quite dry in a hot environment can lead to their death through hyperthermia. Although pigs hyperventilate under hot

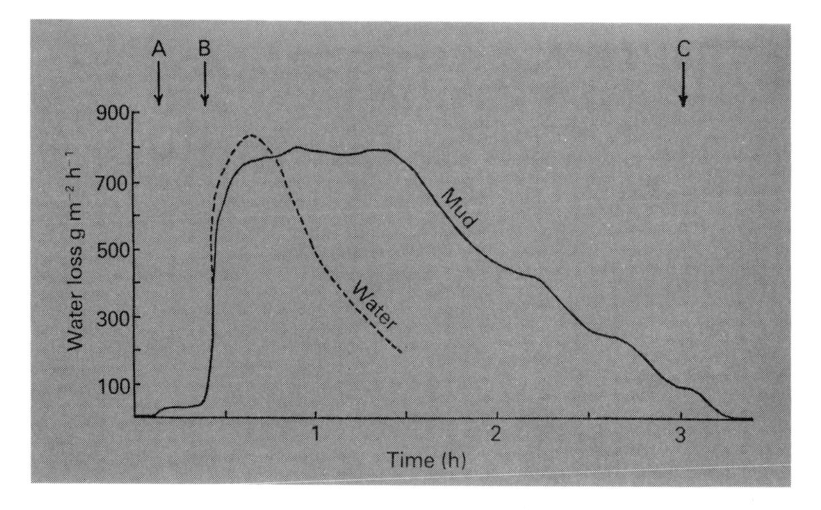

Fig. 8.7 Evaporative water-loss from the skin of a pig measured by the ventilated capsule technique. At A the capsule was placed on the skin. At B mud (———) or water (– – – –) was smeared over the skin. At C the capsule was removed (Ingram, 1965b, by permission of Nature).

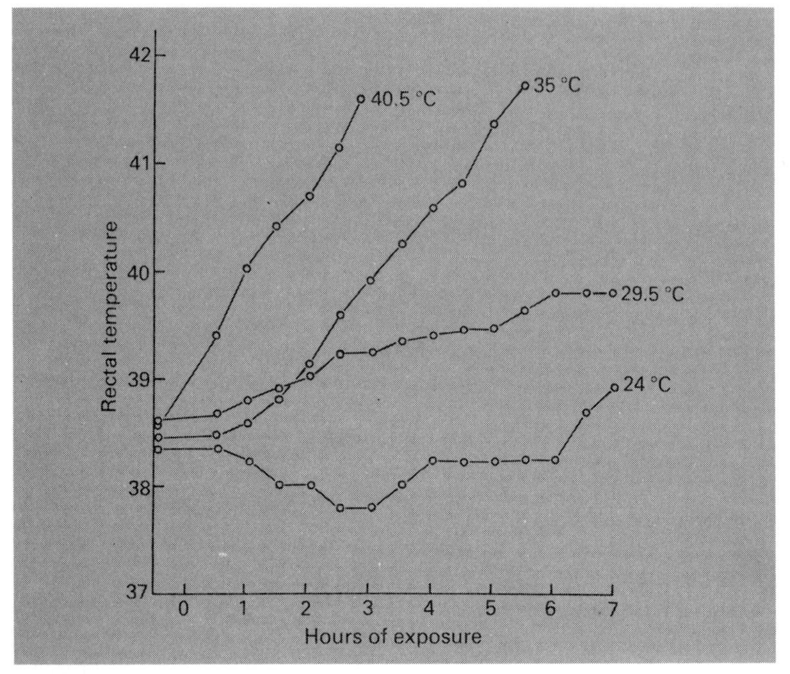

Fig. 8.8 Mean changes in the rectal temperatures of three male Berkshire pigs, each of about 60 kg body weight, exposed to environmental temperatures of 24, 29.5, 35 and 40.5°C, at a constant relative humidity of 65% (Robinson and Lee, 1941, by permission of Royal Society of Queensland).

conditions, the respiratory evaporation is inadequate for the heat loss required to keep down the body temperature when the environmental temperature rises above about 30°C (Fig. 8.8). This is well recognized in pig farming in hot countries, where, in addition to shade, sprays and wallows are provided for the animals to compensate for their limited ability to evaporate water from the respiratory tract.

Radiation

The radiant heat exchange of pigs out-of-doors involves short-wave solar radiation, and is therefore influenced by coat colour, whereas indoors, with only long-wave radiation, there is no effect due to coat colour (see Chapter 3). The reflectances of the skins of white, red and black pigs is shown in Fig. 8.9, and the energy absorptions in Fig. 8.10. Although the different skins have different reflectances for short-wave radiation, they all have a similar low reflectance for long-wave radiation.

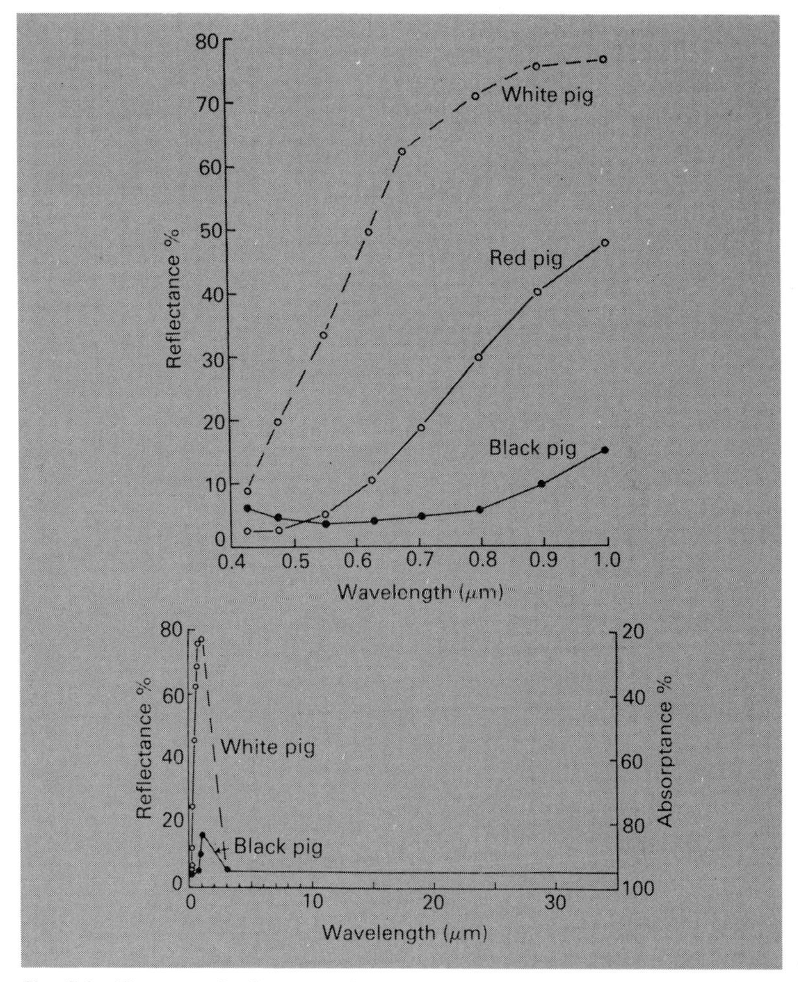

Fig. 8.9 The spectral reflectance of pig skin. Upper diagram: white, red, and black pig skin, between wavelengths of 0.4 and 1.0 μm; lower diagram: reflectance and absorptance of white and black pig skin between wavelengths of 0.4 and 35 μm (Kelly, Bond and Heitman, 1954).

Convection

Convective heat exchange is primarily by forced or natural convection, depending on the skin-air temperature difference, windspeed and body size. Measurements have shown that heat loss from the pig's back increases as windspeed increases, but the rise in heat

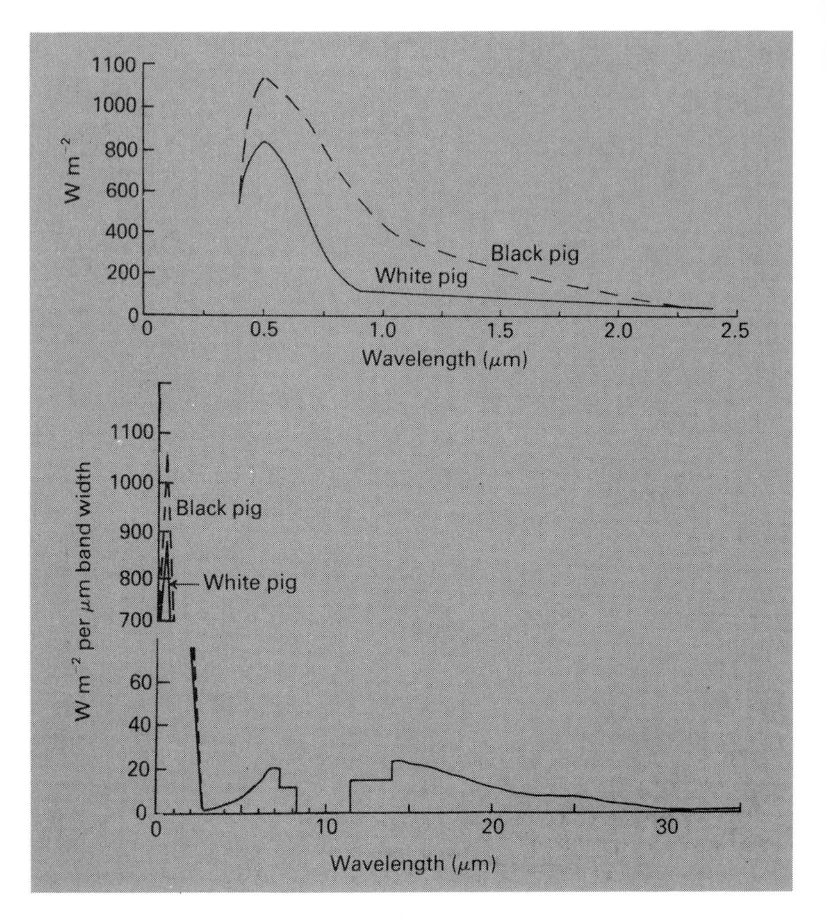

Fig. 8.10 The solar and atmospheric radiant energy absorbed by black and white pigs for typical atmospheric conditions at Davis, California. Upper diagram: absorption over the wavelength range 0.4–2.4 μm; lower diagram: absorption up to 35 μm (Kelly, Bond and Heitman, 1954).

loss is often much less than anticipated owing to changes in the animal's posture (Fig. 8.11). Natural convection predominates under 'still air' conditions; such conditions are provided by windspeeds of $0.1\,m\,s^{-1}$ or less for baby pigs, and by windspeeds up to $0.15–0.2\,m\,s^{-1}$ for 20–60 kg pigs.[396] Natural convection thus occurs in more instances than might be expected, particularly with the larger pig (see Chapter 3).

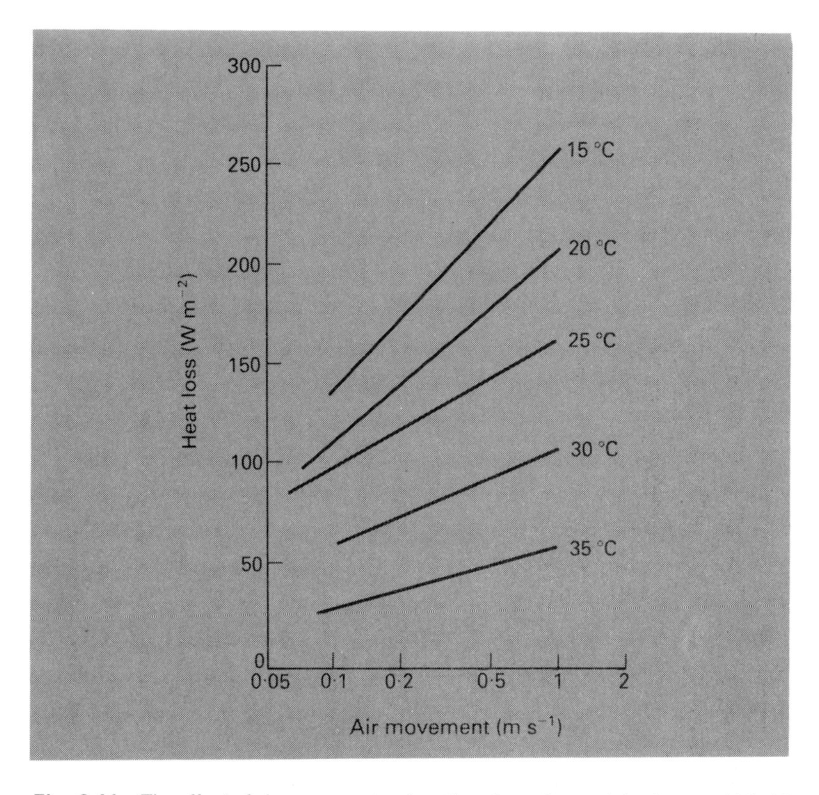

Fig. 8.11 The effect of air movement on heat loss from the trunk in pigs aged 10–12 weeks and weighing 20–25 kg (Mount and Ingram, 1965, by permission of Research in Veterinary Science).

Conduction

Pigs spend much of their time lying down, so that heat exchange by conduction is important. This is influenced considerably by the animal's posture when lying, that is whether the pig is relaxed or is supported on its limbs, as well as by the thermal characteristics of the floor, including temperature, thermal conductance and thermal capacity, and the presence or absence of bedding.[385] As a mean value, approximately 20% of the pig's surface is in contact with the floor during lying. The effective critical temperature of animals weighing 40 kg was found to be 11.5–13°C on straw bedding, 14–15°C on asphalt, and 19–20°C on concrete slats.[515] These figures indicate the correspondingly higher air temperatures required to counter-balance heat losses to the floor. The use of heated floors in pig husbandry leads to heat gain by the animal.

Radiation and convection normally provide the main channels for heat exchange in the pig. The effects of straw bedding extend to providing increased thermal insulation round the animal as well as decreasing direct heat loss through the floor. In the case of the newborn pig, moving the animal at an environmental temperature of 10°C from a bare concrete floor onto straw has the same thermal effect in reducing metabolic rate as raising the environmental temperature to 18°C.[484] A small component of heat transfer, which may be termed conductive, lies in warming ingested food and drink; usually this does not exceed 3% of the total heat loss (see Chapter 3).

Behavioural adaptation to temperature

Pigs normally live in groups, and their behavioural responses to changes in environmental temperature involve reactions with other pigs. An example of this is the huddling that occurs in cool conditions, which has a considerable sparing effect on metabolic heat production (see Figs 8.3, 8.4 and 8.5). Even the individual pig by itself can modify its heat loss by choosing its micro-environment and by postural changes. For example, radiant heat loss from the newborn pig at 20°C is reduced by about 10% by the adoption of a flexed position. Orientation to wind influences convective heat loss, and conductive heat loss is affected by the animal's area of contact with the floor, which in turn depends on the type of posture. Measurements on groups of pigs in a large calorimeter designed like a pig pen have shown that although heat losses at 20 and 30°C might be similar, at 20°C the pigs are in a huddle and at 30°C they are spread out. Another behavioural response to warm conditions that has been noted in the calorimeter is that at 30°C the pigs urinate and defaecate indiscriminately over the floor of the pen, whereas at lower temperatures they confine these activities to one place.[386]

Preferred environments

During the first five weeks following birth, observations on temperature preference have shown that the most common choice of environmental temperature is in the region of 30°C (see Table 6.1). The results from such preference or multiple choice experiments are sometimes unexpected when the animal is subject to competing drives to satisfy different behavioural demands. This was the case with young pigs allowed free movement in an area of woodland and pasture with a hut provided for shelter.[256] When the weather was dry the animals did not begin to take shelter until the environmental temperature was below 5°C, although in the laboratory the critical temperature of pigs similar in age and weight was found to be about 25°C. The behaviour out-of-doors is a consequence of an interaction of drives: for example, the animals stayed in the region of lowest air movement unless food was available *ad libitum* in troughs in a windy area, when they spent more time there. Similarly the exploration of a diverse natural environment presumably took precedence over seeking warm conditions. This example underlines the necessity for proceeding cautiously when laboratory experiments under carefully controlled conditions are to be used to provide information in the more diverse conditions of everyday life.

Operant conditioning (see Chapter 6)

An example of this is illustrated in Fig. 8.12, which gives the results of experiments in which pigs were trained to press a switch with the snout to turn on a battery of infra-red heaters. It was found that the rate of response varied with environmental temperature, decreasing as it became warmer to approach zero at thermal neutrality. When the animals' oxygen consumption rates were measured simultaneously, it was found that animals in the cold, but with access to heating operated by the switch, turned on enough heat to keep their metabolic rates within the range characteristic of thermal neutrality (Figs 8.13a and b). Pigs on a high food intake were less inclined to obtain additional heat than animals kept on a low intake (Fig. 8.14). The operant conditioning results are in good accordance with expectations from measurements of heat production (see also Mount[394]).

Morphological adaptation to temperature

Pigs exposed chronically to different temperatures develop different body shapes and other anatomical features. Litter-mates raised at 35°C and 5°C and maintained at equal body weights showed marked differences in body conformation and hair coat (see Fig. 6.6). The shorter limbs and the denser coat of the pig kept at 5°C are apparent. The pig kept in the cold has a smaller number of blood vessels in the skin than an animal kept in the warm,[261] shorter ears and limbs and more hair[168,254] (see Chapter 6), and deposits more unsaturated fat in its tissues.[169,335]

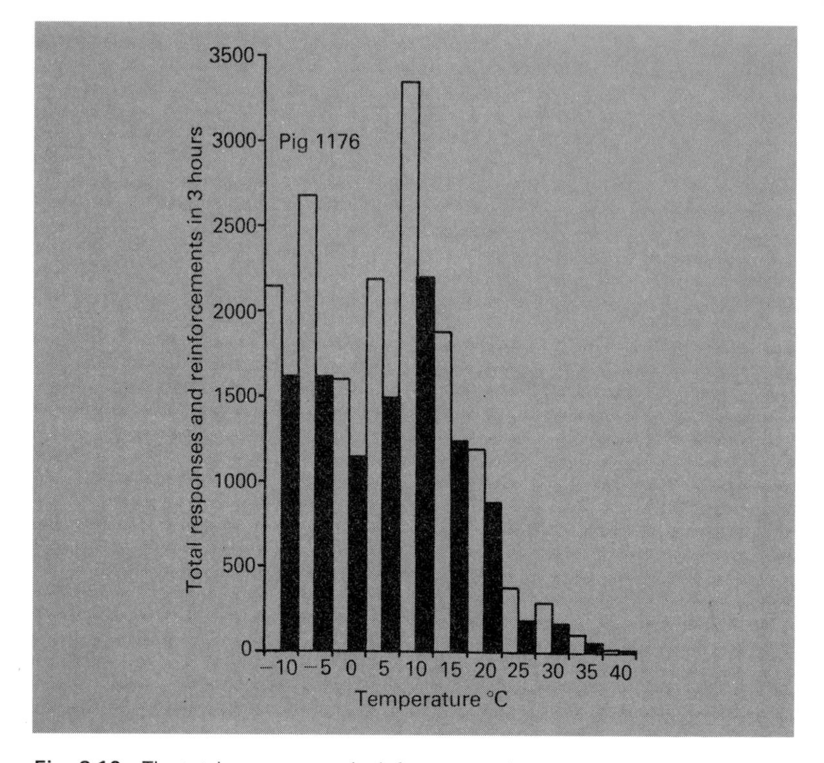

Fig. 8.12 The total responses and reinforcements (rewards in the form of bursts of infra-red heat for 3 seconds) recorded for one pig during three-hour experimental sessions at environmental temperature from −10 to +40°C. Responses, white; reinforcements, black. Responses made during a period of reinforcement had no effect, so that the number of responses exceeds the number of reinforcements (Baldwin and Ingram, 1967, by permission of Physiology and Behaviour).

IMPLICATIONS FOR HUSBANDRY

The susceptibilities of the young pig to cold and of the older pig to heat have important bearings on pig husbandry. In modern intensive pig farming, concern with the environment lies in its effects on feed utilization and growth. The animal's environment sometimes possesses very little diversity, and then the responsibility for the animal's health, productivity and welfare rests on the house designers and users. In more diverse environments the choice of micro-environment by the pig modifies the impact of the macro-environment.

The low thermal insulation of the young pig indicates that the animal should be kept in groups in its accustomed life-style, where huddling in the cold and spreading out in the warm buffer the effects of a fluctuating environment to a marked degree. These reactions are assisted by the presence of bedding.

Maintaining temperatures above the critical level for animals on high planes of nutrition does not lead to an increase in energy retention but instead provides an obstacle to the dissipation of the

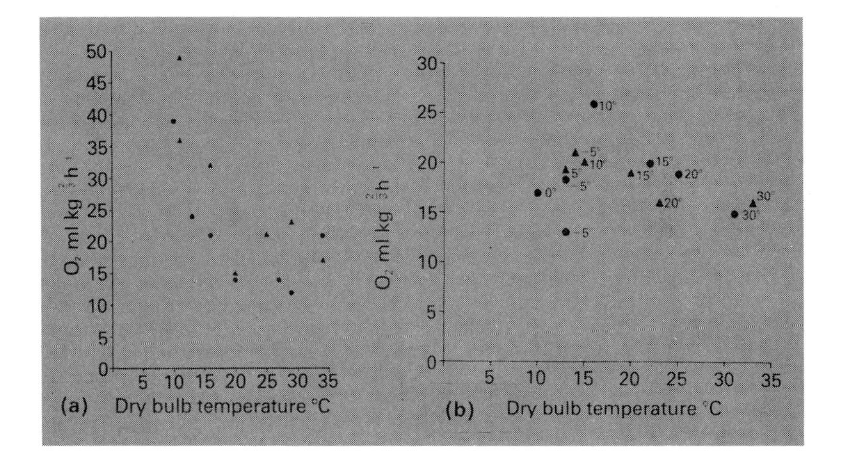

Fig. 8.13 Oxygen consumption of two pigs, (▲ and ●), (**a**) in a chamber without radiant heat; (**b**) in a chamber where the pigs could turn on radiant heaters by pressing a switch. The temperatures written beside the experimental points indicate the air temperature of the chamber at the time the pig was placed in it and before it operated the heaters. The abscissal temperatures were those at which oxygen consumption was measured, and which resulted from the pig operating the heaters (Baldwin and Ingram, 1967, by permission of Physiology and Behaviour).

heat produced on high food intakes, so limiting food intake and reducing productivity.[214, 374a] If pigs are kept too warm then there is a danger that activity or feeding may induce hyperthermia because the animals' evaporative heat loss under dry conditions is inadequate for thermoregulation. Under hot conditions, pigs tend to eat little and often; the effect of this is to avoid large heat increments of feeding at any one time.

Conversely, low feeding levels adopted in the quest for lean carcasses call for somewhat warmer conditions. The interactions of the plane of nutrition and the environmental temperature in relation

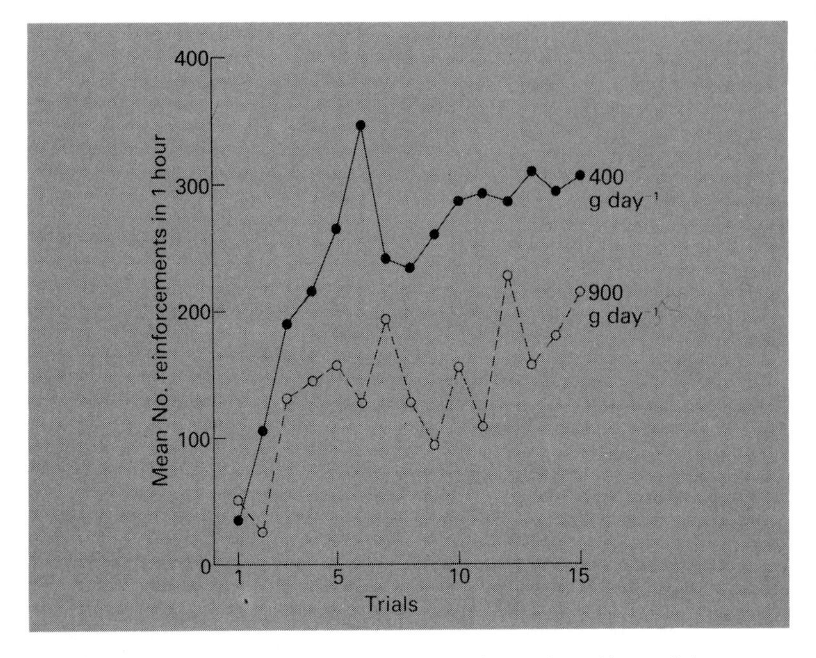

Fig. 8.14 The effect of level of food intake on the number of heat reinforcements produced by pigs operating a switch (Baldwin and Ingram, 1968, by permission of Physiology and Behaviour).

to protein and fat retention are summarized in Fig. 8.15. Both factors, particularly the plane of nutrition, affect fat formation much more than protein deposition. The critical temperature can be estimated by calculating the extra thermoregulatory heat production (ETH) due to the cold by subtracting from the total metabolizable energy (ME) intake both the ME associated with energy retention and the ME required for maintenance at thermal neutrality (see Chapter 2). ETH divided by the appropriate increment in heat loss per °C from Table 8.2 then gives the temperature difference by which the critical temperature exceeds the operating temperature.[393]

Calculated ranges of thermoneutral air temperature for pigs of increasing body weight are given in Table 8.3. The limits were calculated from the minimal metabolic rates and estimates of the core-environment thermal conductances, and show how the range of optimal temperature for productivity is extended to colder levels as the plane of nutrition is increased. Table 8.4 gives mean values for

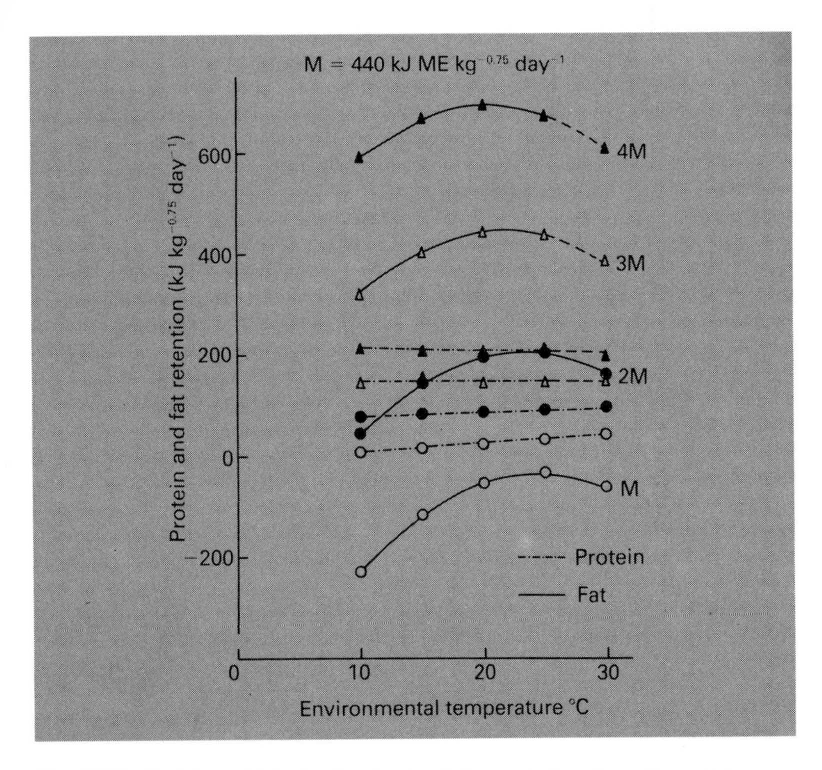

Fig. 8.15 Protein and fat retention in pigs fed once to four times the maintenance requirement (M) at several environmental temperatures. ME, metabolizable energy; temperature, °C (Close, Mount and Brown, 1978, by permission of Cambridge University Press).

heat production at thermal neutrality together with approximate intakes of metabolizable energy, so that the percentage energy retention can be calculated.

The outside air temperature, ventilation, building insulation and the nature of the floor are the primary determinants of the thermal environment in a pig house when taken in conjunction with the stocking density and consequently the heat and water vapour produced by the animals.[80, 93–4, 429, 474] Although an animal's responses may themselves be the best indicator of the suitability of an environment, the difficulty with such observations lies in generalizing the results obtained, which is why physical measurements of different attributes of the environment are usually made instead, or simulated systems employed. Variations between houses and in management do

Table 8.3 The calculated ranges of thermoneutral air temperatures for pigs under standardized conditions. M = 0.42 MJ metabolizable energy intake per $kg^{0.75}$ per day (Holmes and Close, 1977).

	Live weight (kg)	Metabolizable energy intake: M	2M	3M
		Thermoneutral zone		
	2	31–33	29–32	29–31
	20	26–33	21–31	17–30
	60	24–32	20–30	16–29
	100	23–32	19–30	14–28
Pregnant sows 112 days	140			
	Thin	20–30	15–27	11–25
	Fat	19–30	13–27	7–25

not allow conclusions reached for one location necessarily to be valid for another.

One way of overcoming this difficulty is to calculate an equivalent standardized environmental temperature, which is the temperature of a dry standardized environment that would produce the same effect on the animal's energy exchange as the actual environment (see Chapter 5). The critical temperatures, and rates of increased heat loss below the critical temperature, are obtained from laboratory experiments on pigs in which farm conditions have been simulated, so that some behavioural and management factors are incorporated, but which have been carried out under standardized environmental conditions (Table 8.5).[238]

This approach to the problem of making general statements about desired environments for pigs gives an answer in terms of air temperature. For example, for a group of 35 kg pigs fed at the level of 540 kJ metabolizable energy per kg body weight per day, the following air temperatures appear to be required to give minimal heat loss and maximal energy retention under different sets of conditions. The equivalent standardized environmental temperature is in each case 14°C:

Insulated house, insulated floor, no draughts	14°C
Insulation, but draughts present	20°C
Uninsulated house, no draughts, winter	16°C
Uninsulated house, draughts present, winter	22°C
Good straw bed	10°C

Table 8.4 Heat production, under thermoneutral conditions, of pigs of several live weights, fed on different amounts of energy. M = 0.42 MJ ME per $kg^{0.75}$ daily (Holmes and Close, 1977).

Type of pig	Live weight of pig (kg)	Metabolizable energy intake: 0	M	2M	3M
		Heat production (and approximate intake of metabolizable energy in brackets) (MJ per pig daily)			
Milk-Fed					
Newborn	2	0.89	0.96 (0.7)	1.08 (1.4)	1.19 (2.1)
Young	2	1.15	1.40 (1.2)	1.59 (2.4)	1.80 (3.5)
Solid-Fed					
	15		4.94 (3.2)	6.06 (6.4)	7.18 (9.6)
	20	3.76	4.00 (4.0)	5.31 (7.9)	6.61 (11.9)
	60	7.85	8.84 (9.0)	11.37 (18.0)	13.89 (27.1)
	100	11.51	13.00 (13.2)	16.67 (26.5)	20.37 (39.7)
Sows					
	140		16.01 (17.0)	21.63 (34.1)	27.25 (51.1)
Pregnant sows					
60 days	140		18.56	24.18	29.80
112 days	140		20.77	26.56	32.02

Table 8.5 Approximate relations of effective critical temperature (CT), °C, to feeding level and number of pigs in a group for 35 kg pigs, and rates of increase in heat loss below CT. 1 g feed≡12 kJ metabolizable energy (Mount, 1975, by permission of Livestock Production Science).

Feeding level per day		Heat loss above CT per day	CT Group size: number of pigs		
$kJ\ kg^{-1}$	$g\ kg^{-1}$	$kJ\ kg^{-1}$	1	4	9
500	42	270	20	17	16
540	45	290	19	16	14
575	48	310	18	14	12
620	52	330	17	12	10
Rate of increased heat loss below CT ($kJ\ °C^{-1}\ kg^{-1}\ day^{-1}$)			7	5	4
Equivalent increase in feed intake ($g\ °C^{-1}\ kg^{-1}\ day^{-1}$)			0.6	0.4	0.3

Temperature levels recommended in practice for housed pigs are 24°C at weaning, falling to 15°C at 100 kg for growing pigs, 10–15°C for the sow, and 27°C for young piglets.[453] These temperatures can be compared with ranges of thermal neutrality calculated for different levels of food intake in Table 8.4. The increases in food intake calculated to be necessary to offset the effects of lower temperatures are given in Table 8.5.

PIGS IN THE TROPICS

Even the best results on raising pigs in the tropics fall about 15% below performance levels that could be expected in a comparable temperate environment. In animals that are fed *ad libitum*, food intake falls as environmental temperature rises, and food conversion efficiency falls. From a number of different investigations, quoted by Steinbach:[482]

$$F = 145.7 + 1.96T_a - 0.104T_a^2$$

where F = daily food intake, g (kg body weight)$^{-0.72}$; and T_a = environmental temperature, °C.
The mean food intake decreases from 155 g kg$^{-0.72}$ at 10°C to 90 at 35°C. This corresponds to 67 g kg^{-1} at 10°C and 50 g kg^{-1} at 35°C for the 20 kg pig, and 40 g kg^{-1} at 10°C and 30 g kg^{-1} at 35°C for the 60 kg pig.

For each 1°C increase in mean daily temperature above the temperature of maximum gain (24°C for 40 kg pigs, and 16°C for 160 kg), there is a decrease of 30 g in the average daily gain.[374] When the relative humidity increases from 30% to 80% at temperatures above the optimum there is a further decrease in growth rate, of about 15% at 28°C and 30% at 33°C.[372–3] These results refer to pigs fed *ad libitum*; many of the observed variations disappear when the animals are fed on a restricted scale, when food intake is held constant independently of environmental temperature, except where this is so high as to inhibit food intake even on the restricted level. Water intake is increased at high environmental temperatures; whereas the water : feed ratio falls between 2.1 and 2.7 at temperatures between 7 and 22°C, the ratio is 2.8–5.0 at 30 and 33°C under laboratory conditions.[399]

Reproduction

High environmental temperatures tend to depress fertility, but may do so less in pigs that are adapted to hot conditions. The total number of piglets born in a litter and their birth weights do not differ significantly between tropical and temperate sow herds; however, the weights at weaning are considerably less in the tropics.

In the male under hot conditions there are significant decreases in sperm cell concentrations in the ejaculate, and artificial cooling of the boars increases fertility. In the female, there is delay in sexual maturation in hot climates, possibly related to reduced food intake and growth rate. The incidence of temporary anoestrus increases at high temperatures.

The survival of embryos during pregnancy is reduced by heat stress if this occurs within two weeks of mating but not if it occurs later; heat stress in the terminal phases of pregnancy leads to an increase in the number of still-births. During the lactation period, milk yield is reduced at high environmental temperatures, possibly due to low food intake. Many of the effects of high temperatures in reducing reproductive efficiency may be produced indirectly as the result of low food intakes.[483]

COMPARISONS WITH OTHER SPECIES

The high thermal conductance of the newborn pig is comparable with that of the newborn human infant and about double that of the lamb.[29] Between the newborn and mature stages, the pig's environmental temperature requirements change more than man's. The critical temperature is about 34°C for both the newborn pig and for the newborn baby.[384] The mature pig at 34°C without a wallow is distressed and may die in hyperthermia: adult man at 34°C is uncomfortable but he can survive provided that he has water to drink, because he can dissipate heat through his highly effective sweating. The mature pig's critical temperature may fall to 0°C or even lower,[265] whereas adult man's critical temperature is about 28°C. The pig's shift in critical temperature from the newborn to the mature animal is thus more than 30°C, whereas for man it is less than 10°C (see also Ingram and Mount[258]). The corresponding shift for the sheep is even greater than for the pig (see Table 2.1).

The pig's lack of thermoregulatory sweating and absence of highly effective panting lead to a higher thermal dependance on those factors in the environment that influence heat transfer through the non-evaporative or sensible channels. When the effects of a hot environment on body temperature in man, cattle and pigs are related to dry bulb (DB) and wet bulb (WB) temperatures, the weightings are as follows:[249]

$$\begin{aligned}
&\text{Man:} && (\text{DB} \times 0.15) + (\text{WB} \times 0.85)\\
&\text{Cattle:} && (\text{DB} \times 0.35) + (\text{WB} \times 0.65)\\
&\text{Pigs:} && (\text{DB} \times 0.65) + (\text{WB} \times 0.35)
\end{aligned}$$

The high WB coefficient for man is associated with the significance of the ambient humidity for evaporative cooling from sweating, whereas the pig, with its inability to sweat, has a low WB coefficient. Cattle, that sweat more than pigs but less than man, have an intermediate WB coefficient.

9

Cattle

Cattle are of great importance in providing meat, milk and traction power in many parts of the world. They are large animals, naturally free ranging, although they can be raised indoors. Cattle exist in many different climates, but basically they are cold-tolerant; this is the first of their particular climatic characteristics. The large body size of itself confers resistance to cooling (see Chapter 4); in addition, the animal has a coat that can provide considerable thermal insulation. The newborn also share cold tolerance; they are born in a remarkably mature state in respect of locomotion and metabolic responses. The second characteristic relating to climate is that cattle can sweat to such a degree that under hot conditions evaporative heat loss through sweating exceeds that from the respiratory tract. The third characteristic is that cattle have a high rate of water turnover, so that they are highly susceptible to lack of water. There are distinct differences between breeds, particularly in adaptation to high temperatures.

BREEDS

Cattle occur principally as two breed varieties: *Bos taurus*, which includes the European breeds, and *Bos indicus*, the Indian or tropical breeds. Francis[161] uses the terms zebu, Brahman or *Bos indicus* as being practically synonymous in referring to a 'typical zebu', which he defines as a bovine with a well-developed musculo-fatty thoracic hump and dependent dewlap. *Bos indicus* means literally 'Indian ox', and India is regarded as the origin and focus of these zebu cattle, with distribution from China to Africa. The name 'Brahman' is sometimes reserved for a particular type of zebu developed in the U.S.A. There are innumerable gradations and variations between zebu and non-humped cattle.

Zebu cattle thrive better than British breeds in a tropical environment. Animals of the *Bos indicus* variety are better adapted to high temperatures and have a shorter, more glossy and less dense coat than *Bos taurus*. The production of meat and milk has been developed more highly in European cattle than tropical cattle, owing to climate, breed and the application of scientific and technological knowledge.

Modifications of behaviour for temperature adaptation do not occur in European cattle between about 2 and 21°C. The corresponding range for Brahman cattle that does not call for behavioural thermoregulation lies between 10 and 27°C, indicating the higher mean temperature preference of these animals.[189] The differences between lactating and Indian cattle in respect of heat production and rectal temperature in relation to ambient temperature are illustrated in Fig. 9.1. *Bos indicus* have lower production levels and lower metabolic rates than *Bos taurus*, and as the temperature rises their rates of food intake and milk production do not fall off so rapidly. However, a proper comparison of metabolic rates can only be made when the levels of food intake are controlled.[166, 513] The sweating rate of *Bos indicus* is higher than that of *Bos taurus* relative to respiratory evaporation. The appetite of the zebu is not easily depressed by lack of water, and the animal eats a wider range of vegetation than European breeds.[346]

High levels of animal production are associated with high levels of food intake and high rates of heat production that can produce excessive heat stress in hot environments. Some cross-bred animals can maintain productivity coupled with heat tolerance; an example is the Santa Gertrudis, which is a Shorthorn-Brahman cross.[565] The use of cross-breds is of considerable importance, because cattle taken from temperate zones to a tropical environment fall off markedly in productivity so that direct transfer of cattle is self-defeating. An important feature is the inherently higher resistance to disease among cattle indigenous to the tropics compared with imported cattle.

THERMAL NEUTRALITY AND RESPONSE TO COLD

The cold tolerance of cattle is evident from lack of increase in the resting heat production of heifers exposed without shelter to a severe

Canadian winter. These heifers had a thicker coat than other animals kept under warmer conditions, and the associated increased thermal insulation lowered the critical temperature.[524] However, calves fed *ad libitum* and exposed to a mean winter temperature of −20°C did not grow as rapidly when in an open yard as when they were given access to shelter.

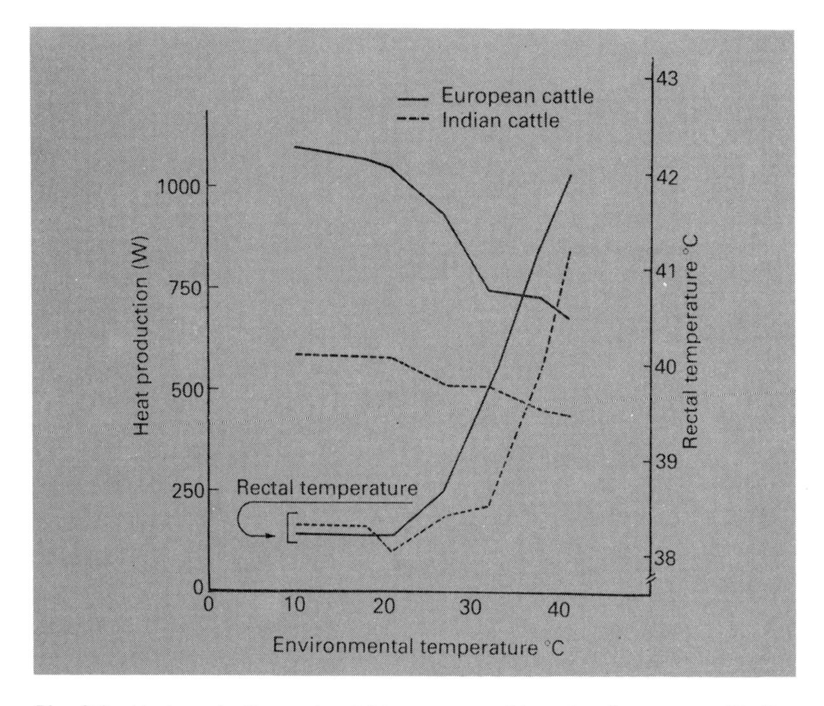

Fig. 9.1 Heat production and rectal temperature of lactating European and Indian cattle as a function of environmental temperature (Yousef, Hahn and Johnson, 1968, after Worstell and Brody, 1953, Mo. Agric. Exp. Sta. Res. Bull. No. 515, by permission of Lea and Febiger).

Fasted cattle at thermal neutrality have a rate of heat production of 70–90 W m^{-2}.[524] When the animal is fed at the maintenance level, metabolic rate is close to 100 W m^{-2}, and this rises to 170 W m^{-2} for the very productive animal on higher food intakes (Table 9.1). At the maintenance level of feeding heat production may be increased by 20% to 50% by activity in grazing out of doors. Below the critical

Table 9.1 Heat production of cattle confined at thermoneutrality and at different stages of production (Webster, 1974).

	Body weight (kg)	Surface area (m^2)	Heat production	
			W	$W m^{-2}$
Fasting metabolism				
Calf 1 month old	50	1.24	109	88
1 year old	300	4.11	384	93
Steer 2 years old	450	5.39	401	74
Growing cattle				
Veal calf 1.5 kg gain day^{-1}	100	1.97	304	154
Baby beef 1.0 kg gain day^{-1}	150	2.58	359	139
	350	4.55	656	144
1.3 kg gain day^{-1}	150	2.58	387	150
	350	4.55	707	155
Store cattle maintenance	250	3.64	448	123
0.4 kg gain day^{-1}	250	3.64	576	158
Fat stock 0.8 kg gain day^{-1}	450	5.39	845	157
1.5 kg gain day^{-1}	450	5.39	943	175
Beef cow maintenance	450	5.39	578	107
Dairy cattle				
Dry, pregnant	500	5.79	604	104
2 gallons day^{-1}	500	5.79	747	129
5 gallons day^{-1}	500	5.79	891	154
8 gallons day^{-1}	500	5.79	1034	178

temperature, heat production increases by about 2.4 $W m^{-2}$ for every 1°C fall in environmental temperature.

The resting heat production of the young 40 kg Ayrshire bull calf, given 4 litres of milk per day, is near 100 $W m^{-2}$. On the third day after birth its critical temperature is 13°C, and later, on the twentieth day, this falls to 8°C. The evaporative heat loss at low temperatures is 17 $W m^{-2}$, independently of the quantity of feed.[184] In other experiments the critical temperature for both Friesian and Hereford-Friesian calves has been found to be about 8°C when the animals are fed 950 kJ metabolizable energy per $kg^{0.75}$ per day. The calves' heat losses are found to range from 560–668 kJ $kg^{-0.75}$ per day (6.5–7.7 W $kg^{-0.75}$) between 15°C and 5°C environmental temperatures.[527]

From several different estimates, it can be said in general that at temperatures below thermal neutrality the overall insulation of cattle is near 0.42 °C $m^2 W^{-1}$, a value that falls to about 0.33 in the absence of vasoconstriction and pilo-erection under warm conditions. The calf in the cold has an overall insulation of about 0.33 °C $m^2 W^{-1}$ on the

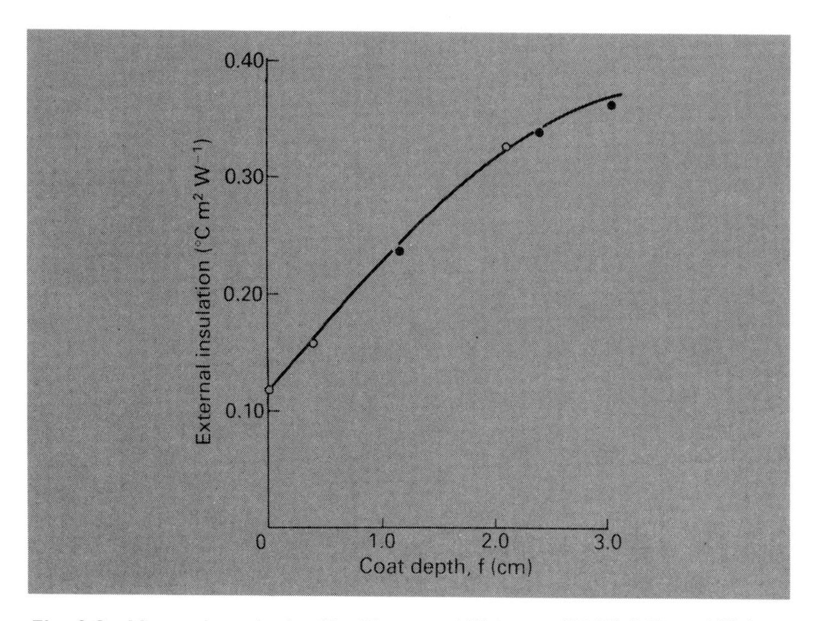

Fig. 9.2 Mean values obtained by Blaxter and Wainman (1964) (○); and Webster and Young (1970) (●) for the external insulation of cattle having different coat depths. The curve is

$$I_e = 10^{-3}\,(118 + 132f - 16.4f^2)$$

where I_e is external insulation, and f = mean coat depth, cm (Webster, 1974).

day of birth, rising to the adult level of 0.42 by 20 days of age, high values that are associated with the cold tolerance that is characteristic of cattle. The overall insulation consists of three additive insulations in series: the tissue insulation, the coat insulation, and the air-ambient insulation; the last has an average value of 0.12 °C m^2 W^{-1} in still air (see Chapter 4). Tissue insulation in the calf is 0.09 °C m^2 W^{-1} when vasoconstricted and 0.04 °C m^2 W^{-1} vasodilated, with corresponding values for the adult steer of 0.14 and 0.04 °C m^2 W^{-1}. Local cold vasodilatation (see Chapter 4) reduces tissue insulation in those parts of the appendages where it occurs. The insulation of the hair coat in the calf is 0.11 °C m^2 W^{-1} when unerected and 0.16 °C m^2 W^{-1} when erected;[184] the insulation of the hair coat increases with depth of coat, although the rate of increase falls off for longer coats (Fig. 9.2).

Using estimates of internal and external insulation, Webster[524] has calculated critical temperatures for cattle of different body weights, growing at different rates, lactating and non-lactating (Table 9.2).

Table 9.2 Critical temperatures of cattle housed in conditions of very low air movement (wind velocity 0.16 m s^{-1}). I_t, tissue insulation; I_e, external insulation (Webster, 1974).

	Body weight (kg)	Coat depth (cm)	Thermal insulation (°C m^2 W^{-1})		Heat production at critical temperature (W m^{-2})	Critical temperature (°C)
			I_t	I_e		
Calves						
Newborn	35	1.2	0.09	0.25	100	+9
1 month old	50	1.4	0.10	0.26	120	0
Veal calf	100	1.2	0.12	0.25	154	−14
Beef cattle						
Baby beef 1 kg gain day^{-1}	150	1.4	0.14	0.26	139	−12
	350	1.4	0.19	0.26	144	−12
1.3 kg gain day^{-1}	150	1.4	0.13	0.26	150	−15
	350	1.4	0.19	0.26	155	−26
Store cattle, maintenance	250	2.2	0.16	0.33	123	−16
0.4 kg gain day^{-1}	250	2.2	0.14	0.33	158	−30
Fat stock, 0.8 kg gain day^{-1}	450	2.0	0.19	0.32	157	−36
1.5 kg gain day^{-1}	450	1.4	0.19	0.26	175	−36
Beef cow, maintenance	450	2.9	0.26	0.36	107	−21
Dairy cattle						
Dry, pregnant	500	1.6	0.27	0.29	104	−14
2 gallons day^{-1}	500	1.4	0.26	0.26	129	−24
5 gallons day^{-1}	500	1.2	0.24	0.25	154	−32
8 gallons day^{-1}	500	1.2	0.22	0.25	178	−40

These values apply to animal shelters that are unheated but dry and free from draughts. With the exception of young stock, the critical temperatures are all below −10°C, and highly productive stock have critical temperatures below −30°C. The importance of the level of feeding is underlined by results obtained on two steers, 2 years old and weighing 500 kg, which gave the critical temperature as 6°C when the animals were on the maintenance level of feeding, and as high as 18°C when fasting, although each steer had an 8 mm thick coat.[64] Although draughts and wet conditions would have to be allowed for in practice, these estimates indicate metabolic indifference to a large range of environmental temperature. Low air temperatures are consequently unlikely to have a great effect on the performance of cattle on those parts of the earth where they are husbanded.

Wind, radiant exchange and rain affect the critical temperature of well-fed cattle when this is expressed as the air temperature of a non-standardized environment (see Chapter 5). In Table 9.3, the effects of

Table 9.3 Estimated critical temperature of a well-fed beef cow in different cold environments, I_t, tissue insulation; I_e, external insulation (Webster, 1974).

Environment	Net radiation ($W m^{-2}$)	Thermal insulation (°C $m^2 W^{-1}$) I_t	I_e[1]	Critical temperature (°C)
Dry, calm ($V = 0.4 m s^{-1}$), overcast	−10	0.25	0.29	−13
Dry, calm, 8 h direct sunshine	+63	0.24	0.29	−21
Dry, calm, 4 h sunshine, 16 h cloudless night	−68	0.26	0.29	−6
Dry, windspeed 4.5 $m s^{-1}$ (10 mph)[2], overcast	−10	0.26	0.17	−3
Overcast, raining, coat wet, windspeed 4.5 $m s^{-1}$	−10	0.28	0.09[3]	+2[3]

[1] uncorrected for radiation exchanges
[2] windspeed at an elevation of 1 m
[3] extrapolated from measurements made with sheep

radiation are apparent in the first three examples. Eight hours of direct sunshine lower the mean 24-hourly critical temperature from −13°C to −21°C; 4 hours of sunshine combined with the low radiant temperature of the sky on a cloudless night raise the critical temperature from −13°C to −6°C, the rise in air temperature being required to compensate for the net radiation loss that occurs in contrast to the overcast condition. Wind raises the critical air temperature to −3°C, and rain raises it further to +2°C. Driving rain, therefore, needs a compensatory rise of 15°C at these temperature levels. These values are probably maximum values for cold tolerance in animals of good condition. An additional factor, for which estimations have not been made, is that cows lying in cold mud probably lose a considerable amount of heat by conduction.[524]

Lactation

Exposure to cold depresses lactation, and three possible mechanisms have been suggested.[277,501] First, energy is channelled into the increased heat production demanded by the cold environment. Second, reduction in blood flow to the gland occurs on cooling. Third, Holmes[237] found that cooling one half of the udder by clipping hair from its surface and exposing it to wind reduced milk supply from that half compared with the other half that was covered, indicating a possible direct local effect of cold on milk production in the gland.

Adaptation to cold

Cattle acclimatized to 10°C for one year were not affected by exposure to 2°C, but were stressed by 35°C. Cattle acclimatized to 26°C were stressed by 2°C but not by 35°C.[287] The coat is important in adaptation to cold, and observations have been made on the coats of cattle living out-of-doors or indoors during the Canadian autumn and winter. The animals outside had twice the amount of hair as those indoors, suggesting that exposure to cold leads to reduced shedding of the summer coat in the autumn. Insulative acclimatization in the form of a thicker coat is more marked in cattle fed at maintenance than in those fed on a high plane of nutrition; the effect of the thicker coat is to compensate for the lower rate of heat production that occurs in animals that are fed only at the maintenance level, which leads to a rise in the critical temperature (see Chapter 2). Calculations suggest that dairy type cattle kept outside during the winter are less cold tolerant than beef cattle.[524]

RESPONSES TO HIGH TEMPERATURES

The tolerance of cattle to cold is matched by their susceptibility to high temperatures, which has prompted considerable efforts in breed selection and in assessing the performance of cattle in hot climates.

The rectal temperature of cattle is usually close to 39°C.[46] It rises when the animal feeds or is active, and it falls when cold water is ingested in large quantities. When cattle have been exposed to heat for a long period, there is acclimatization in that a given acute heat stress produces smaller effects on rectal temperature, respiratory rate and heart rate than before the prolonged exposure. When three calves were exposed to 45°C for 5 hours on each of 21 successive days, acclimatization became evident during the first 10 days, with the initial rectal temperature before the heat exposure falling progressively. A fall in food intake and lower levels of heat production and thyroid activity were observed, and the coat became thinner.[49]

Evaporative heat loss

Measured rates of evaporative heat loss from cattle have ranged from 17 $W m^{-2}$ below the critical temperature[65] up to 116 $W m^{-2}$ in the hyperthermic zone above thermal neutrality. When humidity is low, evaporative heat loss from cattle allows the thermally neutral zone to extend up to environmental temperatures equal to body temperature, as for example at 40°C dry bulb and 21°C wet bulb temperatures.[500] In one series of experiments on Jersey cattle weighing about 400 kg evaporativc heat loss was 18% of total heat loss at 15°C and 84% at 35°C.[342] Brahman cattle have lower vaporization rates than European below 32°C, in accordance with their lower metabolic rates, and reach maximum vaporization at 35°C; European cattle reach their maximum at 27°C. Under hot conditions, evaporative heat loss from the skin increases more than that from the upper respiratory tract (Fig. 9.3).

Sweating

When cattle are exposed to heat there are periodic increases in evaporative loss instead of a continuous increase, with the stepped increases taking place at the same time on different parts of the skin surface.[48, 500] The sweat glands of *Bos indicus* are larger and more numerous than those of *Bos taurus* (Table 9.4).

The evaporation of sweat normally takes place from the skin surface, where its cooling power is most effective, although under hot humid conditions there may be some evaporation from the hairs. It has been observed that the sweat collects in drops on the hairs, which are water-repellant, and spreads rapidly on the skin surface. This combination of repulsion and attraction tends to prevent the transfer of sweat from the skin to the hairs, so making evaporative heat transfer from the body more effective. In one series of measurements in Australia the sweating rates of Jersey, zebu × Jersey and Hereford

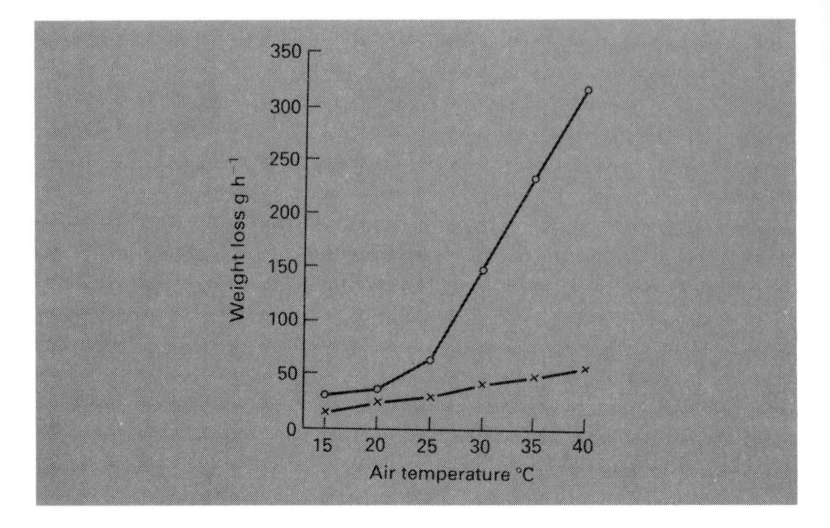

Fig. 9.3 Respiratory (X) and cutaneous (○) evaporative losses in Ayrshire bull calves in relation to environmental temperature (from McLean, 1963, by permission of Journal of Physiology).

heifers were found to range from 28 to 438 $g\,m^{-2}\,h^{-1}$ [11] (see also Table 3.2).

Indian cattle have long limbs and dewlaps, and it has been suggested that these provide an increased surface area for the dissipation of heat. However, at high environmental temperatures such dissipation would not be through the non-evaporative channels, and for evaporative heat loss the concentration of sweat glands on the

Table 9.4 Average dimensions and density of sweat glands on the mid-flank of adult cattle* (Macfarlane, 1964, by permission of American Physiological Society).

	Bos indicus	*Bos taurus*
Length, μm	936	724
Diameter, μm	173	129
Volume, $\mu m^3 \times 10^{-6}$	20–25	8–12
Number per cm^2	1507	1005

* Based on Nay and Hayman[405]

The *Bos indicus* breeds are Sindhi and Sahiwal, the *Bos taurus,* Jersey, Friesian, and Red Poll. Shorthorns have even smaller glands.

dewlap appears to be no greater than elsewhere on the body surface, although there have been some indications that the glands are larger. Surgical removal of the dewlap has not produced any marked differences in response to high temperature, and its possible role in heat dissipation remains open to question.[334]

Respiratory evaporation

The respiratory rate gives an indication of a cow's thermal state: above 80 per min, the animal is very warm, at 20 per min the cow is cool and near or below the critical temperature. Cattle at 35°C have been found to have respiratory rates of 100 and 160 per min at relative humidities of 35 and 75%.[342] In Brahman cattle, with their more highly reflecting coats, the rises in respiratory rate and in deep body temperature following exposure to sun are less than they are in European cattle. Brahman cattle also lose only 12% of their evaporation from the respiratory tract at an air temperature of 38°C, compared with 24% for the Shorthorn at the same temperature.

Panting in cattle is not as effective as sweating in providing evaporative cooling. The respiratory ventiliation may increase 10-fold in cattle, 12-fold in sheep, 15-fold in the rabbit, and in the dog, where panting is highly effective, as much as 23-fold. In the first phase of rapid shallow panting, under mild heat stress, the dead space ventilation increases more than the alveolar ventiliation, so that there is little effect on the partial pressure of CO_2 in the arterial blood and the acid-base balance. Under more severe heat stress, when the deep body temperature in cattle approaches 41°C, 'second phase' breathing occurs, in which the respiratory frequency falls and the tidal volume increases (Figs 9.4 and 9.5). 'Second phase' breathing increases the alveolar ventilation up to five-fold and leads to increased loss of CO_2 and respiratory alkalosis, when the pH of the arterial blood may exceed 7.7.[195, 500, 540]

Radiation

Exposure to intense solar radiation can add considerably to the heat load on an animal. Coat colour is important for the absorption of solar radiation (see Chapter 3); in the visual region of the spectrum, white coats have a low absorptance and a correspondingly high reflectance, and black coats have a high absorptance, with correspondingly smaller and greater radiant heat loads respectively.[79] The smooth-coated *Bos indicus* has a higher reflectance and a lower surface temperature than *Bos taurus*.

Kelly, Bond and Heitman[294] made measurements with a directional

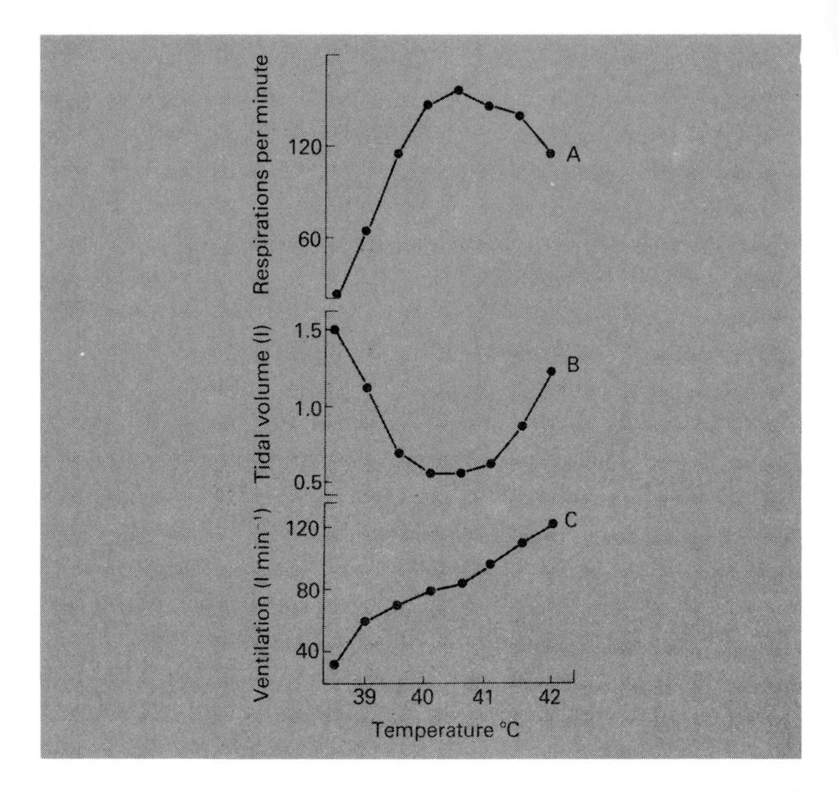

Fig. 9.4 Changes in respiratory rate (A), tidal volume (B), and minute volume (C) of the ox during hyperthermia. Temperature is rectal temperature. Mean result from 9 animals (Bianca and Findlay, 1962, by permission of Research in Veterinary Science, redrawn by Whittow, 1971, by permission of Academic Press).

radiometer of the radiant energy expected to fall on cattle from the earth and sky hemispheres both under a shade and when exposed to the sun. The reduction in radiant intensity due to standard cattle shades in California was found to be 50–65% on the animal's back from the sky hemisphere and 25–30% from the earth hemisphere.

Endocrine and cardiovascular effects

Exposure to heat and an increase in body temperature reduce thyroid activity in cattle, the effect of exposure taking about 60 hours to develop.[48, 500] Starvation also brings about a reduction, and *ad libitum* feeding at low temperatures increases thyroid activity.

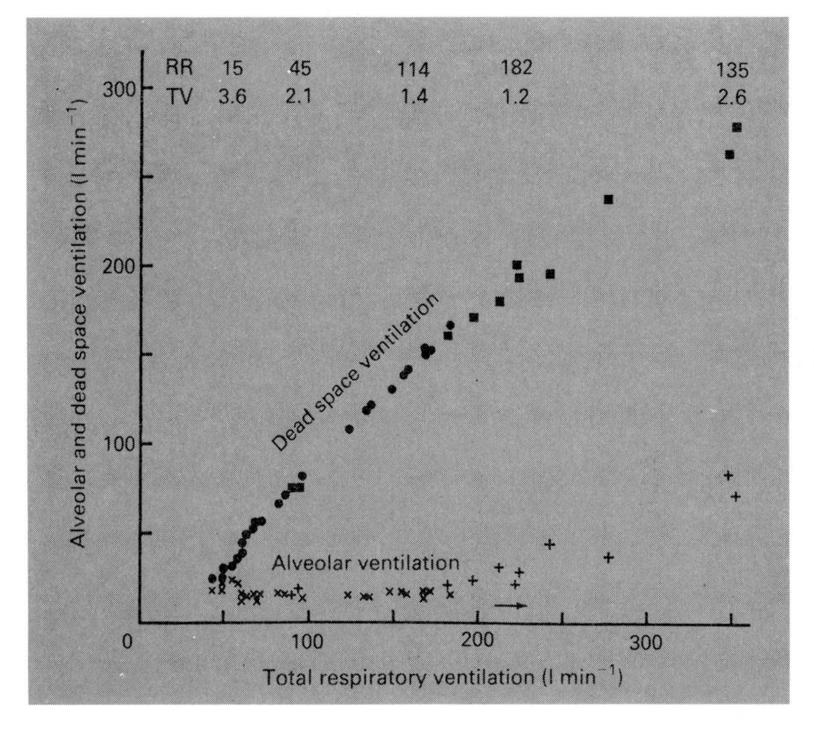

Fig. 9.5 The changes in alveolar and dead space ventilation with increasing total respiratory ventilation of the 280 kg ox as the dry bulb/wet bulb temperatures rise from 16°C/14°C to 40°C/38°C. The arrow indicates the transition from rapid shallow to slower deeper breathing. RR, average respiration rate (breaths per min); TV, average tidal volume (litres) (Hales, 1967, by permission of Journal of Physiology).

As in other animals, exposure to moderate heat leads to peripheral vasodilatation with a fall in the peripheral circulatory resistance and a fall in blood pressure; there is at the same time an increase in blood volume and an increased cardiac output. Large increases may occur in the skin temperature on the shank and the ear when the cattle are fed (Fig. 9.6) and when the animals are exposed to high temperatures.

Food intake

Cattle exposed to moderate heat have a reduced food intake. The food consumption of lactating Holstein cattle falls by 20% at 32°C, and declines to zero with a decrease in rumination at 40°C, with an accompanying decline in heat production. Daytime grazing is reduced

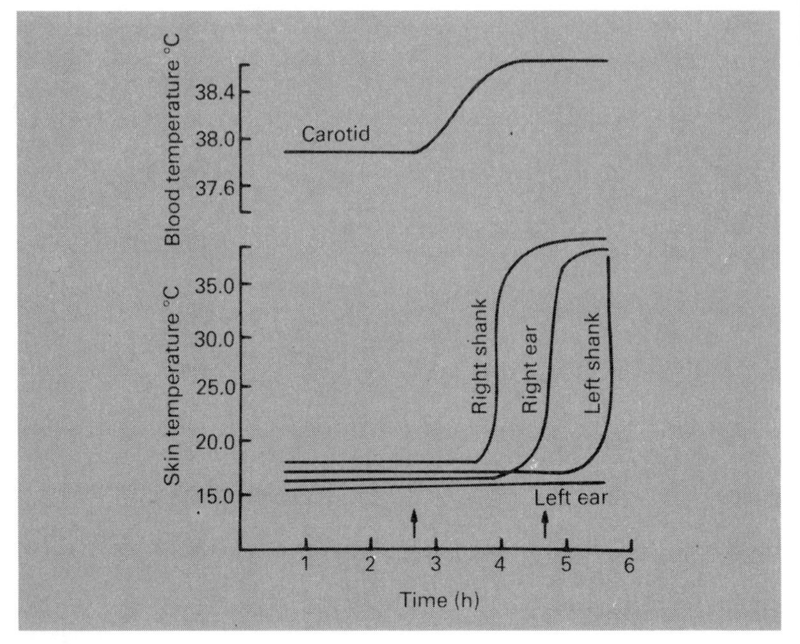

Fig. 9.6 The effect of feeding on skin temperatures on the ear and shank of the leg of the ox (Ingram and Mount, 1975a, redrawn from Ingram and Whittow, 1962, by permission of Springer–Verlag).

under tropical conditions,[307] and the animals spend much time in the shade;[565] for example, 67% of the grazing time of cows in Fiji was found to be at night.[417]

Feed consumption begins to decline when the environmental temperature rises to 24–26°C for Holstein cattle, 26–29°C for Jerseys, above 29.5°C for Brown Swiss and 32–35°C for Brahmans.[556] Growth rate is reduced under hot conditions, but the reduction in growth is masked to some degree by the increased retention of water, so that the reduction in body weight increase is not as marked as would otherwise be the case. Water turnover in cattle is discussed in Chapter 11.

REPRODUCTION

In the bull, high environmental temperatures depress sperm concentration and mobility and reduce male fertility. Internal scrotal temperature varies little with exposure to environmental temperatures from −3 to 31°C, but at 40°C there is reduced spermatogenesis, an increased number of abnormal spermatozoa, and histological evidence of damage to the seminiferous tubules. In the cow, exposure to heat may shorten oestrus or temporarily stop the oestrous cycle; the baseline concentration of plasma luteinizing hormone is reduced in the heat, and the peak concentration at oestrus is reduced, with longer time intervals between peaks. In the pregnant animal under hot

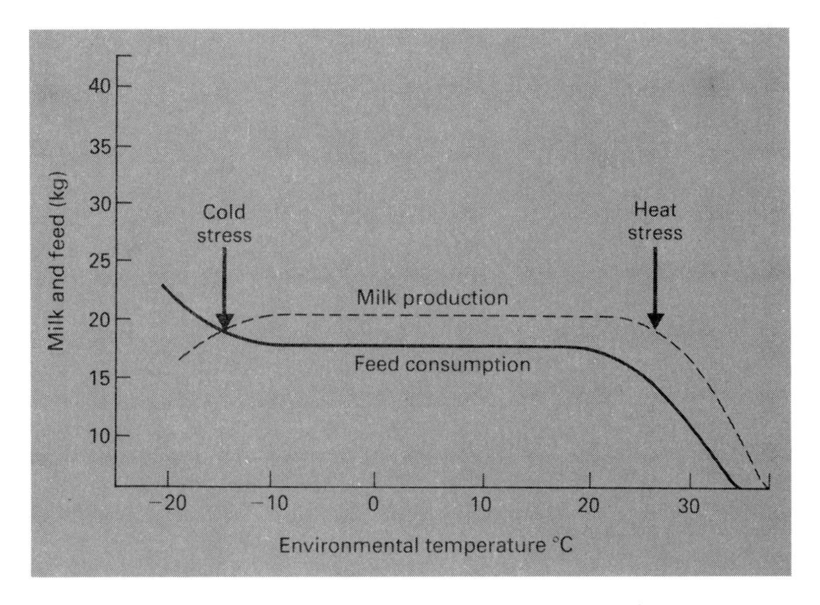

Fig. 9.7 Diagrammatic illustration of the effect of environmental temperature on milk production and feed consumption of cattle (By H. D. Johnson, from Hafez, 1968b, by permission of Lea and Febiger).

conditions there is a high rate of embryonic mortality, with the critical time occurring soon after mating.[500] Pregnant cattle exposed to high temperatures produce small calves.

Lactation

Heat stress reduces milk yield, with a more marked effect in mid-lactation than at other stages (Fig. 9.7), milk yield falling by about one kg for each °C rise in rectal temperature. The general level of milk production in Brahman cows is less than in European breeds, but when the environmental temperature rises milk production does not fall off in the Brahman until 35°C is reached, whereas the corresponding decline in European cattle occurs at 29°C. A high rate of milk production is accompanied by a high water intake. The decline in lactation at high temperatures is associated with a fall in water consumption, but the consumption then rises as temperature rises further, a thermal effect seen in non-lactating animals (Fig. 9.8).

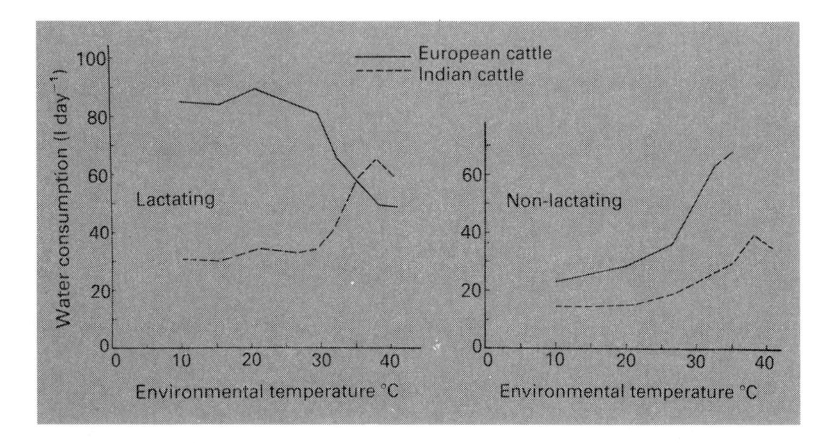

Fig. 9.8 Effect of environmental temperature on water consumption of lactating and non-lactating European and Indian cattle (Yousef, Hahn and Johnson, 1968, from Worstell and Brody, 1953, Mo. Agric. Exp. Sta. Res. Bull. No. 515; and Johnson, Ragsdale, Berry and Shanklin, 1963, Mo. Agric. Exp. Sta. Res. Bull. No. 846, by permission of Lea and Febiger).

The optimal temperature for milk production has been suggested as 10°C. A 'temperature-humidity index' has been used instead of the dry bulb temperature by itself, to assess the thermal environment for cattle under hot conditions where humidity plays a large part in thermoregulation; above a dry bulb temperature of 18°C with 50% relative humidity, milk production is adversely affected in most cows.[279] After 3–5 weeks' exposure to 29°C in one series of experiments most of the obtainable acclimatization was achieved in lactating Holstein cattle, but at a level below the animal's potential for milk production at 18°C; however, in dry cows exposed to 29°C, while the animals were still at the high temperature there was a return to the 18°C values in respect of rectal temperature, feed consumption, water consumption and body weights.[278] The effects of high humidity on milk yield become pronounced only above 24°C.

Growth is depressed by high temperatures, but because under these conditions the consumption and digestibility of feed are higher in zebu than European cattle, rates of gain are higher in zebu.[155] However, in temperate regions European cattle have higher growth rates than zebu.

IMPLICATIONS FOR HUSBANDRY

As cold-tolerant animals, cattle need only shelter that is adequate to allow them to avoid extremes of cold, wind and precipitation. Tables 9.2 and 9.3 indicate that in a temperate climate cattle need shelter from wind and rain but little else. Housed cattle need high rates of ventilation if relative overheating is to be avoided, and ventilation is important in reducing the risk of pneumonia. It is hot conditions that are inimical to cattle; shelter is needed from solar radiation, and in some areas cooling devices such as showers are useful. The animal's high rate of turnover of water demands an ample water supply.

Webster[524] states that the main reason for housing cattle is one of convenience, and he remarks: 'supplementary heating for cattle is not necessary in the United Kingdom and in most areas of cattle production, but shelter from wind and rain may be beneficial as may be a confinement system that allows the animal to select its own most comfortable environment'. The direct benefits of winter housing are likely to be small, and the water vapour produced by increased respiratory activity in an environment that is warmer than the animals need tends to produce condensation and requires a high ventilation rate for its removal. Shelter from wind, rain and summer sun is all that is required.

It is useful to attempt an assessment of animal-environment interactions in relation to productivity under husbandry conditions. This approach has been used by Smith[475] in assessing environmental effects on steers kept during the winter in simple shelter, roofed or unroofed, exposed to the weather, and fed on poor to medium quality hay; the results which he used were taken from McCarrick and Drennan.[333] Smith summarized the weather over the period of the observations, and took as a reference base the mean air temperature when the animals showed no weight change, that is they were feeding at maintenance level. In a subsequent experiment the mean air temperatures in the intervals between animal weighings, when related to the reference temperature, accorded with the loss or gain in body weight. Smith has used this information in conjunction with temperature charts of the United Kingdom to predict mean weight losses over the winter season: these amount to about 12 kg for a 200 kg animal under these conditions of housing and feeding in Eastern England.

Susceptibility to heat means that variations in heat tolerance between the different breeds become important in determining the choice of breed for meat and milk production under given conditions. The selection of cattle for hot environments is best achieved by measuring productivity in the particular environment, because the complex interaction of factors makes selection based on isolated criteria difficult, and there is much to support this view.[70] When suitable animals have been found it is then useful to measure various characteristics of sweating and body temperature, cardiovascular responses and hormonal variations to act as sets of predictors for making further selection. Heat tolerance of cattle is largely judged on rectal temperature measurements. If thermostability is required, in some areas it then becomes necessary to engineer an environment in which the animals can flourish, although this is expensive. Performance testing may indicate that under some conditions high productivity is associated with some degree of thermolability. Physiological indices so far developed, based on rectal temperature, respiratory rate and evaporative loss, indicate those animals that can maintain heat balance under hot conditions but the indices do not predict their productive performance.[286]

The upper limits for optimal production by dairy cows probably lie between 20 and 25°C for *Bos taurus* and 30 and 35°C for *Bos indicus*. In the case of beef cattle, 27°C has been found to be too hot for optimal production of Shorthorns, although in all these instances the level of production and acclimatization must be allowed for.[524] Resistance to disease is a most important factor in the selection of cattle for hot climates, and *Bos indicus* cattle exhibit relative resistance to a number of parasites in addition to being more heat-tolerant and more efficient in feed utilization under hot conditions than *Bos taurus*.[79, 359]

Selection for hot environments

For production in the tropics animals should have high efficiency of feed utilization, efficient heat loss mechanisms and ability to withstand some rise in body temperature and some dehydration, together with resistance to local diseases. The three approaches to this problem are to select within native breeds, to upgrade by crossing with improved breeds, and to develop new breeds.[565]

McDowell[334], in considering the improvement of livestock production in warm climates, is of the view that the genetic potential of many breeds of livestock that are native to hot countries is so limited that they cannot respond by increased productivity to improved environment and nutrition. He concludes that much wider use could be made of cross-breeding as a means of overcoming deficiencies in the native stock.

Variation in cattle breeds

Highly adapted cattle can be obtained by cross-breeding tropical and temperate species, followed by selection. The difference between the breed types lies in the markedly different air temperatures that are required to produce a similar rise in rectal temperature. The air temperatures required to raise the rectal temperature by 1.3°C were 40.5°C for a Brahman × Shorthorn × Hereford cross, 40°C for an Afrikander × Shorthorn × Hereford cross, and 32°C for a Shorthorn × Hereford cross. The differences in response of the rectal temperature were related to the different capacities for evaporative water loss. One basis for selection is the maintenance of a normal rectal temperature without a reduction in food intake.[512]

10

Sheep, goat and deer

Domestic sheep (*Ovis aries*) and goats (*Capra hircus*) have many similarities. Together with cattle, they belong to the family Bovidae of the order Artiodactyla, the even-toed ungulates (see classification in Chapter 11). Deer are members of the family Cervidae, also of the order Artiodactyla. Sheep and goats will be discussed together, and deer separately.

There are more than 200 breeds of sheep widely distributed in the world, from the Arctic Circle to the southern parts of South America and New Zealand. The largest numbers are in the warmer parts of the temperate zones; wild sheep are found particularly in mountainous contry in the Northern hemisphere. Sheep occur in many climates, hot and cold, wet and dry; they also live in semi-arid regions that experience extremes of heat during the summer and of cold during the winter. However, few occur in hot wet climates on account of fleece rot, foot-rot and problems with parasites. The world total of sheep is about 1000 million, a number reached following a rapid increase during the last 100 years. They are used for the production of wool, hides and meat, and in some areas for milk and even as pack animals. Terrill[497] reproduces Mason's[358] table listing types of sheep with their countries of origin, environmental humidity and temperature, distribution, types of coat, uses and feed conditions.

The high level of tolerance of sheep to both hot and cold environments, covering most inhabited areas, constitutes their chief climatic physiological characteristic. Their adaptability is due primarily to the highly insulating fleece that gives protection both from the heat load due to solar radiation and from the effects of cold. The second physiological characteristic that is important in the animal's adaptation to the thermal environment is its economy in the use of water, and its ability to produce a highly concentrated urine.

Goats are most numerous in India, China and Turkey, although they also exist in large numbers in Northern Africa, the Middle East, and some countries of Central and South America. They are kept for the production of milk, meat and fibre (mohair and cashmere). Goats do not occur at the higher latitudes and are therefore not as widely distributed as sheep, although they co-exist in many places with sheep, when they compete successfully for available forage. Goats are more concentrated than sheep in the drier tropical and subtropical areas; their ecological range is very wide, from tropical rain forests to the dry desert, including higher rugged country.[497] Goats are important in subsistence agriculture, and the Saanen goat of Switzerland has contributed greatly to milking herds throughout the world through its introduction into breeding herds of indigenous animals.[375] The total number of goats in the world is about 400 millions.[157]

RESPONSES TO THERMAL ENVIRONMENT

The range of bodyweight amongst the mature ewes of the commoner British breeds extends from 32 kg for the Welsh Mountain to 63 kg for the Clun Forest and 86 kg for the Border Leicester. This body size, coupled with the highly insulating fleece, makes the mature sheep less susceptible than either man or the pig to variations in the thermal environment. Some sheep have coats that are loosely packed, like the Awassi, and some have tightly packed coats, like the Merino, with other intermediate types. The deep body temperature is close to 39°C, with a normal range of ± 1.5°C.

Exposure to cold

The high thermal insulation for sheep, particularly for the Merino where the fleece insulation is $0.023\,°C\,m^2\,W^{-1}$ per mm depth, is noteworthy.[43] The insulation of Down, Blackface and Down × Cheviot fleeces is rather less at $0.015\,°C\,m^2\,W^{-1}$ per mm. The efficiency of the coat as a thermal insulator depends on the degree to which still air is trapped and on how much solar radiation can penetrate (see Chapter 4). The combination of a high thermal insulation and a high cold-induced maximum metabolic rate[42] of

Table 10.1 Calculated environmental temperatures (°C, dry bulb) at which heat loss would equal summit metabolism in lambs and adult sheep, from Alexander, 1962 and Bennett, 1972 (Alexander, 1974).

	Wind-speed ($m\,s^{-1}$) State of coat		
	0.1 Dry	~7 Dry	~7 Thoroughly wetted
Adult 7 mm fleece, av. summit	−56	−11	+14
7 mm fleece, low summit	−26	+4	+25
100 mm fleece, av. summit	(below abs. zero)	−120	−70(?)
Lamb short coat 5 kg	−69	−25	−7 (+4)[1]
short coat 2 kg	−32	−4	+13 (+23)[1]
long coat 5 kg	−105	−54	−35 (−19)[1]

[1] Naturally wet newborn

$10\,W\,kg^{-1}$ or $25\,W\,kg^{-0.75}$ leads to a very considerable degree of resistance to cold (Table 10.1).

Wind reduces the thermal insulation of the fleece, and in Fig. 10.1 fleece insulation is shown relative to fleece length and wind. An approximate formula for the effect of wind on fleece insulation, I_f(°C $m^2\,W^{-1}$), is

$$I_f = f(0.0122 - 0.0028\,V^{0.5})$$

where f = fleece depth, mm; and V = wind speed, $m\,s^{-1}$.[7, 283]

The effects on metabolic rate of a fall in environmental temperature depend in part on conditions that affect the thermal insulation of the coat, including wetting and wind. In Fig. 10.2, metabolic rate for different fleece lengths is related to windspeed and wetting. These results show that a cold, wet, windy environment, with a temperature close to 0°C, can provide demanding conditions even for the mature sheep with a full 100 mm fleece, leading to a doubling of metabolic rate. A sheep with only a 7 mm fleece exposed to the same conditions would have a very high heat loss in excess of $300\,W\,m^{-2}$.

Alexander[7] remarks that heavy vertical rain produces much smaller effects than thorough wetting by hosing; rain wets the wool only along the dorsal midline of the sheep so that that there is little effect on the insulation of the fleece as a whole. The effects of wind and wetting appear to be additive.[62] Conductive heat loss can play an important part; a sheep living on cold, poorly insulated ground may

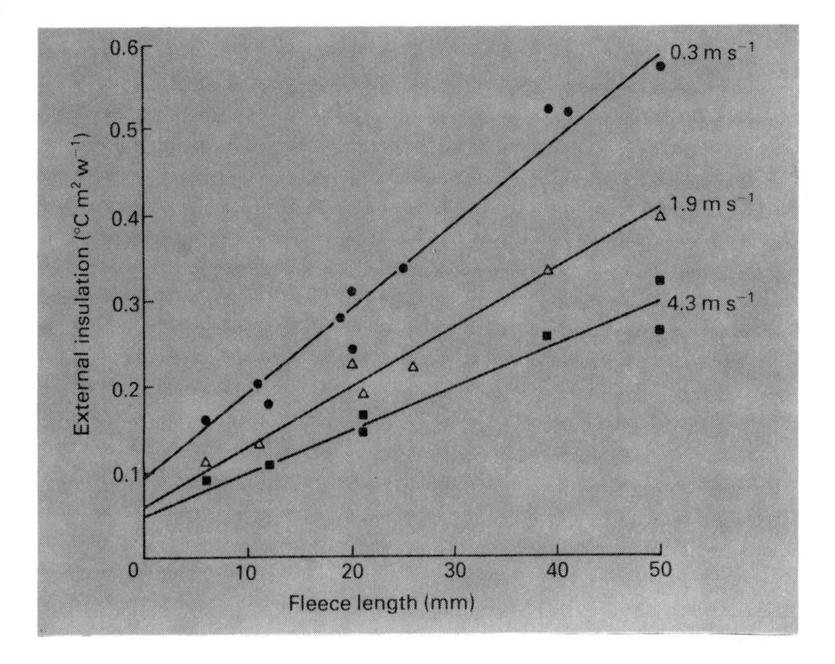

Fig. 10.1 Effect of fleece length on external insulation of sheep exposed to 3 windspeeds: ●, 0.3 m s^{-1}; △, 1.9 m s^{-1}, ■, 4.3 m s^{-1}. The relations were little affected by whether the wind was directed at the mid-side or at the hind quarters (Joyce and Blaxter, 1964, redrawn by Alexander, 1974, by permission of Cambridge University Press and Butterworths).

dissipate up to 30% of its minimum heat production by conduction.[178]

Shearing

Wool does not grow at a uniform rate throughout the year, and there is correspondingly a seasonal variation in the coat covering. The rate of wool growth is related to the plane of nutrition, the environmental temperature and the photoperiodicity; there is a much greater growth in the autumn than in the spring. In the United Kingdom, sheep are normally shorn in the late spring; in some parts of the world they are shorn twice a year.[478]

The safeguards against hypothermia that are provided by the coat are removed when the sheep is shorn, and body temperature can then fall more rapidly in a cold environment. Shearing in cold weather

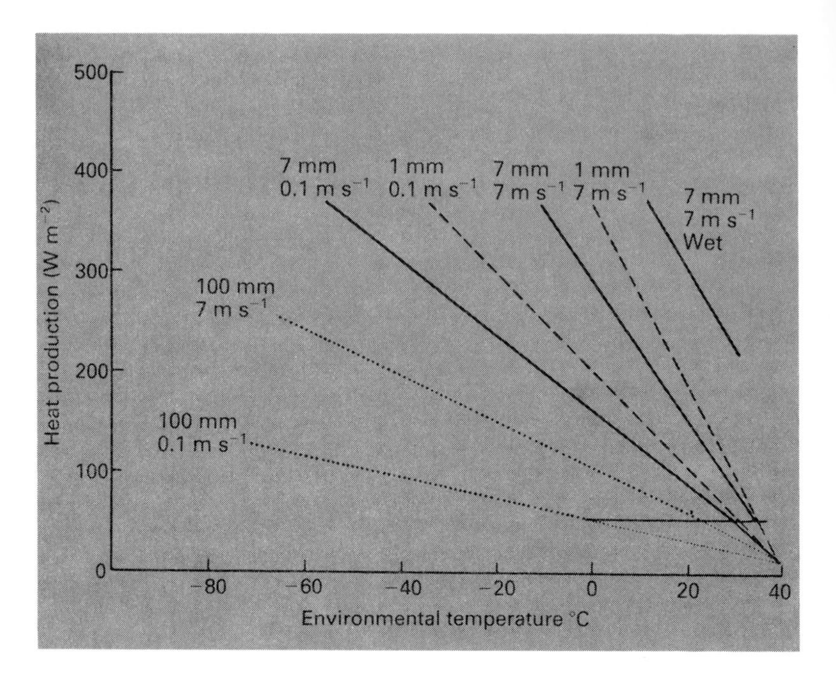

Fig. 10.2 Metabolic rate related to environmental temperature in British bred sheep with different fleece depths (1 mm, 7 mm, 100 mm) and exposed to 'still air' (wind 0.1 m s^{-1}) or a wind of 7 m s^{-1} (25 km h^{-1}). The relations were calculated from formulae of Joyce, Blaxter and Park (1966), assuming no solar radiation, and the sheep are considered as cylinders, radius 120 mm. The metabolic rate of wet sheep is taken from Panaretto, Hutchinson and Bennett (1968) (Alexander, 1974).

raises metabolic rate several fold above the resting level, and, apart from the cold stress produced by shearing, the animals can be effectively starved because the food intake that is available becomes inadequate, although shorn sheep increase their food intake for each hour of grazing.[15] There is then the depletion of fat reserves;[350] on exposure of the closely shorn sheep to cold, the metabolism of fat is increased (Fig. 10.3), and carbohydrate may also be used in the short term. Sheep can die following shearing; animals that are in poor condition are particularly susceptible.[79]

When shorn sheep are exposed to acute cold stress (for example −20°C and 1.8 m s^{-1} wind) for several hours, or to moderate cold (+8°C) for two weeks, acclimatization develops. The acclimatization takes the form of increased resistance to cooling and increased skin

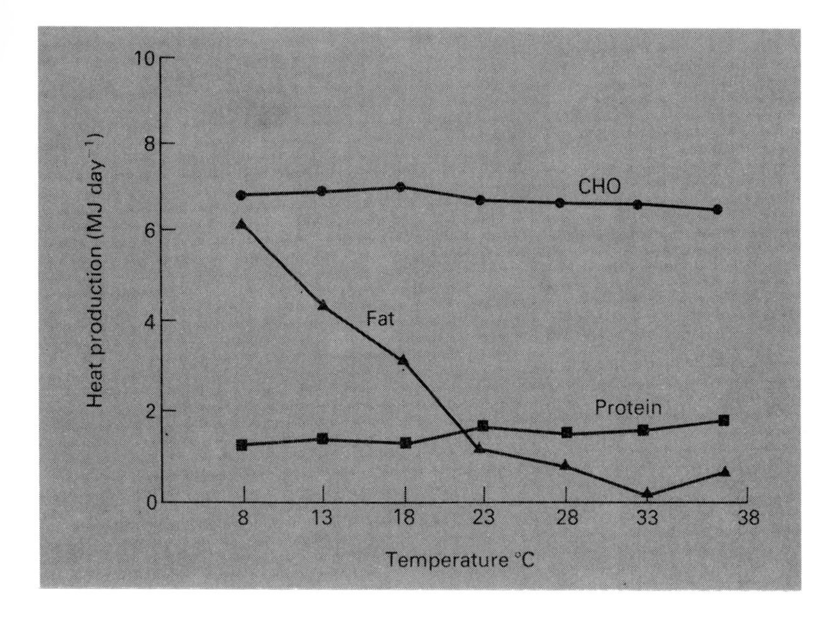

Fig. 10.3 The production of heat from the metabolism of carbohydrate, fat and protein by two sheep in relation to environmental temperature (Graham *et al.*, 1959, by permission of Cambridge University Press).

temperatures on the extremities, apparently related to an increased metabolic rate.[143, 473]

Critical temperature

A consequence of the high insulation of the fleece is the low critical temperature of −20°C or even lower in well fleeced animals.[7] When the sheep is shorn to a fleece length of about 7 mm (normal machine shearing), the critical temperature can rise to 28°C; when the animal is closely clipped, to 1 mm fleece length, the critical temperature can be even higher, at 30°C. Consequently the shorn sheep is as cold-susceptible as naked man (see Chapter 7). The actual critical temperature depends on the level of feeding, being higher on a low plane of nutrition than on a high plane (see Fig. 2.5). Webster[526] has calculated critical temperatures for mature sheep under different conditions (Table 10.2), showing the lowering of critical temperature produced by a high metabolic rate and the rise in critical temperature that follows shearing or saturation of the fleece with water.

Table 10.2 Approximate critical temperatures of mature sheep under different conditions of management and environment. I_t, tissue insulation; I_e, external insulation; M^*, thermoneutral metabolic rate (Webster, 1976b, by permission of Swets and Zeitlinger).

Management	Fleece depth (mm)	Wind-speed ($m\ s^{-1}$)	M^* ($W\ m^{-2}$)	I_t ($°C\ m^2\ W^{-1}$)	I_e ($°C\ m^2\ W^{-1}$)	Critical temperature (°C)
Fasting, confined	10	0.2	50	0.14	0.20	+25
	60	0.2	50	0.14	0.67	+9
Maintenance						
confined	60	0.2	70	0.13	0.67	−7
outside, shorn	10	0.9	90	0.14	0.17	+13
windy	10	4.3	90	0.14	0.10	+19
windy, dry fleece	60	4.3	90	0.14	0.39	−3
windy, wet fleece	60	4.3	90	0.14	0.20	+12
Full feed, confined	40	4.3	150	0.08	0.50	−40

Goats

There are many fewer measurements on goats than on sheep, but it is evident that goats are sensitive to cold.[276] Dairy goats begin to shiver when they are exposed to $-5°C$; the Angoran goat in Texas is particularly susceptible to chilling in cold wet winds and losses may be high.[497] Measurements on a 32 kg Alaskan mountain goat showed a minimal metabolic rate of $260\ ml\ O_2\ kg^{-1}\ h^{-1}$ and a critical temperature close to $-20°C$. At $-50°C$, the metabolic rate rose to $600\ ml\ O_2\ kg^{-1}\ h^{-1}$.[303] Exposure to cold that is sufficient to raise the total body oxygen consumption of the lactating goat by 46% has been shown to reduce mammary blood flow and oxygen consumption to two thirds of their thermoneutral values and to reduce milk secretion to 40%.[503]

The 15 kg Sinai goat is adapted to its hot environment in having a high hyperthermic point (see Chapter 1) and a low minimal metabolism compared with the neighbouring mountain goat weighing 30 kg (Fig. 10.4). For similar body weights, the minimal metabolic rates of the two animals would be even further apart (see Chapter 2).[466] Bianca and Kunz[51] have described the changes in body temperatures and respiratory rates that occur in three breeds of goats in response to cold and heat.

ADAPTATION TO HEAT

Evaporation

Respiratory and cutaneous evaporation in sheep have been calculated to share approximately equally about 25% of the total heat loss at thermal neutrality or below.[280] Above thermal neutrality, respiratory evaporative heat loss increases relative to cutaneous evaporation, and accounts for an increasing part of total heat loss (Fig. 10.5). Panting in mature sheep occurs as rapid shallow breathing mainly through the nose at rates up to about 300 per min; 20 per min is the minimum resting rate.[346] The maximum evaporative rate is 9 mg water per respiration, falling to 4 mg per respiration at very high rates of panting. 'Second phase breathing' (see Chapter 9) may develop with continued exposure to heat, with deeper slower respirations with the mouth open, leading to increased alveolar ventilation and respiratory alkalosis which do not occur in rapid panting.[196] A particular feature of the sheep's respiratory response to high temperatures is that polypnoea is readily evoked by heating the scrotum of the ram, and the high respiratory rate is maintained in spite of the concomitant fall in rectal temperature. The scrotal skin is well endowed with thermal receptors.[516]

Sweating plays a secondary role in evaporative cooling in sheep, in contrast to its important role in cattle (see Table 3.2 and Fig. 9.3).

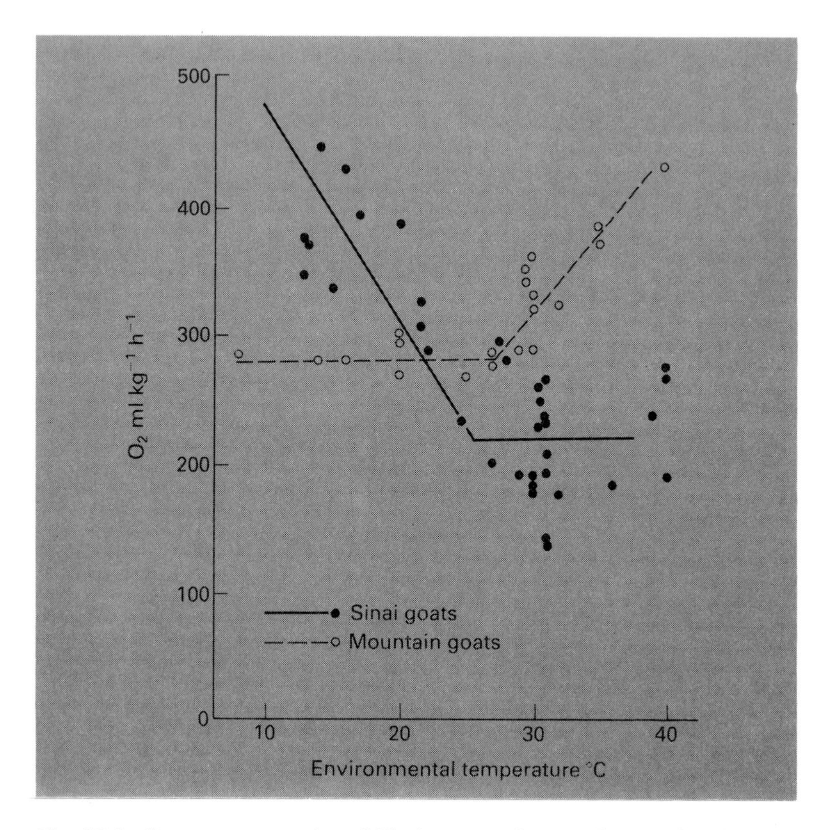

Fig. 10.4 Oxygen consumption of Sinai goats and mountain goats in relation to environmental temperature (Shkolnik *et al.*, 1972, by permission of Zoological Society of London).

Sheep sweat more than goats and characteristically display a synchronous discharge of sweat glands over the body surface.[67] The minimum loss of water from the skin of sheep is about 8 g m^{-2} h^{-1}, and the maximum is from the scrotal skin that can produce sweat at 200 g m^{-2} h^{-1}; the local cooling so produced favours normal spermatogenesis.[518–9] In this connection, panting can be induced more easily by warming the scrotal skin than other areas of skin.[516–7] This is also found in the pig, where simultaneous cooling of the hypothalamus diminishes the panting response.[257]

Radiation

Fig. 10.6 gives values for fleece and skin temperatures in the Awassi and Merino sheep exposed to intense solar radiation. Solar radiation

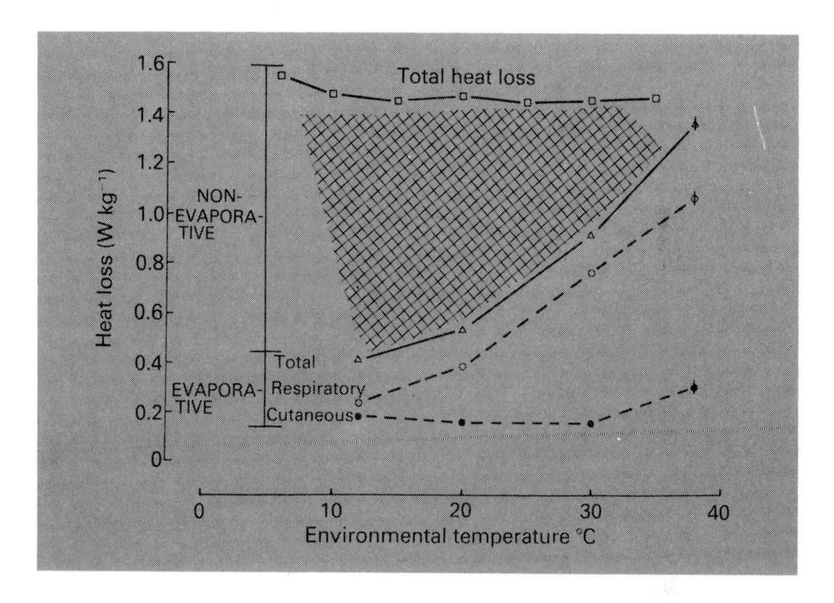

Fig. 10.5 The evaporative and non-evaporative components of total heat loss from sheep. Values at 12–35°C are for animals carrying 3–8 cm wool (from Brockway *et al.*, 1965) and at 38°C for shorn animals (from Hofmeyr *et al.*, 1969) (Johnson K. G., 1976, by permission of Swets and Zeitlinger).

(short-wave) penetrates the loose coat of the Awassi and produces a high skin temperature; however, because the coat is loose and open a large part of the absorbed radiant energy is lost by convection. Radiation is absorbed in the outer part of the fleece, but the surface may be cooler than the deeper fleece because convective loss occurs most readily at the surface. The tightly packed coat of the Merino sheep allows very little penetration by sun's rays, with the result that the skin temperature is not as high as in the Awassi, although the surface temperature of the fleece can be very high, up to 85°C. The reduction in heat flow through the fleece as wool length increases is shown in Fig. 10.7. Heat is lost from the hot surface of the Merino by infra-red radiation (long-wave) to any cooler surroundings, such as shaded surfaces or sky. Solar radiation is reflected from the smooth white coats of some animals, such as Ogaden sheep

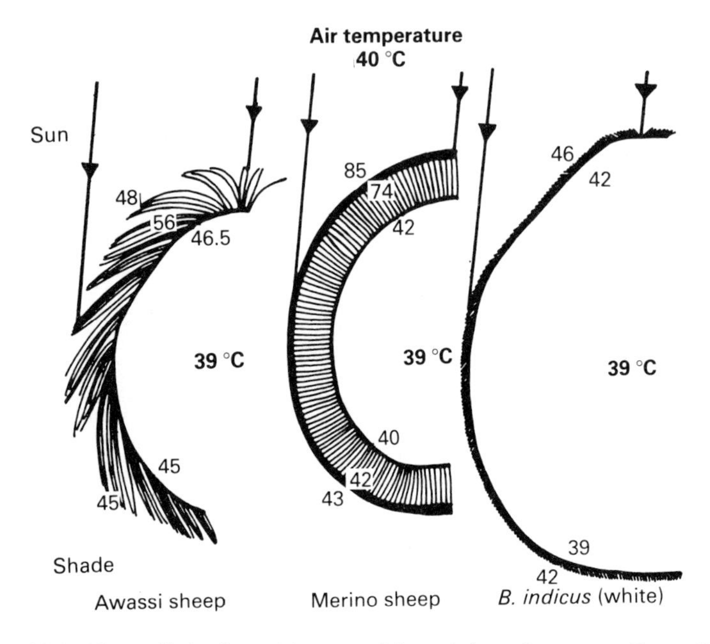

Fig. 10.6 Thermal behaviour of the coats of Awassi sheep (loose open fleece of hair and wool), Merino sheep (dense fleece) and Brahman cattle (short, shiny hair coat). The Awassi depends on convection for most of the heat loss, the Merino loses much of the incoming solar radiation by long-wave radiation and the Brahman reflects solar energy (Macfarlane, 1968a, by permission of Lea and Febiger).

(Persian) and goats, so further reducing the heat load. In Fig. 10.6, the inclusion of the surface and skin temperatures of *Bos indicus* illustrates the value of a reflecting surface exposed to intense solar radiation. Convection, radiation and reflection from the animal serve to diminish the heat load in these different adaptations.

A heat balance can be drawn up for the Merino sheep in a hot environment (Fig. 10.8).[346,423]

The possible net gain of 165 W under still air conditions could be largely dissipated by wind, but the residual net heat gain must be dissipated by evaporation, primarily from the upper respiratory tract

Input	Watts	Output	Watts
Metabolic heat	45	Free convection in still air	70
Solar radiation	280	Long-wave radiation	220
Sky radiation	70		
Ground radiation	60		
Total heat load	455	Non-evaporative heat loss	290

(see Fig. 10.5). There is also an increased non-evaporative heat loss from the relatively poorly insulated limbs under such conditions (see Chapter 3).

When the Merino sheep is shorn, the albedo increases from about 0.4 to 0.7. The albedo is the ratio of the total luminous flux reflected

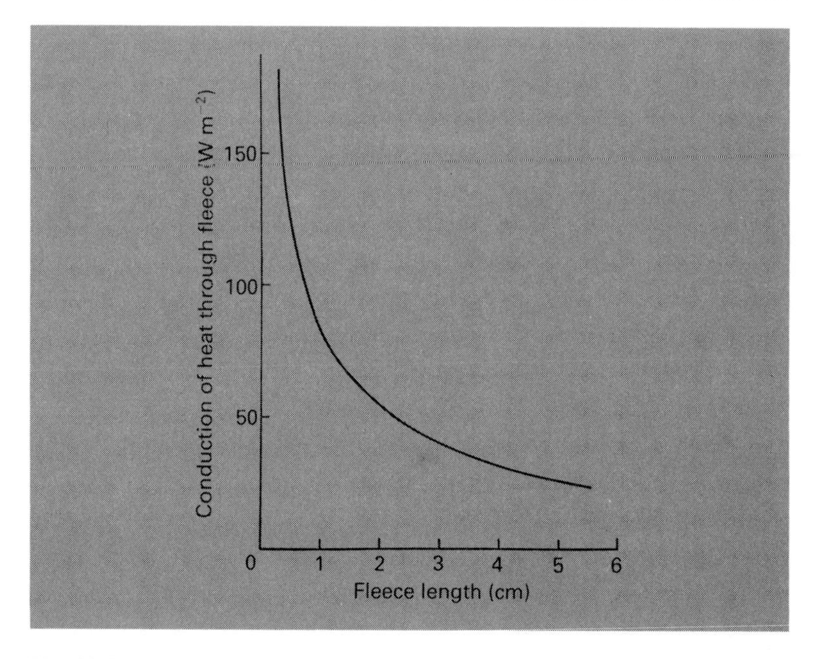

Fig. 10.7 The relation between fleece length and heat flow through the fleece in Merino sheep exposed to simulated radiant heating at an air temperature of 36°C (Parer, 1963, by permission of Cambridge University Press).

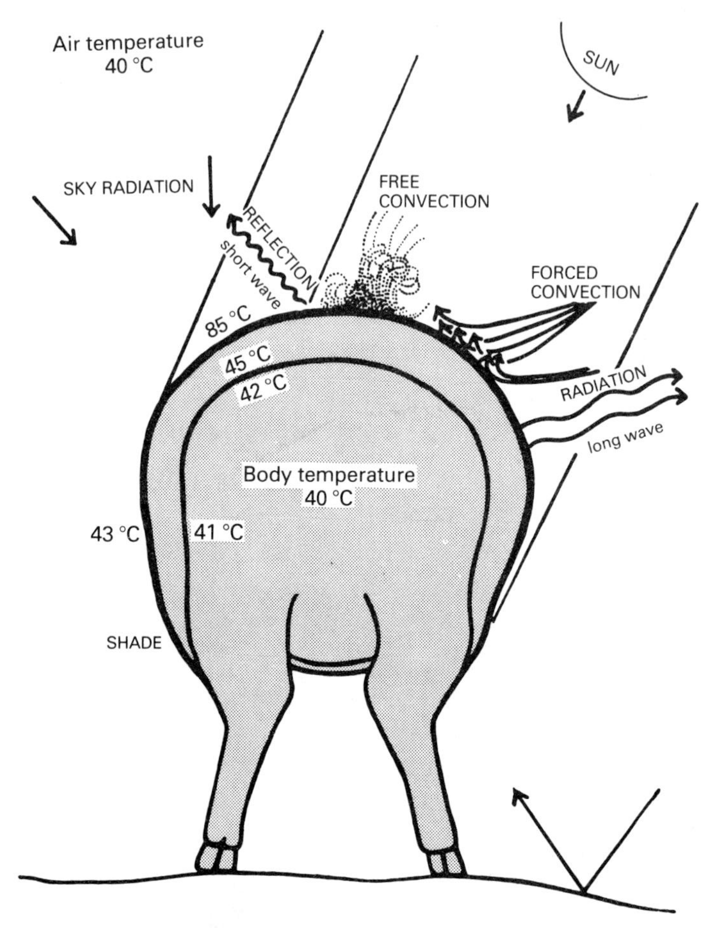

Fig. 10.8 Physical aspects of heat regulation by the fleece of a Merino sheep. Convection takes place both by the upward passage of heated air and by wind action. Radiation is to some extent reflected, but energy absorbed in the outer 1 cm of fleece induces temperatures of 85°–92°C in still air. This energy is lost by long-wave radiation to the sky (Macfarlane, 1968a, by permission of Lea and Febiger).

to that received, and the rise in albedo helps to decrease the heat load that would otherwise arrive at the skin after shearing. However, for a sheep standing in the open the energy absorbed is still two to three times the amount that is absorbed when there is a 5 cm fleece. When the animal is in the shade, convective heat loss and cooling are greater from a shorn than an unshorn sheep.

Shearing in a hot summer leads to loss of protection of sheep from solar radiation with a resulting increase in the absorption of heat. The animals pant and double their water intake. Shearing in various parts of the world is most appropriately done in the absence of extremes of temperature, that is in spring or autumn, particularly avoiding either cold wind and rain or high temperatures and strong sunlight.[350]

Convection

The effect of wind is pronounced. A surface temperature of the fleece of 90°C in still air is reduced to 60°C within a matter of minutes by a wind of $5\,m\,s^{-1}$; a desert wind of this speed is common in the heat of the day. The effect of wind in reducing the insulation of the fleece has been illustrated in Fig. 10.1.

Upper limiting temperatures

The upper limit of environmental temperature that can be withstood by the sheep depends partly on the animal's metabolic rate. Animals on a high plane of nutrition have lower temperatures for their hot limits than animals on lower planes: pregnancy, lactation and exercise also lower the hot limiting temperatures. The range of lethal limits probably lies within 40–60°C.

Goats tolerate hot-dry environments, but respond differently from Ogaden sheep in the heat in that the goat skin temperature rises less than that of the sheep, due to the greater reflection of solar radiation from the goat's shiny hair surface. Tropical and desert sheep nearly always have hair coats with good reflecting qualities, rather like the goat's coat.[347–8]

NEWBORN LAMB

The newborn lamb's body weight is in the region of 3–4 kg. Thermal neutrality lies in the range 25–30°C, with a heat production of $60–70\,W\,m^{-2}$. As the temperature falls below thermal neutrality, the increment in heat production in the long-coated lamb is less than it is in the short-coated, corresponding to the thermal insulation and rates of heat loss (Fig. 10.9).

There is often a fall in deep body temperature immediately after birth, with recovery within a few hours; heat loss from the newborn lamb depends on the environmental conditions at the time, and it is enhanced by the lamb being born wet. There are considerable differences between breeds in the extent to which rectal temperature falls when the newborn lamb is exposed to cold. Scottish Blackface,

Cheviot and Soay lambs suffer only small falls in body temperature in the cold, compared with warm conditions, but the indications are that some Merinos and Finnish Landrace show significant reductions soon after birth and Southdowns also tend to be hypothermic (Table 10.3).

The maximum metabolic rate of which the animal is capable when

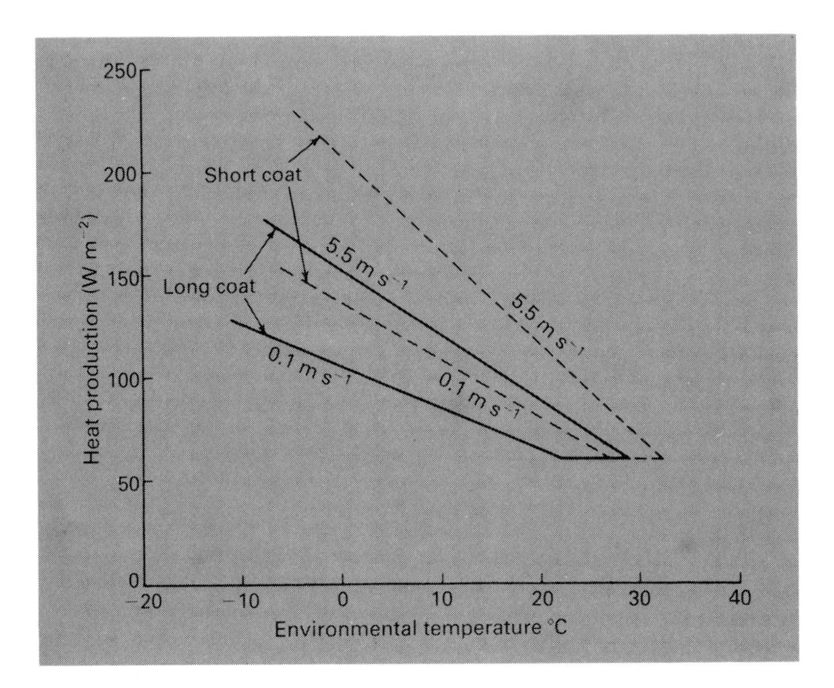

Fig. 10.9 Relation between metabolic rate and environmental temperature in Merino lambs (less than 1 day old) with short curly coats and long straight coats, in still air (wind 0.1 m s^{-1}) and a wind of 5.5 m s^{-1} (20 km h^{-1}) (Alexander, 1962, by permission of Australian Journal of Agricultural Research).

it is faced with the demand of a cold environment determines the cold limit that it can withstand. Alexander[7] gives the maximum metabolic rate of the lamb as 20 W kg^{-1} (28 W kg$^{-0.75}$), irrespective of the weight of the lamb. Calculated values for cold limits, including wind and wetting, are given for both adult sheep and lambs in Table 10.1. It is clear that the mature sheep could survive for limited periods under remarkably extreme conditions, and the lamb, particularly the long-coated, is also very resistant to cold. However, these are limiting values for temperature and continued existence in any given

Table 10.3 Rectal temperature measured 20 min after birth, and birth weight in lambs of six breeds in various weather conditions. EAT, effective ambient temperature from air temperature and windspeed (Sykes *et al.*, 1976, by permission of Animal Production).

Breed	No. of sheep	Rectal temperature (°C) Mean	SD	Birth weight (kg) Mean	SD	EAT (°C) Mean	SD
Weather—cold; EAT −5°C to 1°C							
Blackface	24	39.1	1.08	3.6	0.79	−1.4	1.52
Cheviot	17	39.4	1.06	3.7	0.65	−1.5	1.44
Soay	9	39.3	0.93	2.2	0.34	−0.4	0.91
Merino	26	34.9	2.75	3.7	0.57	−1.2	1.13
Southdown	20	37.1	2.04	3.6	0.61	−1.6	1.47
Finnish Landrace	24	34.5	3.26	2.2	0.65	−1.4	0.57
Weather—warm; EAT 1°C to 10°C							
Blackface	32	39.3	0.80	3.7	0.54	4.8	3.64
Cheviot	9	39.6	0.37	3.7	1.03	2.5	0.98
Soay	26	39.5	0.75	2.0	0.54	3.9	2.40
Merino	13	37.3	1.67	3.7	0.64	4.2	2.84
Southdown	4	38.1	1.57	3.0	0.58	2.0	0.89
Finnish Landrace	4	38.2	0.13	2.5	0.85	6.9	2.50

environment would depend not only on temperature but also on food supply, activity and shelter from wind, rain and snow.

The rise in heat production that occurs in the newborn lamb exposed to cold is brought about both by shivering and by increased metabolism in brown fat (see Chapter 2). Dissection of newborn lambs of several breeds has shown that brown adipose tissue constitutes about 1.5% of the body weight, with the largest amount in the perirenal-abdominal and prescapular-cervical regions. These are the principal sites of the non-shivering thermogenesis that has been estimated to contribute about 40% of the maximum metabolic response to cold.[9]

Under hot conditions, panting is the major route of heat loss. The respiratory water loss is approximately proportional to the respiratory rate; at low humidities the water loss is about 1 mg per breath (Fig. 10.10). The respiratory rate can reach 400 per min at an environmental temperature of 45°C, with a deep body temperature of 41–43°C. Evaporation from the skin at the minimum level is about the same as the minimum respiratory loss, $10\,g\,m^{-2}\,h^{-1}$, but the maximum rate from the skin, $40\,g\,m^{-2}\,h^{-1}$ in the Merino lamb, is less than half that from panting. Despite the limited sweating capacity of

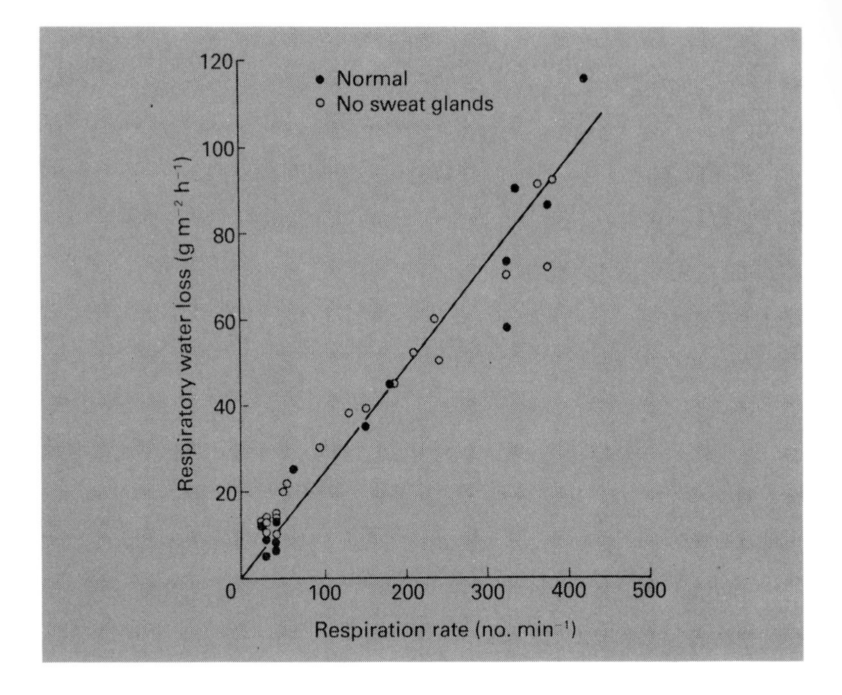

Fig. 10.10 Relation between respiratory rate and respiratory water loss in Merino lambs. Water vapour pressure gradients from lungs to air were high throughout (Alexander and Williams, 1962, by permission of Australian Journal of Agricultural Research).

normal lambs, lambs without sweat glands are less heat-tolerant than normal lambs, and reach their maximum evaporative capacity at an environmental temperature that is 6°C lower than normal lambs.[7]

Animal productivity

Although sheep can produce meat, milk and wool under a wide range of climatic conditions, the efficiency of production is highest in temperate climates. Some animals are well adapted to high temperatures; for example, the Australian Merino has marked ability to produce wool under hot conditions, although wool production is less in hot regions than in temperate, largely on account of decreased food intake. Another effect of high temperatures is to reduce fertility, and pregnant sheep exposed to high temperatures tend to produce small lambs. It is important to use sheep that are suitable for the intended environment.

DEER

Deer are widely distributed in Europe, Asia, North and South America, and North Africa. They live mainly in forests, but some live in open country, like the barren ground caribou in North America.

Deer range in size from that of a small dog to that of a large horse. The Lesser Mouse Deer (*Tragulus Javanicus*) has the distinction of being the smallest ungulate, with body weights recorded from 1.37 to 1.85 kg. It maintains homeothermy between environmental temperatures of 15 and 30°C; above 30°C it becomes hyperthermic, with evaporative heat loss limited to 54% of its heat production.[541] At the other end of the body weight range, the American moose and the European elk, with body weights up to 800 kg, are both sub-arctic species, extending southwards to the Rockie Mountains and Northern Europe. Of the various types of deer, those that have received some investigation include in particular red deer and caribou.

Red Deer

Interest in the red deer (*Cervus elaphus*) has increased considerably on account of its potential for meat production:[26,199] 'the intriguing possibility that the deer might in some situations be an alternative to the sheep as a meat animal'.[59] The animal evolved in temperate woodland; it occupies roughly the same ecological niche as hill sheep. The red deer can grow to a considerable size, with the two-year-old stag reaching 180 kg body weight. Numbers in the Scottish Highlands were estimated as 200 000 in 1973 by the Red Deer Commission[430]; a more recent estimate gives 270 000.[368]

There is little difference in digestibility, metabolizability, or efficiency of utilization of feed nitrogen and energy between red deer and sheep, although deer retain proportionally more of their metabolizable energy as protein whereas sheep retain a greater proportion as fat. However, the deer may be less adapted to cold than sheep or cattle, particularly when on a low feed intake. The animals have a characteristically lean carcass, containing approximately 2.5% fat in the whole live weight, much less than in domestic livestock, with the fat deposits being intermuscular rather than intramuscular, and with little or no subcutaneous fat.[290,470]

A high metabolic rate has been reported for red deer,[470] and for other species of deer, with a fasting heat production of 4.5 $W\,kg^{-0.75}$

that is 30% higher than sheep and 10% higher than cattle (see Chapter 2), with correspondingly higher maintenance requirements.[76, 412, 442] The critical temperature is close to 13°C on a low plane of nutrition, falling to 8°C on a high plane, under conditions approaching those of still air.

The rectal temperature of deer of 40–50 kg is close to 39°C, with skin temperatures on the trunk of 33°C at 16°C air temperature and 29°C at 8°C. Evaporative heat loss as a proportion of total heat loss ranges from 32% at 16°C (that is, at thermal neutrality) down to 17% at 4°C air temperature; in the cold it remains at about 19 $\mathrm{W\,m^{-2}}$. The mean external insulation (coat plus air-ambient; see Chapter 4) is about 0.28 $\mathrm{^\circ C\,m^2\,W^{-1}}$ and the tissue insulation is close to 0.09 $\mathrm{^\circ C\,m^2\,W^{-1}}$, lower than in adult sheep (0.14), thin cattle (0.14), and mature fat cattle (0.30).[470] Inadequate food supply reduces the resistance of the animals to cold, with mortalities of 10–60% amongst calves in winter.[364]

Caribou calves

The caribou and the reindeer are probably sub-species of the same species, *Rangifer tarandus.* They are domesticated and managed like domestic cattle in northernmost parts of Europe, and undergo seasonal migrations between winter and summer feeding grounds. Reindeer feed on lichen (reindeer moss) in the winter and in summer they also have grasses and saplings. They provide draught, meat, milk and hide.

Caribou calves follow the mother about in spite of severe weather conditions involving cold, wind and precipitation. Their temperature regulation is remarkably well established at birth, and the calves have a strong metabolic response to cold. In one series of observations, the metabolic rate of 5 kg calves, soon after birth, was doubled when the environmental temperature fell to 0°C, and when wind and precipitation were added the metabolic rate increased to over five times the resting value. The newborn calves can therefore withstand a very demanding environment as a result of their metabolic response, although they do suffer mortality if they cannot avoid extreme conditions.

In contrast to newborn calves, the 9-month-old calf shows no increase in metabolic rate between 25°C and −55°C, and there is no evidence of shivering. This indicates a very high overall insulation that allows these older calves to remain near thermal neutrality even under severe winter conditons, so that a metabolic response to cold is largely unnecessary.[208]

11

Comparative aspects of ungulates in hot climates

Ungulates are of great importance in providing man with food, drink, transport and clothing. Many of them are native to hot climates, and it is useful to consider some comparative aspects of their climatic physiology, particularly their water requirements and the role of evaporative cooling, with reference to animal production in hot regions of the world.

The ungulates include, amongst other species, horses, asses, pigs, camels, deer, cattle, sheep, goats, antelopes and gazelles. The importance of the ungulates has been described by Fraser[162] in the following terms: 'The ungulates stock man's larder and, even today, supply a great deal of his draft power... The ungulates, indeed, are the principal means whereby man can live off the herbal covering of the globe.' Harlan[202] has estimated from statistics from the Food and Agriculture Organization of the United Nations that of the 110 million metric tons of meat produced in the world in 1974, 42.5 came from pigs, 42 from beef, 20.7 from poultry, 5.4 from lamb, and 3.2 from goat, buffalo and horse combined.

Table 11.1 provides a classification of ungulates, divided into the Perissodactyla, or odd-toed ungulates, and the Artiodactyla, or even-toed. The Perissodactyla have no bony outgrowths from the skull (the horns of the rhinoceros are dermal in origin). The Artiodactyla do have such outgrowths; in the bovids thay are permanent, unforked, persistently horn-covered, and they are present in both sexes; in the cervids, they are branched, seasonal and present only in the male except for the reindeer, *Rangifer tarandus* (see Chapter 10). Ungulates are widely distributed in the arctic, desert, grasslands and tropics. The surviving wild Perissodactyla are limited to Africa, Asia, Central and South America; the numerous Artiodactyla are widely distributed on all continents except Australia.

ANIMAL PRODUCTION

Animal production and activity in a hot environment are hampered by the decrease in feed intake that occurs at high temperatures, and by the animal's need to dissipate the additional heat production that is associated with productivity (see Chapter 2). The reduced animal productivity at high temperatures has induced man to search for heat-adapted animals and for means of enhancing adaptation to heat. It has also led to means of combating high temperatures by environmental control, using shades and wallows to counteract the effect of hot surroundings. Other difficulties facing animal production in hot regions of the world lie in the lack of suitable forage during some seasons, and the high fibre content and low protein value of much of the available forage.

A factor that has caused unsatisfactory results in transferring livestock from a temperate to a hot climate is the high level of disease and parasite infestation in hot climates. Growth rates, milk yields, egg production and fertility all decline as a result of disease and lack of heat tolerance. Native breeds of animals are better adapted to the environment and they are more resistant to disease; for example, resistance to parasite infestation is higher in *Bos indicus* than *Bos taurus*. However, the levels of productivity of breeds that are indigenous in hot countries are lower than those of selcted animals raised in temperate climates.[258]

The combination of uniformly warm environments and high humidities does not favour animal production. Unchanging day-length in equatorial regions means that coat changes, reproductive cycles and metabolism are not linked to photoperiodic seasonality as they are in temperate climates. However, wet tropical zones, with their high rainfall and vegetation, do provide less direct solar heat load than the desert, and support buffalo and beef cattle as domestic animals used productively.[375] Indian cattle have been used extensively as meat-producing animals in Brazil and Australia; water buffalo are indigenous to India, Burma and Malaya, and there are many in China and

Table 11.1 Summary of classification of odd-toed and even-toed ungulates (Yousef, 1976, by permission of Swets and Zeitlinger).

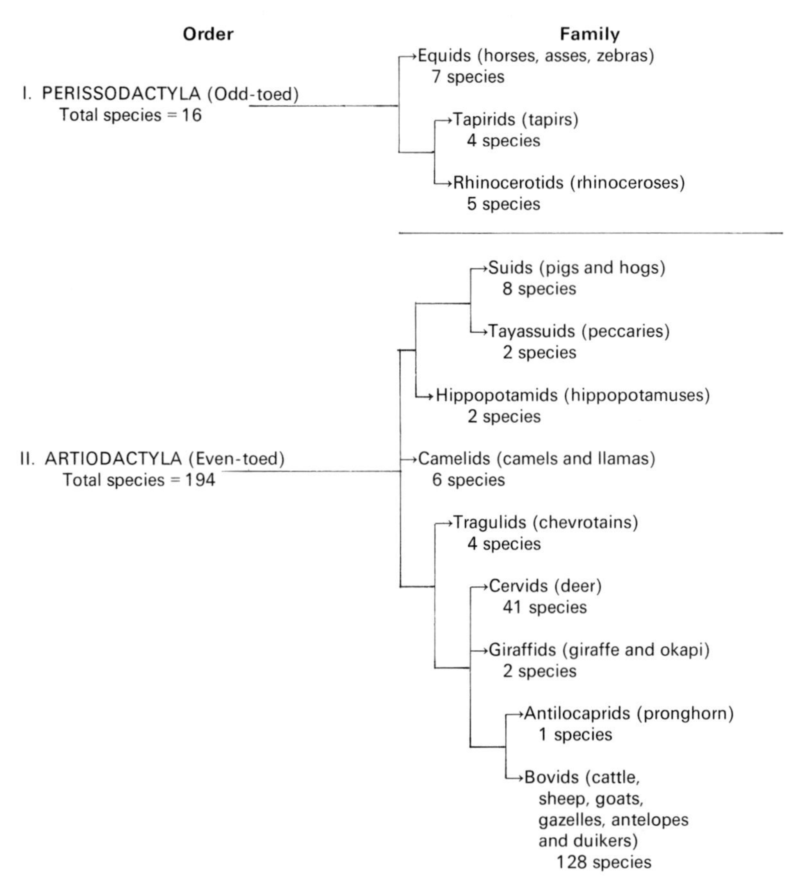

Africa, where they are used for draught purposes and sometimes for milk.

Deserts, with their lack of water and plants, are clearly not compatible with considerable animal production, but many semi-desert areas in Africa, America, the Middle East, Asia, and Australia are used for animal production. These areas merge with semi-arid scrubland which maintains numbers of ruminants. Sheep in particular exist in a very wide range of climates and they are important in hot, semi-arid regions in China, northern Africa, the Middle East and Australia.

Types of climate have been described in Chapter 5, and adaptation of animals to heat has been discussed in general terms in Chapter 6, for man in Chapter 7, pig in Chapter 8, cattle in Chapter 9 and sheep in Chapter 10. Some of this discussion, and other aspects of animals' reactions to hot environments, can be found elsewhere.[258]

ADAPTATION TO HOT CLIMATES

Various morphological adaptations are evident from an animal's appearance, whereas many of the physiological adaptations are not so apparent. For example, pigs kept in warm environments develop longer limbs and larger ears than animals of the same breed kept in the cold (see Chapter 8). Sub-tropical cattle have dewlaps and long limbs (see Chapter 9). Desert sheep also have long limbs, long ears and long tails, and typically they also have local stores of fat, chiefly either as a fat tail or as a fat rump. There are local deposits of fat in the neck and dewlap in Masai and Blackhead sheep, and the humps of the camel contain a fat store that varies with the food supply. The metabolic and morphological adaptations of ungulates to hot environments have been extensively reviewed by Macfarlane.[346–51]

Behavioural adaptation is also important. Under hot conditions animals tend to become nocturnal in habit; this is the case in the tropics and sub-tropics, where cattle graze for considerable periods at night. Animals seek shade from hot sun; if shade is not available they turn themselves to present the smallest bodily profile area to the sun. The pig seeks out shaded wet spots in hot weather; wallowing in mud effectively compensates for the animal's inability to sweat.

Thermoregulation in wild ungulates

The rectal temperatures for 20 species of wild ungulates at rest are given in Table 11.2, and the 24-hourly variations in rectal temperature for some of them in Table 11.3. The rectal temperature appears to be independent of climate, and both arctic and desert species can tolerate varying degrees of hyperthermia.

EVAPORATIVE COOLING

In hot conditions, evaporative heat loss may provide the only channel through which the animal can dissipate its heat, and the supply of water and mode of evaporative cooling are then of great importance (see Chapter 3).

Table 11.2 Rectal temperature, T_{re}, and respiratory frequency, RF, of 20 species of wild ungulates at rest. T_a, air temperature (Yousef, 1976, by permission of Swets and Zeitlinger).

Species	Common name	RF min^{-1}	T_a, °C	T_{re}, °C	References
Even-toed ungulates					
Tayassu tajacu	Collared peccary		Summer	37.5–40.0	Zervanos and Hadley[567]
			Winter	37.5–40.0	
Gifaffa camelopardalis	Giraffe		15	38.8	Bligh and Robinson[70a]
Ovibos moschatus	Muskox			40.0	Irving and Krog[266a]
Oryx beisa	Oryx	17	30	39.4	Taylor[489a]
Taurotragus oryx	Eland		30	39.8	Taylor[489a]
Gazella thomsonii	Thomson's gazelle	25	30	39.4	Taylor[490]
Gazella granti	Grant's gazelle	42	30	39.8	Taylor[490]
Cervus elaphus	Red deer	15–70	20	38.5–39.6	Johnson *et al.*[280a]
Connochaetes taurinus	Wildebeest	34	30	39.0	Taylor *et al.*[494]
Aepyceros melampus	Impala	45	30	39.2	Mailoy and Hopcraft[355]
Alcelaphus buscelaphus	Hartebeest	28	30	38.7	Mailoy and Hopcraft[355]
Rangifer tarandus stonei	Stone caribou		−12 to −26	39.0	Irving and Krog[266a]
Rangifer tarandus	Reindeer	50	30	38.6	Yousef and Luick[566]
Kobus defassa	Waterbuck	25	30	39.6	Taylor *et al.*[494a]
Lama guanico	Guanaco	25	30	38.7	Rosenmann and Morrison[450a]
Camelus dromedaris	Camel	12	30	36.7	Maloiy[352a]
Syncerus caffer	Buffalo		18	38.5	Bligh and Robinson[70a]
Hippopotamus amphibus	Hippopotamus		32	35.4	Whittow[537]
Odd-toed ungulates					
Ceratotherium	White rhinoceros		25	35.2	Whittow[537]
Equus assinus	Ass, Burro, Donkey	15	30	37.1	Maloiy[352a]

Table 11.3 Range of body temperature (T_b) and its 24-hourly variation (ΔT_b) in hydrated wild ungulates in the field or in a simulated desert environment in the laboratory (air temperature 22–40°C) (Yousef, 1976, by permission of Swets and Zeitlinger).

Species	Range in T_b °C	Type of study and air temperature, °C	ΔT_b	References
Collared peccary	37.5–40.0	Field	2.5	Zervanos and Hadley[567]
Giraffe	37.7–39.0	Summer and winter	1.3	Bligh and Robinson[70a]
		28–42 5–18		
		10–25		
Camel	35.2–37.8	22–35	2.6	Bligh and Robinson[70a]
Buffalo	36.9–40.2	10–25	3.3	Bligh and Robinson[70a]
Burro (donkey, ass)	36.5–39.0	18–35	2.5	Yousef and Dill[564a]
Eland	33.9–41.2	22–40	7.3	Taylor[489a]
Oryx	35.7–42.1	22–40	6.4	Taylor[489a]
Impala	38.8–40.4	22–40	1.6	Maloiy and Hopcraft[355]
Hartebeest	38.8–40.4	22–40	1.6	Maloiy and Hopcraft[355]
Grant's gazelle	39.4–40.4	22–40	1.0	Taylor[490]
Thomson's gazelle	39.0–40.0	22–40	1.0	Taylor[490]

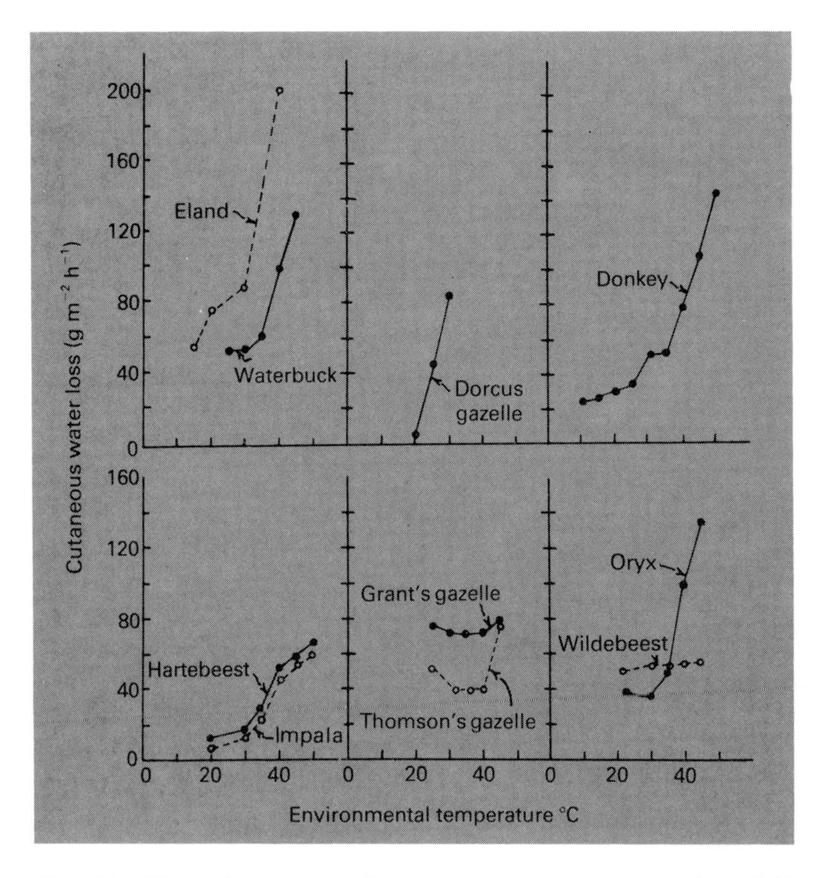

Fig. 11.1 Effects of environmental temperature on cutaneous water loss of 10 species of ungulates (Yousef, 1976; from results of Maloiy, 1970; Maloiy and Hopcraft, 1971; Taylor, 1970, 1972; Ghobrial, 1970; by permission of Swets and Zeitlinger).

There is great variation in the relative importance of cutaneous and respiratory evaporative heat loss amongst the ungulates.[537] When the environmental temperature rises to 40°C or more, some ungulates, such as the wildebeest and Grant's gazelle, show no change in cutaneous water loss, whereas others, such as eland, oryx and donkey, show a considerable increase (Fig. 11.1). The wildebeest and Grant's gazelle pant and do not sweat; some others, like the camel, sweat and do not pant; between these extremes there are species that both pant and sweat to different degrees. Ungulates that sweat under hot

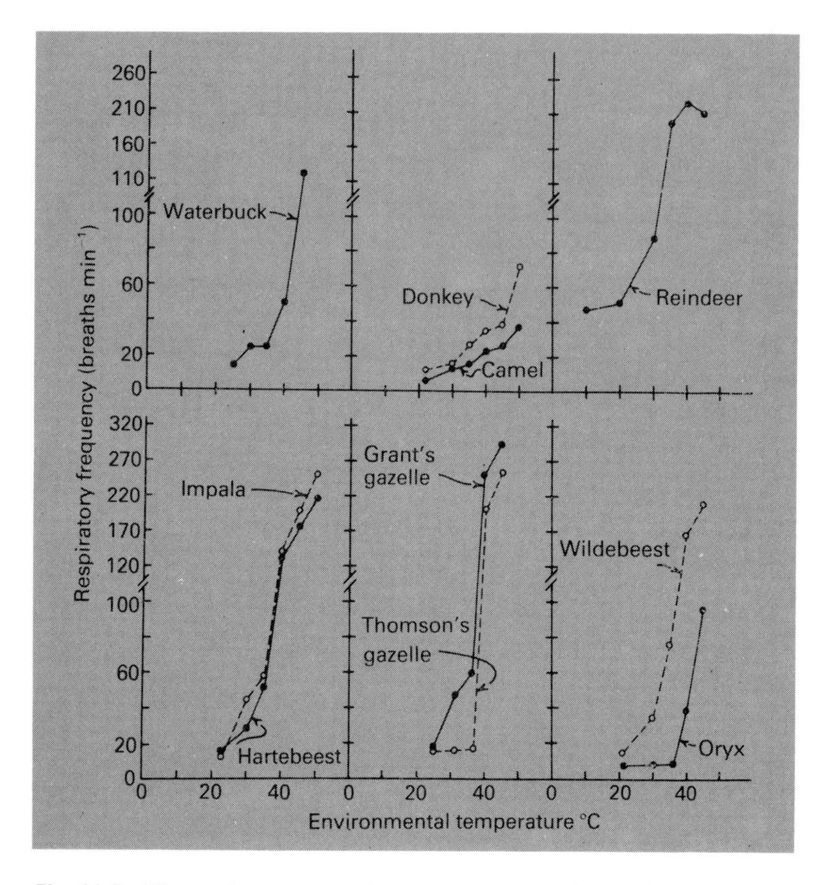

Fig. 11.2 Effects of environmental temperature on respiratory frequency of 10 species of ungulates (Yousef, 1976; including results from Maloiy and Hopcraft, 1971; Taylor, Robertshaw and Hofmann, 1969; Taylor, 1972; by permission of Swets and Zeitlinger).

conditions show a much smaller increase in respiratory frequency when they are exposed to high air temperatures than is the case with panting ungulates. For example, the respiratory frequency observed at a high air temperature was 38 min^{-1} in the camel, which sweats, whereas the respiratory rate was 295 min^{-1} in Grant's gazelle, which pants (Fig. 11.2). Small ruminants up to about 20 kg seem to be panting animals, large ungulates, like camels and horses, sweat, and those of intermediate size both pant and sweat.

Sweating

Cattle show a step-wise increase in cutaneous moisture loss until a continuous output is reached; in these animals cutaneous evaporative loss is the major avenue of evaporative cooling (see Fig. 9.3). Sheep and goats show discrete discharges of water on to the skin, with low levels of cutaneous evaporative cooling (see Fig. 10.5). Whereas in cattle most evaporative cooling at high temperatures takes place through sweat gland activity, in sheep the roles of the sweat glands and respiratory tract are reversed. The partition of evaporative water loss between the sweat glands and respiratory tract in cattle and sheep is given in Table 11.4.

Table 11.4 Partition of evaporative water loss between the sweat glands and respiratory tract in adult cattle and sheep (Macfarlane, 1968a, by permission of Lea and Febiger).

	Sheep	Cattle
Density of sweat glands/cm^2	260–300	800–1500
Sweat gland volumes, mm^3	0.001–0.008	0.006–0.015
Maximum sweat secretion rate, g m^{-2} h^{-1}	32	230
Maximum respiratory evaporation, g m^{-2} h^{-1}	95	41
Maximum evaporative cooling, W m^{-2}	86	180
Respiratory rate/min (air temperature 40°C)	260	170

Exposure to 40°C produces cutaneous moisture loss with considerable variation between a number of East African bovids (Fig. 11.3). Duiker, Grant's gazelle, Thomson's gazelle and oryx show periodic discharges of moisture on to the skin surface, Defassa waterbuck and eland show a gradual rise, and buffalo show an immediate large increase that is sustained during the heat exposure. The sweat glands of all the species are found to be under adrenergic nervous control, as in domestic bovids, without correlation between the pattern of sweating and the phylogenetic position of the species. Water loss per m^2 per hour at 40°C is correlated with adult size; the smaller animals dissipate their excess heat through the respiratory tract, and the larger animals through cutaneous evaporation.[445]

Rates for cutaneous evaporation in other species that can be compared with those in Fig. 11.3 have been obtained under hot conditions by a number of workers: the pig at 35°C environmental temperature has an evaporative rate from the skin of 24 g m^{-2} h^{-1},[251] sheep at over 30°C produce 63 g m^{-2} h^{-1},[77a] and at 40°C *Bos taurus*

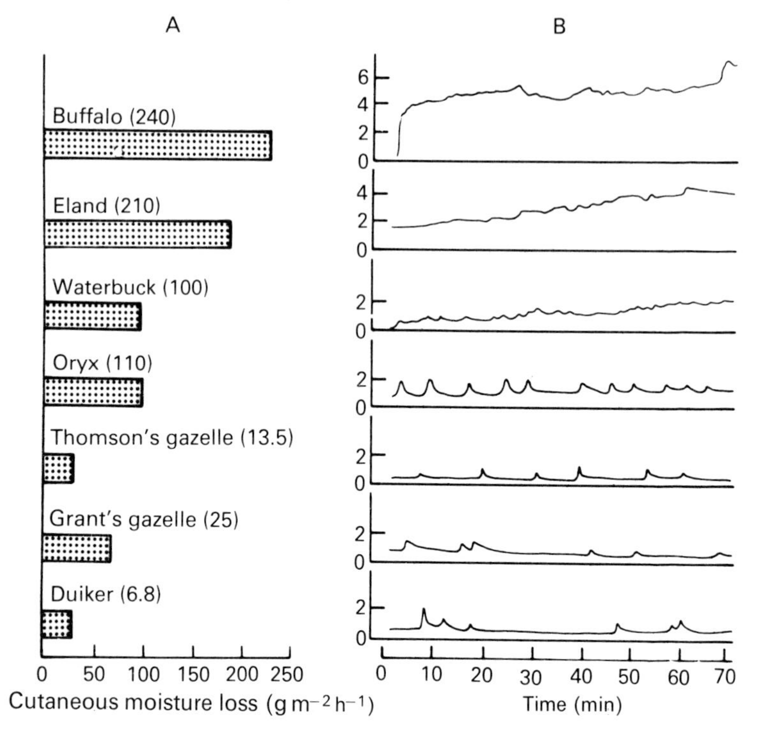

Fig. 11.3 (A) The water loss from the skin in several East African mammals at 40°C dry bulb and 23°C wet bulb, with the animal's weight in kg in parentheses. (B) The pattern of sweat gland activity; the ordinate in each case is the rise in water vapour content of air passing through a capsule on the skin, used as a measure of evaporative heat loss (Robertshaw and Taylor, 1969, by permission of Journal of Physiology, redrawn by Whittow, 1973, by permission of Academic Press, copyright).

produces 145,[338] goat 50, horse 100, donkey 166, and llama 250 $g\,m^{-2}\,h^{-1}$.[12]

In general, three types of sweating can be discerned in ungulates.[280] There is first the intermittent synchronous discharge of sweat glands as observed in the sheep and goat.[67, 443] The second type is a stepwise increase, seen in the ox.[338] The third type takes the form of a steady increase in sweating on exposure to hotter environments, and this occurs in the camel[444] and the llama.[12]

Respiratory evaporation

The rapid movement of air to and fro over the turbinate region in the nasopharynx during panting leads to evaporative cooling of the blood flowing through the surface tissues. The turbinate system is well developed, for example in the sheep's nasopharynx, with the result that in the summer desert the sheep which has a facial skin temperature of 42°C has a surface temperature of 34°C over the turbinate bones as a consequence of the movement of dry air over the wet mucosa.

The increased respiratory frequency that occurs at high temperatures is combined with a small tidal volume, so that there is considerable ventilation of the respiratory dead space (where the heat exchange occurs) but little change in the alveolar ventilation (where the exchange of gases occurs). However, if the alveolar ventilation is increased, carbon dioxide is removed from the blood at an increased rate and a respiratory alkalosis results. This can occur in both cattle and sheep exposed to very hot conditions. There is a fall in the respiratory rate from the peak level that is reached at high temperatures, together with an increase in tidal volume (see Figs, 9.4 and 9.5).

WATER ECONOMY

Macfarlane[349] refers to the jungle syndrome of high energy and water turnovers, with low salt tolerance and poor kidney concentration. This is exhibited by cattle, banteng and buffalo. This can be contrasted with the adaptation to water restriction shown by sheep, goat and camel. The rates at which energy and water are used by the different genera vary by a factor of three, and the rates of use of energy and water are closely linked. As a generalization, the larger the mammal, the lower the metabolic rate and the water turnover.

The following table summarizes some of the relative degrees of adaptation shown by jungle, intermediate and desert types of ruminant:

	Cattle	Sheep	Camel
Energy turnover	+++	++	+
Water turnover	+++	++	+
Renal concentration	+	++	++
Vasopressin sensitivity	+	++	+++
Salt tolerance	+	++	++++

The half-time for the turnover of water, measured with daily temperatures reaching 40°C, is 2.5 days for cattle, 4 days for sheep, and as long as 8 days for the camel. Water turnover can be defined as

Table 11.5 Comparison of the heat input and evaporative cooling of sheep and cattle in adaptation to hot environments (Macfarlane, 1968a, by permission of Lea and Febiger).

	Hair sheep	Wool sheep	*Bos taurus*	*Bos indicus*
Surface temperature in sun, °C	56	92	58	46
Reflectance of coat	++	+	+	+++
Long-wave emission	+	+++	++	+
Skin colour	White	White	White	Black
Water turnover ml $kg^{-0.82}$ 24 h^{-1}	180	220	530	400
Evaporative cooling $\frac{\text{sweating}}{\text{respiratory}}$ ratio	0.2	0.12	4	6
Water reabsorption colon	+++	+++	+	++
kidney	+++	+++	+	++
Loops of Henle	Long	Long	Short	Short

the amount of water passing through an animal per unit time. A diuresis readily occurs in cattle after drinking, but sheep need to take in a lot of water, up to 3% of body weight, to produce a diuresis. The larger daily needs of cattle for water are in keeping with their less concentrated urine and their lower capability to reabsorb water in the colon and kidney when compared with the sheep.

The kidney contains both long and short loops of Henle, the long loop being associated with more water reabsorption and a more concentrated urine, and the short loop with less reabsorption and more dilute urine. Cattle have mainly short loops, and sheep mainly long loops (Table 11.5). Cattle urine can reach a concentration of 2.6 osmoles per litre, whereas Merino sheep can reach 3.5 to 3.8 osmoles per litre.[347] These values are reached when the animals are deprived of water under hot conditions; normally, values over 1.5 osmoles per litre are rare. Cattle and sheep normally have glomerular filtration rates of 90 to 150 ml min^{-1} per kg, falling to about 40 during dehydration; the value for the camel is normally 60, falling to 15 ml min^{-1} per kg when water is restricted.

Bos indicus cattle drink less than *Bos taurus* when the two species are exposed to similar conditions.[48] However, *Bos indicus* still needs twice as much as the sheep and camel per unit of metabolic body size; in actual amounts cattle may take in 50 litres, and sheep about 3 litres. Sheep and goats can live in temperate climates without drinking, if the pasture is not dry (50% water content), but cattle must have water even under cool conditions. Amongst cattle, *Bos indicus* has lower rates of water turnover than *Bos taurus*, and, amongst sheep, Merinos have lower rates than Leicesters. In sheep, there is a high water turnover in the wet season, and a low water turnover in the dry season; the total body water increases when water turnover increases, as occurs on more succulent pastures. When sheep are shorn, the increased heat absorption in a hot climate leads to an increased water turnover. Goats need rather less water than sheep, and have a lower water turnover.[349]

ADAPTATION TO HOT ARID CONDITIONS

Ungulates in the desert have to deal with both heat and lack of water. Dehydration and exercise increase rectal temperature; 46.5°C has been observed in a Grant's gazelle maintained for 6 hours with no apparent ill effect. The brain temperature is probably several degrees below this, because the carotid artery that supplies the brain divides at the base of the brain into a rete mirabile of hundreds of small parallel arteries that pass through the cavernous sinus, and the blood passing through them is cooled. The cooling results from heat transfer from the arterial blood to cooled blood reaching the cavernous sinus from the nasal passages, where it was cooled previously by heat transfer through evaporative loss during breathing[22, 493] (see Chapter 4).

Cattle are animals that have a wet tropical origin, and they are basically suitable for wet grazing. If they are deprived of water they die in a few days on account of their rapid water turnover. Sheep and goats are more suited than cattle to arid environments, because they have lower water requirements, higher renal concentrating power and lower metabolic rates. Sheep and goats can survive with 1.3–1.5% NaCl in the drinking water, whereas for cattle and pigs 1% NaCl in the drinking water is near the physiological limit. Goats survive in the desert better than most sheep and the camel is well adapted to an arid environment.[350]

Of the antelopes, the eland, which is the largest antelope, corresponds to cattle in respect of water and energy turnover rates, the wildebeest to sheep and goats, and the oryx to the camel. The oryx appears to have turnover rates even lower than those of the camel. The maximum urinary osmolar concentrations that have been reported for several East African mammals are given in Fig. 11.4.

Dehydration

During dehydration, ungulates show a fall in heat production and evaporative loss, and the importance of evaporative cooling in thermoregulation in the heat draws attention to the availability of water and to the consequences of dehydration. The total body water of mammals living in hot climates usually exceeds 70% of body weight; the animals are lean and have large extracellular volumes. In temperate regions where animals are fatter the percentage water content is less than this. Cattle in the wet tropics have as much as 85% of body weight as water, equatorial sheep 75%, and Merino sheep in the hottest part of Australia 74%. By contrast, fat sheep in the temperate zone may have below 50% of body weight as water.

Large domestic animals such as cattle and sheep present a complication with respect to dehydration owing to the large capacity of the rumen, which in a cow can hold 100 litres of water. In the ruminant the gut contents, which may form 20% of the body weight, are 85% water. Cows deprived of water for 2 days at 40°C lost 12% of body weight, 95% of which was water loss.[49] There was evidence of fluid loss from the blood, but the total solids in the serum increased by only half as much as in man, probably on account of the large

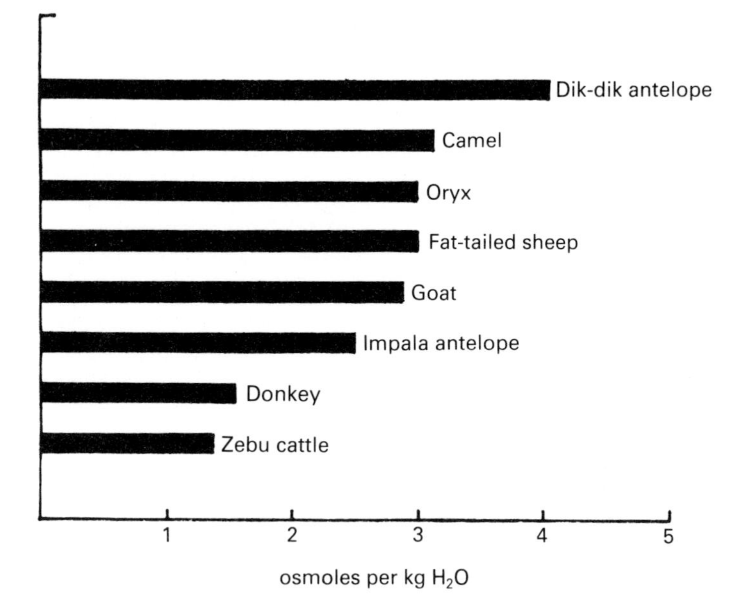

Fig. 11.4 The maximal urinary osmolar concentrations of several East African mammals. Urinary osmolalities were measured after the animals had been severely dehydrated (Maloiy, 1972, by permission of Zoological Society of London).

reservoir of water present in the rumen. During dehydration the body temperature at which sweating began was slightly higher than in control animals.

Merino sheep are intermediate between camels and cattle in their ability to survive water deprivation under hot conditions. During the first day at 40°C day temperature and 25°C night temperature, cattle lose 10% of body weight, sheep 8% and camels 4%; sheep can withstand a weight loss of 30% following deprivation of water for 5 days, some of the water being lost from the rumen (Fig. 11.5). The faecal water content falls only to 60% in cattle after 3 days without water, whereas in sheep faecal water is 45% after 6 days and in camels about 40% after 5 days. Table 11.6 gives some indication of relative losses of water through different channels in ungulates adapted to a dry existence, the relation of body fluids to the survival of ruminants without water in desert environments is given in Table 11.7, and the Merino sheep, *Bos indicus* and *Bos taurus* are compared in Table 11.8 in respect of water balance and metabolism.[348]

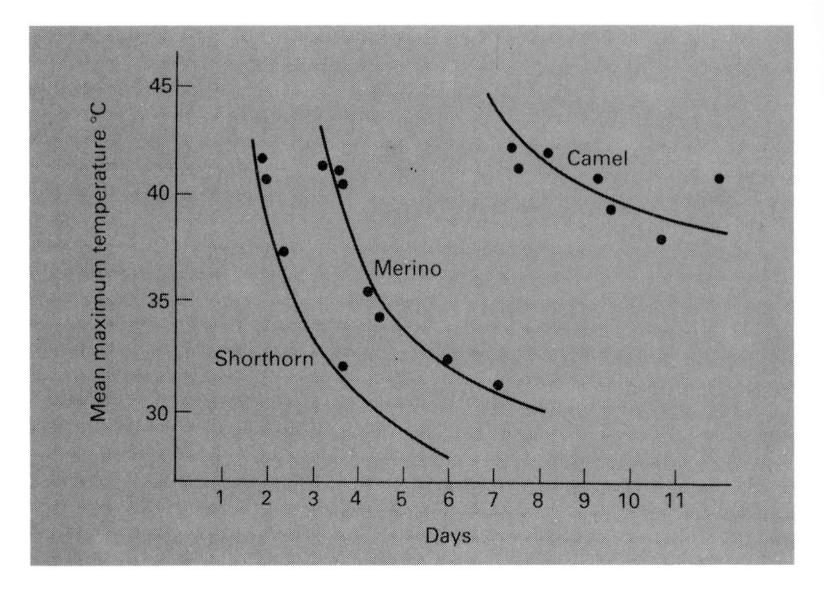

Fig. 11.5 20% weight loss without water in summer. Evaporative water loss is greater as air temperature rises. A loss of 20% of body weight occurs in 7–10 days in camels, 4–5 days in sheep and 2–3 days in cattle, at air temperature maxima over 40°C. Death from loss of 28–32% of body weight would occur in about 15 days in camels, 7 days in sheep and 4 days in cattle (Macfarlane, 1968b, by permission of Lea and Febiger).

The small black Sinai goat can withstand dehydration even to 40% of body weight, and continue to eat even at a loss of 30% of weight. When given water, it takes in the volume required for rehydration in 2–3 minutes. The goat's red blood cells are significantly more fragile in hypotonic solution than those of the camel; 50% haemolysis occurs in goats' red blood cells at 110 mM NaCl, but a solution as hypotonic as 50 mM is needed for 50% haemolysis of camels' red blood cells. However, the rumen in the goat plays a major role in avoiding excessive dilution of the plasma, retaining the water that is drunk to release it only gradually later on.[104]

In a study of Thomson's and Grant's gazelles, Taylor[491] has shown that in these animals body temperature increases during dehydration and the threshold temperature for panting is increased, particularly in Grant's gazelle, which extends its range into the hot arid regions of East Africa. This change in threshold can be demonstrated quite dramatically if the animals are exposed to 40°C,

Table 11.6 Relative loss of water by various routes among ungulates adapted to dry country. Quantitative data are not standardized so that only relative amounts are shown. Water elimination takes place in different proportions through the various routes. Camels and llamas do not increase respiratory water loss to any extent when heated, whereas Zebu cattle show some increase. European cattle have moderate respiratory cooling, whereas sheep evaporate most of the water for heat regulation from the nasal cavities (Macfarlane, 1964, by permission of American Physiological Society).

	Respiratory	Percutaneous	Renal	Faecal
Camel	±	+++	+	+
Horse	±	++++	++	++
Cattle	++	+++	+++	+++
Sheep	++++	+	++	++

when normally hydrated animals pant but dehydrated ones do not; they simply allow the body temperature to rise. Taylor has also shown that Grant's gazelle can thrive without water in places that have had no rain for two years. The oryx and gazelle survive while eating dry grasses and shrubs, but it has been demonstrated that these plants collect water from the desert air at night (Fig. 11.6), when the animals are grazing, so mitigating some of the effects of lack of water in the forage.[564] Several species of antelope can survive prolonged drought; for example, the eland can obtain enough water from moist leaves.

Camel

The camel is the most celebrated of the desert mammals, and some aspects of its water economy that have already been discussed lend

Table 11.7 Relation of body fluids to survival of ruminants without water in desert environments with daily maximum temperature 40°C (Macfarlane, 1968b, by permission of Lea and Febiger).

Phenomena	Camel	Merino sheep	Shorthorn cattle
Rate of weight loss % per day	2.0	4.5	7.0
Percentage of fluid lost from plasma	4.5	8.0	10.0
Days survival at maximum of 40°C	12–15	6–8	3–4
Maximum urine concentration osmoles/litre	3.8	3.1	2.6
Maximum faecal dehydration % water	38	45	60
Maximum plasma sodium osmoles/litre	0.202	0.185	0.170
Water loss as % of weight lost	85	74	66
Initial water replacement as % of weight lost	60	72	84

Table 11.8 Comparison of features of tropical or desert types of *Bos indicus* with Shorthorn or Hereford, *Bos taurus*, bred for temperate climates and with Merino sheep (Macfarlane, 1968b, by permission of Lea and Febiger).

Function in hot environments	*Ovis aries* (Merino)	*Bos indicus* (Brahman)	*Bos taurus* (Shorthorn)
Behaviour of coat in sun	wool (longwave radiation)	short, shiny (reflecting)	longer (heat absorbing)
Sweat glands	small +	large +++	smaller ++
Respiratory cooling	important ++	slight +	moderate ++
Foraging in the sun	common	common	rare
Range of foodstuff eaten	restricted	wide	restricted
Rumen yield of metabolites	+++	+++	++
Intestinal absorption of metabolites	+++	+++	++
of water	+++	++	+
Efficiency of conversion	+++	+++	++
Metabolic rate and water turnover	low	higher	highest
Response to vasopressin	+++	++	+

+++ major activity ++ moderate + minor

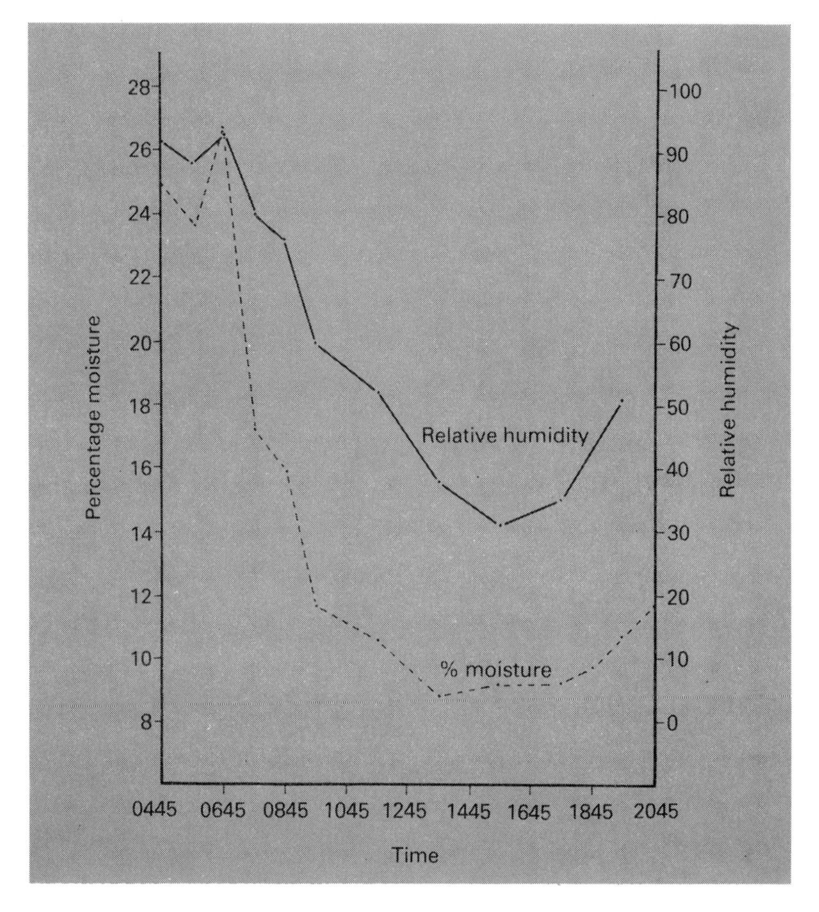

Fig. 11.6 The relation between the relative humidity of the air at various times of the day and the moisture content of a perennial desert grass *Stipagrostis uniplumus* (Louw, 1972, by permission of Zoological Society of London).

credence to the stories of its capacity to survive for long periods without water. The camel is a ruminant and its normal capacity for water is therefore large. The camel's kidney is capable of producing urine (3.1 osmoles per litre) that is three times as concentrated as sea-water (1.1 osmoles per litre), and it is therefore possible for the animal to drink water that is more saline than other animals can tolerate and to eat plants with a high salt content. The faeces are also very low in water content when water intake is low (see Table 11.7).

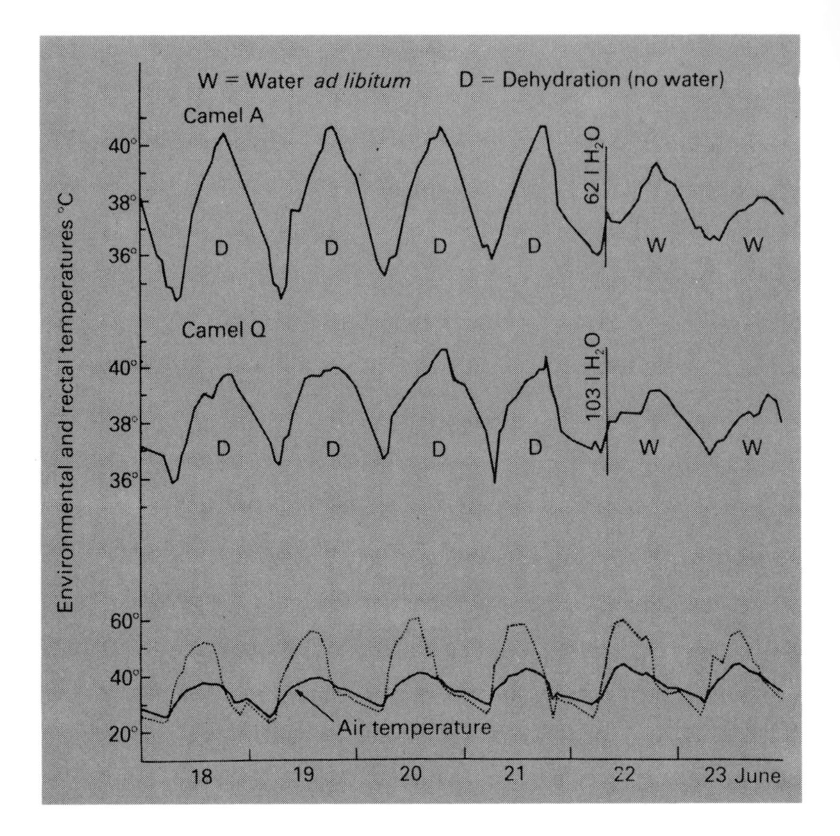

Fig. 11.7 The rectal temperatures of two camels in relation to environmental temperature over several successive 24-hourly cycles. The 24-hourly variation in rectal temperature is considerably larger in the dehydrated camel than in the animal that is allowed to drink freely. Black bulb temperature (· · · ·), which depends on air and mean radiant temperatures and solar radiation (see Chapter 5) (Schmidt–Nielsen, Jarnum and Houpt, 1957, by permission of the American Physiological Society).

The camel is physiologically adapted to withstand the arid conditions of the desert; it has a low rate of water turnover and a low metabolic rate. The camel sweats, and it cannot be independent of drinking water, like the kangaroo rat (see Chapter 6), but under conditions where water is restricted its water loss in the urine, faeces and through evaporation is considerably reduced, nearly halved.[354] Schmidt–Nielsen[455] made observations on camels deprived of water and found them not only exceptionally able to withstand dehydration of up to 25% of body weight, but also that in the desert winter it took 17 days to lose 16% of body weight, and even in summer 7 days were required to lose 25%.

An adaptation exhibited by the camel that helps it to reduce water loss is its ability to allow fluctuations in its body temperature (Fig. 11.7). Even under conditions where water is freely available the 24-hourly variation in body temperature is 2–3°C, but under conditions of dehydration the temperature not only rises to a higher level during the day but also falls to a lower value at night. This increased fluctuation is associated with an increase in the body temperature at which sweating begins, and a decrease in the threshold temperature for increased metabolism. The net result is that the camel can store much of the heat load it receives during the hot desert day and lose it by convection and radiation during the cool desert night. The day-time hyperthermia has the effect of sparing water that would otherwise be used for evaporative cooling, and it may be regarded as an adaptation to high temperatures that takes place particularly when dehydration is threatened. This adaptation is of special value in larger animals, which have a larger heat capacity per unit surface area and consequently a higher ratio of heat storage potential to heat exchange with the environment than in smaller animals. It is also seen, for example in the eland and the burro.

The camel's resistance to dehydration is increased by a smaller reduction in plasma volume on dehydration than is seen in most animals.[348] Camel tissues are also more resistant to high osmotic pressures in the plasma; it is this tolerance by the tissues, including red blood cells, that is particularly important in the animal's resistance to dehydration rather than its renal concentrating power.

Behavioural adaptations contribute to the water economy of camels. These include seeking shade, lengthwise orientation to the sun, sitting with the legs folded under the body, and remaining sitting on the same ground so that contact with a fresh hot surface is avoided.[354]

The donkey is another animal frequently used as a beast of burden under desert conditions. To some extent it also allows its body temperature to fluctuate under conditions of dehydration, although the daily variations are not as great as in the camel. Tolerance to dehydration is similar to that observed in the camel (over 25% of body weight), without the relatively large reduction in plasma volume seen in man. However, the donkey reduces food intake and loses weight while drinking 0.75–1.00% NaCl and its renal concentrating power is not high.[354]

Comparison of the camel and donkey under simulated desert conditions, but allowed water *ad libitum*, shows that the donkey's

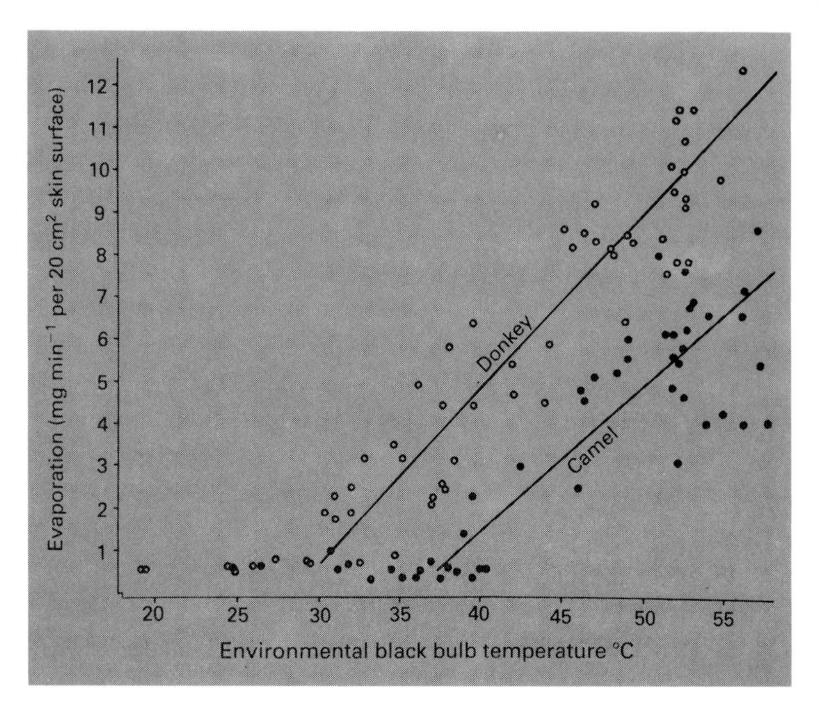

Fig. 11.8 Measurements of the evaporative water loss from shoulder skin in a donkey and a camel in relation to black bulb temperature. The increase in evaporation occurred at a lower black bulb temperature in the donkey than in the camel (Schmidt–Nielsen, Jarnum and Houpt, 1957, by permission of the American Physiological Society).

water turnover is much higher per unit body weight, although in both animals evaporation accounts for the largest part of the water loss. The camel's threshold temperature for sweating is close to 40°C, whereas for the donkey it is near to 30°C (Fig. 11.8).

12

Birds

As with the other species that have been discussed in earlier chapters, the thermal physiology of birds is characterized by a number of special features. The first characteristic is that birds are homeotherms that have relatively high deep body temperatures, usually above 40°C, higher than the usual deep body temperatures of mammals. The second characteristic is that they have insulating layers of feathers that offset the tendency to heat loss that accompanies their relatively small body size. The third characteristic is that under hot conditions body temperature is controlled by evaporative heat loss from the respiratory tract, with only a minor contribution from cutaneous evaporative loss. The fourth characteristic, which applies to birds when they are flying, is that they are exposed to potentially high levels of forced convective heat transfer.

TEMPERATURE REGULATION AND METABOLIC RATE

Body temperatures of inactive birds at moderate environmental temperatures range from 39 to 44°C during the daily waking period (Table 12.1), compared with 36 to 40°C for eutherian mammals. Deep body temperatures of the common domesticated species, that is pigeons, ducks, geese, turkeys and chickens, range from 41.2 to 42.2°C, whereas the range for the common domesticated mammals is 36.4 to 39°C.[536] The body temperature of birds increases as the body weight falls, except in very small birds (Fig. 12.1); this variation does not occur in mammals.

Table 12.1 Body temperatures for birds of various orders (Dawson and Hudson, 1970, by permission of Academic Press).

Order	No. of species sampled	Range (°C)
Sphenisciformes (penguins)	6	37.0–38.9
Struthioniformes (ostrich)	1	39.2
Casuariiformes (casuaries and emu)	4	38.8–39.2
Apterygiformes (kiwis)	3	37.8–39.0
Tinamiformes (tinamous)	1	40.5
Gaviiformes (loons)	1	39.0
Podicipediformes (grebes)	4	38.5–40.2
Procellariiformes (albatrosses, shearwaters, petrels, and allies)	13	37.5–41.0
Pelicaniformes (tropic birds, pelicans, frigate-birds, and allies)	9	39.0–41.3
Ciconiiformes (herons, storks, ibises, flamingos, and allies)	12	39.5–42.3
Anseriformes (screamers, swans, geese, and ducks)	28	40.1–43.0
Falconiformes (vultures, hawks, and falcons)	12	39.7–42.8
Galliformes (megapodes, curassows, pheasants, and hoatzins)	22	40.0–42.4
Gruiformes (cranes, rails, and allies)	7	40.1–41.5
Charadriiformes (shorebirds, gulls, auks, and allies)	39	38.3–42.4
Columbiformes (sand-grouse, pigeons, and doves)	5	40.0–43.3
Cuculiformes (cukoos and plantain eaters)	2	41.9–42.3
Strigiformes (owls)	9	39.2–41.2
Caprimulgiformes (goatsuckers, oilbirds, and allies)	5	37.6–42.4
Apodiformes (swifts and hummingbirds)	25	35.6–44.6
Coraciiformes (kingfishers, motmots, rollers, bee-eaters, and hornbills)	1	40.0
Piciformes (woodpeckers, jacamars, toucans, and barbets)	10	39.0–43.0
Passeriformes (perching birds)	101	39.2–43.8

Birds that are native to hot environments have body temperatures that are not significantly different from those of birds in the temperate zones. The upper lethal body temperatures are similar for both desert and non-desert birds, at about 47°C.

Small birds spend much time foraging to obtain food required for maintenance and thermoregulation. Birds migrate in temperate regions and so avoid severe climatic conditions and the difficulties of

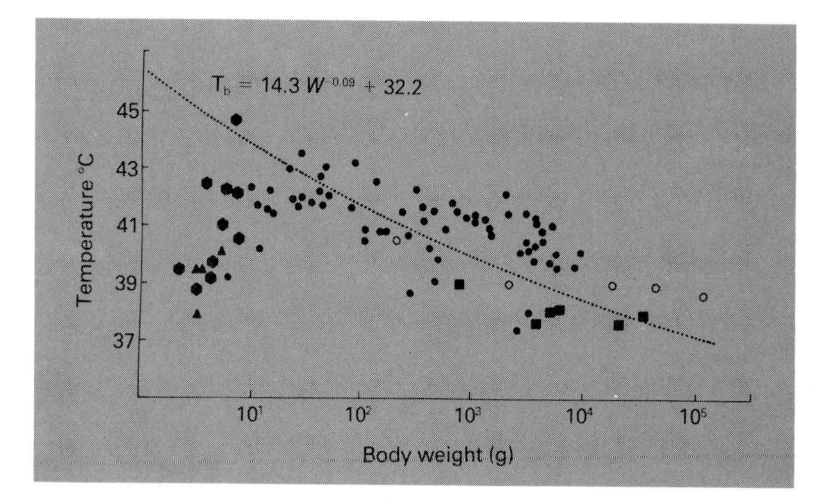

Fig. 12.1 Relation between deep body temperature, T_b, and body weight, W, in birds. ○ = Ratites; ■ = penguins; ▲ = calculated for hummingbirds; ⬢ = hummingbirds; ● = other birds (McNab, 1966, by permission of Condor, redrawn by Whittow, 1973, by permission of Academic Press).

finding food in the winter. Rising[441] comments that migration appears to be an adaptation to avoid food shortage rather than simply to avoid cold, and can be regarded as analogous to hibernation in mammals.

Metabolic rate

The relation of metabolic rate to environmental temperature is shown for both well-feathered and poorly-feathered laying hens in Fig. 12.2. The minimal metabolism of 50 W m^{-2} for these particular birds corresponds to approximately 4 W kg^{-1} (see Tables 2.3 and 2.4 for effects of body size). A similar value for birds of two strains, weighing close to 1.5 and 2.2 kg, can be calculated from the results of Lundy,

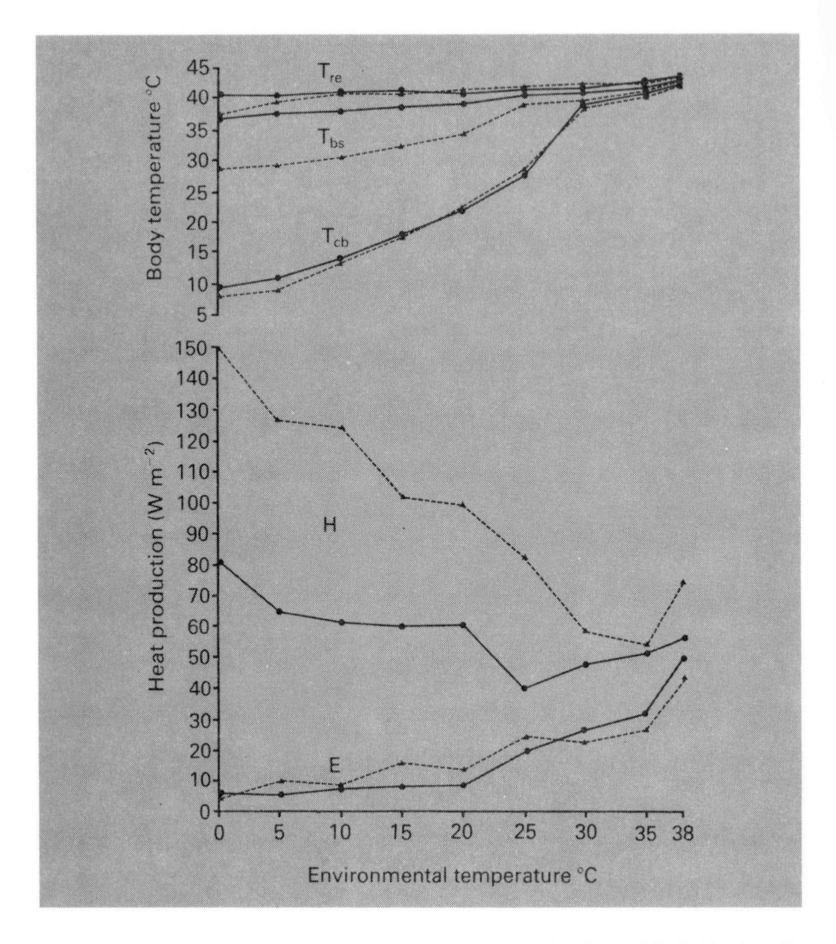

Fig. 12.2 Body temperatures and rates of heat production of individual well-feathered (●—●) and poorly feathered (▲—▲) birds in relation to environmental temperature. T_{re}, rectal temperature; T_{cb}, comb temperature; T_{bs}, temperature of back between scapulae; H, heat production; E, rate of evaporative heat loss (Richards, 1977, by permission of Cambridge University Press).

MacLeod and Jewitt[327], who also found a pronounced light-entrained 24-hourly variation in metabolic rate in both fed and fasted birds with rates about 30% higher in the light than in the dark. Measurements on black duck and mallard using radio-telemetry have also shown that birds weighing about 1 kg have mean daily rates of heat production that approach 4 W kg^{-1};[555] these values correspond to those for well-feathered chickens at 25°C in Fig. 12.2.

The newly hatched chick

The range of temperature for optimum development of the egg during incubation is narrow; for domestic chicken it is 37–38°C, and the percentage of hatching falls off rapidly on either side of this. However, eggs can be stored at 11–13°C for several weeks without affecting hatchability.[144]

The newly hatched chick has a deep body temperature that is about 2.5°C below that of the adult bird; the adult level is reached about 6 days after hatching. The increase is related to a rise in metabolic rate and increased feather insulation.[467]

At hatching, the domestic fowl has a limited ability to regulate its metabolic rate, although it can thermoregulate over only a narrow range of environmental temperature. Thermoregulation improves with age, to become mature after two weeks. Shivering can be elicited even in the newly hatched chick.[165]

EXPOSURE TO COLD

Birds have a homeothermic response to cold, that is, a rise in metabolic rate. There is in addition some evidence of acclimatization to cold (see Chapter 6) in the form of increased metabolism in some species, for example in the chicken (*Gallus domesticus*), the pigeon (*Columbia livia*) and the house sparrow (*Passer domesticus*).[101, 204, 441] Exposure of the bantam chick to cold during the first three weeks of life has produced cold adaptation with a 35% higher maximum metabolic rate than that of warm-adapted chicks.[20] The variation in metabolic rate that occurs over the yearly cycle in the house sparrow approaches two-fold, with the spring and summer rate being rather more than half the winter rate.[441]

There is evidence that the increased metabolic rate in response to cold and in cold acclimatization is due mainly to shivering.[88, 165, 204, 534] However, there is also the suggestion that non-shivering thermogenesis occurs in chickens and some other birds, but its site of origin is uncertain.[441] Brown adipose tissue (see Chapter 2) has not been found in birds, and Freeman[163] was unable to find any thermogenic effect of catecholamines either in adult or in newly hatched chickens. Freeman[164] suggests that newly hatched chicks may use carbohydrate in muscle for non-shivering thermogenesis. The rise

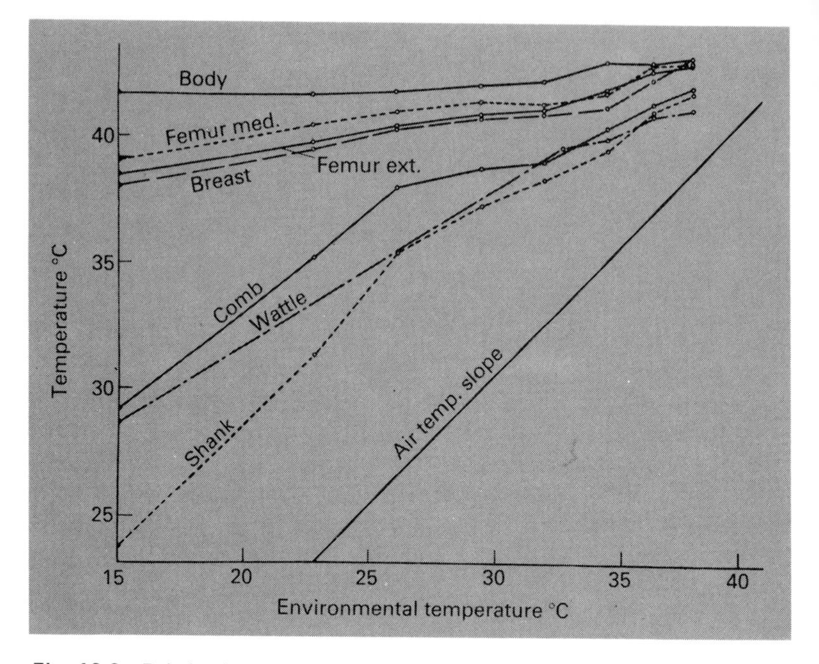

Fig. 12.3 Relation between environmental temperature and body surface temperatures in the hen. The air temperature slope represents the isotherm between skin and air temperature (Wilson, Hillerman and Edwards, 1952, by permission of Poultry Science).

in plasma free fatty acids and the fall in RQ in the newly hatched chick show that lipids are being used although there is no brown fat.

Siegel[468] has reviewed the effects of cold on energy metabolism in birds, and remarks that birds living in polar regions are remarkably tolerant of cold; this also applies to the smaller species. Pigeons with a critical temperature of 20°C had their metabolic rates multiplied by more than three when they were exposed to −30°C, with a correspondingly increased food intake. House sparrows can increase their metabolic rates to face the demands of −30°C if they have been acclimatized to winter conditions, but in the summer they are unable to cope with even 0°C.

Emperor penguins in polar regions spend up to 4 months fasting while breeding at rookeries 80 km or more from the sea. Between 20 and −10°C, the metabolic rate remains almost constant, about 46 W for a 24 kg bird. Below −10°C, the metabolic rate increases, and at −47°C it is 70% above standard, with a mean thermal conduct-

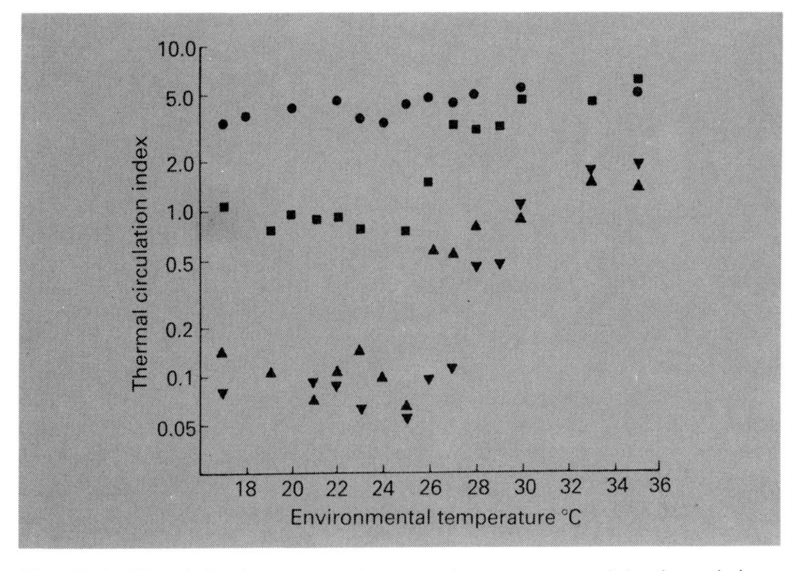

Fig. 12.4 The relation between environmental temperature and the thermal circulation index. Mean results from 31 experiments on 14 birds. ●, mean feathered skin; ■, shank; ▲, toe; ▼, comb (Richards, 1971, by permission of Journal of Physiology).

tance of $1.57\,W\,m^{-2}\,°C^{-1}$ (corresponding to a mean insulation of $0.64\,°C\,m^2\,W^{-1}$ or 4.1 clo: see Chapter 4). Walking 200 km to the sea and back requires less than 15% of the energy reserves of a breeding male emperor initially weighing 35 kg. Collective thermoregulation through huddling is important for energy consevation, because otherwise the total energy reserves might be exceeded.[320, 422]

The feathers of birds are fluffed out in the cold, and provide thermal insulation by trapping air in layers round the body. When a bird moults, or is poorly feathered, it has a higher metabolic rate in the cold[467] (Fig. 12.2). The temperature of skin covered by feathers does not change very much when the environmental temperature changes. In the breast area, for example, there is a rise of only 2.5°C in skin temperature when the temperature of the surroundings rises from 10 to 30°C, although the surface temperature of the shank increases by 15°C (Fig. 12.3).

Measurement of the thermal circulation index (see Chapter 4) for the shank, toe, comb and feathered skin gives values for the index for feathered skin that are much higher than those for the uninsulated surfaces at temperatures below 25°C. However, above 26–27°C the index rises rapidly at the naked skin sites (Fig. 12.4), indicating vasodilatation and increased blood flow. The index for feathered skin remains unchanged, suggesting a relative lack of peripheral vasomotor control over the well insulated body as contrasted with the legs and comb. Blood vessels in the unfeathered legs act as counter-current heat exchangers (see Chapter 4), and so act to conserve heat in the cold and to allow increased heat dissipation in the warm.

Critical temperature

Birds' extraordinary tolerance of cold is referred to by Irving[265] with reference to domestic pigeons that lived at −40°C without food or water for from 48 to 144 hours, and a pigeon that kept its normal body temperature for one hour when exposed to −90°C. The critical temperature (see Chapter 2) of the pigeon has been variously estimated to lie between 14 and 28°C in different measurements, so that exposure to low temperatures is metabolically costly. Some birds have very low critical temperatures; for example, the large arctic glaucous gull, of body weight 1.5 kg, appears to have such a low critical temperature that it can live in the arctic winter at no great metabolic cost.

In the domestic fowl, during the first week after hatching the critical temperature is about 34°C,[437] at a body weight of 35–40 g. This falls to 32°C at 5 weeks old and a weight of 260 g, and to 18°C in the adult.[487] However, when the environmental temperature rises the food intake falls and there is an accompanying fall in heat production (see Figs 12.7, 12.8, and Chapter 2). This makes the demonstration of the critical temperature difficult unless food intake is controlled.

Hypothermia

If body temperature falls considerably, either as a result of food deprivation or of adaptation, torpor results. Entry into torpor is largely passive, and once the bird is inactive, metabolic rates may fall to as low as 2% of the resting levels, with a marked fall in heart rate and a slowing of respiratory rate. Body temperatures during deep torpor vary with species, but may fall towards 8°C. Spontaneous arousal from torpor is accompanied by a rapid rise in metabolic rate, with shivering: heart rates may rise to 1000 min^{-1} in humming birds. Some small birds, such as the hummingbird (*Trochilidae*), pass through a period of regulated hypothermia during the night,[310–11] an adaptation that leads to considerable energy saving in a bird that has to spend most of its waking time seeking food to satisfy metabolic demands that are very large relative to its body mass. There is little evidence of seasonal dormancy of the type observed in some mammals[136] (see account by Siegel[468] for further references).

EXPOSURE TO HEAT

Many species of birds can tolerate a temporary hyperthermia of 2–4°C. When birds are exposed to a heat load, the body temperature that is reached tends to approach that of the environment in small passerines, but in non-passerines the hyperthermia is not so marked. This may be related to higher metabolic rates in the small passerines:

passerines: $M = 6.3\,W^{0.72}$
non-passerines: $M = 3.8\,W^{0.72}$ (Lasiewski and Dawson[312])

where M is metabolic rate in watts and W = body weight in kg. The more marked temporary hyperthermia in passerines may be of adaptational significance since these birds are less able to dissipate heat by evaporation.[469] The metabolic rates of birds are higher than those of mammals.[456, 538]

Chickens reared in high environmental temperatures show morphological adaptation in the form of larger combs than those found in chickens raised in the cold, comparable with the larger ears and longer tails of warm-reared pigs (see Chapter 8). As a behavioural form of adaptation, under very hot conditions in the middle of the day some birds seek refuge in shade or rock crevices, and soaring birds climb to great heights.[136, 258]

Evaporative heat loss

Many birds are able to dissipate more heat by evaporation than can be produced by metabolism. Evaporation occurs principally from the respiratory tract; there are no sweat glands. However, van Kampen[288] finds that more than half the total evaporation at environmental temperatures below 20°C takes place from the skin. Richards[437] remarks that non-sweating skin is not impermeable to water vapour, and that it can play an important part in water loss and thermoregulation. Richards concludes that the 2 kg fowl at 0–5°C loses between 10 and 15% of its metabolic heat by unavoidable water loss and convective transfer to the respired air. Then, as the temperatures rises above 25–27°C, evaporative heat loss moves into a major role[450] with a marked increase in the respiratory loss (Fig. 12.5).

At 35°C, evaporative heat loss can account for over half the heat produced, and for all of it at 40°C, although King and Farner[298] give examples where evaporative heat loss accounts for no more than 50% of heat production. At environmental temperatures above 30°C the respiratory tract is the major site for evaporative cooling. The lungs of birds are relatively smaller and less elastic than those of mammals, and the air sacs are a potential area for evaporation in addition to the

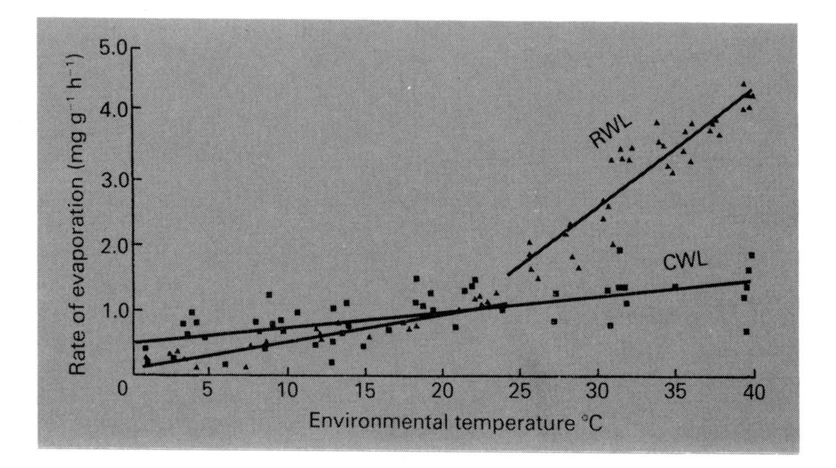

Fig. 12.5 The rates of cutaneous (CWL) and respiratory (RWL) water losses (y) as functions of environmental temperature (T_e; x). The points represent the mean of duplicate measurements on individual birds weighing 2 kg. Regression line for CWL (■): $y = 0.52 + 0.02x$, $r = 0.67$, $P < 0.001$. Those for RWL (▲): $y = 0.15 + 0.04x$, $r = 0.91$, $P < 0.001$ (T_e 0 to 23°C): and $y = -2.86 + 0.18x$, $r = 0.93$, $P < 0.001$ ($T_e > 23$°C) (Richards, 1976, by permission of Cambridge University Press).

areas involved in the mammal.[258, 437] It should be pointed out that the posterior air sacs receive fresh outside air, whereas the anterior sacs receive air that has passed through the tertiary bronchi of the lung (Fig. 12.6). As additional areas for vaporization, therefore, the posterior sacs are the likely sites, although their blood supply is poor. In fact, Menuam and Richards[361] provide evidence that evaporative loss in hot environments occurs mainly from the upper respiratory tract during panting and the air sacs are unlikely to be involved. In the mammal, evaporation occurs from the nasal, buccal and tracheal surfaces, but it is not considered to occur from the lungs themselves (see Chapter 3).

Panting

When the deep body temperature reaches 41–43°C, varying with the species, panting begins. There is an increased respiratory frequency with diminished tidal volume, but in spite of the reduction in tidal volume a respiratory alkalosis develops.[89] The maximum respiratory rate that is reached in the chicken can exceed 200 per minute, at a deep body temperature of 44°C. Panting can be prevented

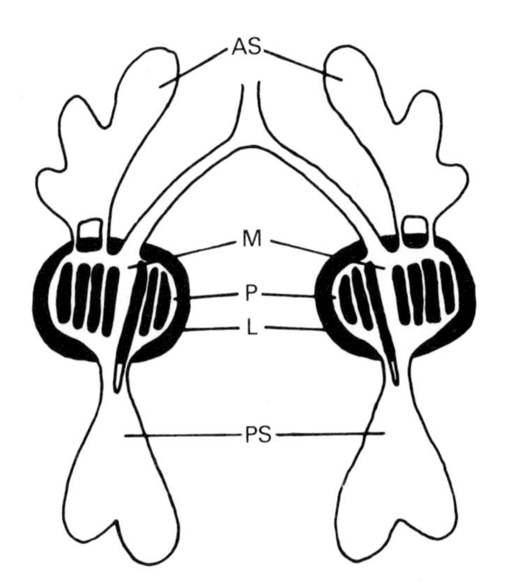

Fig. 12.6 Diagram of lungs and air sacs in the bird. AS, anterior sacs; P, parabronchi; PS, posterior sacs; M, mesobronchus; L, lung tissue (Ingram and Mount, 1975a, by permission of Springer–Verlag).

by cooling the hypothalamus.[258, 437] Some birds increase the effect of panting by 'gular flutter', a rapid fluttering of the gular area produced by flexing of the hyoid apparatus. Gular flutter may be more effective than panting, because it involves less skeletal musculature and less air movement.[298] Gular flutter increases evaporative heat transfer and convective heat loss by forced convection, and does so at low metabolic cost.[137]

Birds, with few exceptions, do not appear to have special mechanisms for dealing with hot and dry conditions. Their tolerance of some degree of hyperthermia of as much as 4°C above normal body temperatures, and their behavioural patterns that reduce heat stress, together with panting and gular flutter, help survival in the hot day time in the desert.[137, 298]

WATER REQUIREMENTS

Amongst domestic chicken, a week-old chick consumes about 28 ml of water per day, and an 8-month laying pullet about 500 ml per day. Most of this intake is from water actually drunk, and not from succulent foods. The daily water loss through all channels is 20 ml for the chick, and 410 ml for the pullet, less than the intakes because the growing fowl is in positive water balance.

Dehydration

Survival time without water is 6–8 days in hens that are producing eggs and about 23 days in non-producers. Some survive longer; at 9°C, White Rock hens survive 60 days and White Leghorns 22 days. At 30°C, White Rocks survive 21 days and Leghorns 11 days, so that environmental temperature is important. Deprivation of water reduces food consumption, which falls to zero with total deprivation. High mortality can occur if water is ingested rapidly after periods of deprivation of 24–48 hours.[467]

Smaller birds have a relatively greater need for access to water than larger birds. At least one large bird, the ostrich, reacts like the camel and donkey (see Chapter 11) when it is deprived of water, by allowing its body temperature to rise.[101] Counter-current exchange of heat and water (see Chapter 4) is important in the water economy of birds.[457]

Birds produce uric acid instead of urea from the degradation of protein, and this leads to the production of 20% more water from protein metabolism than in mammals. Uric acid is also highly insoluble, leading to further saving of water, and water is re-absorbed in the cloaca.

There is a nasal gland in the bird that secretes a concentrated solution of sodium chloride. This is well developed in marine birds, where it acts in addition to the kidney in maintaining the water-salt balance of the animal's body fluids.[418]

FLIGHT

During flight, the rise in metabolic rate can be as much as seven times the resting level (in humming birds[310]). About 18% of the metabolic heat is lost by evaporation and the remainder by convection into the air stream and by radiation.[209–10] Budgerigars flying at 35 km h^{-1} were found at 20°C to lose 85% of their heat by convection and radiation, whereas at 37°C evaporative cooling accounted for 47% of heat loss, and most of the remainder of the heat produced was stored. The net water loss during migration would be 11 mg $g^{-1} h^{-1}$, corresponding to 1.1% of body weight per hour. If the bird could tolerate a water loss equal to 15% of body weight, it could fly for 14 h at 35 km h^{-1}, a distance of 490 km.[508] The primary energy source is fat.[509–10] There may be a problem of hyperthermia during flight, because only moderate activity even at a relatively low environmental temperature leads to the production of more heat than is required for the maintenance of body temperature.[298]

However, the naked feet of many birds can act as heat-regulating devices, and measurements on the herring gull have shown how effective this can be.[35] Both when the birds were at rest and when they were flying in a wind tunnel, heat loss from the feet was considerable, when estimated from measurements of blood flow and arterio-venous temperature differences. Heat production at rest was 8 W, and 37–56% of this was lost from the feet over an environmental temperature range of 10–35°C. During flight, heat production was about 57 W, and 46 W of this, or 80%, was lost from the feet. The results suggest that no evaporative loss is necessary for heat dissipation below an air temperature of 35°C.

HUSBANDRY

Temperature and growth in the domestic chicken

Barrot and Pringle[30–2] found that the optimum temperatures for maximum growth in the domestic chicken were 34°C at one day old, falling by about 0.5°C per day to 19°C at 32 days old. The intake of metabolizable energy and the rate of heat production for 1.5 kg birds is shown in Fig. 12.7.

Poorly-feathered birds are at a disadvantage in the cold, particularly because the fowl possesses little of the subcutaneous fat that occurs in many mammals and it has to rely on its feathers for most of its thermal insulation. Normally-feathered laying hens suffer from hypothermia when exposed to temperatures between 0 and 5°C, but in poorly-feathered birds the core temperature begins to fall when the environmental temperature is between 15 and 20°C, and their relative lack of thermal insulation is also evident in their higher rates of oxygen consumption at environmental temperatures of 30°C and below (Fig. 12.2).

The feed intake of birds fed *ad libitum* falls as the temperature rises (Fig. 12.8). The birds' protein requirement does not change very much with temperature. Egg production can be maintained even at 30°C

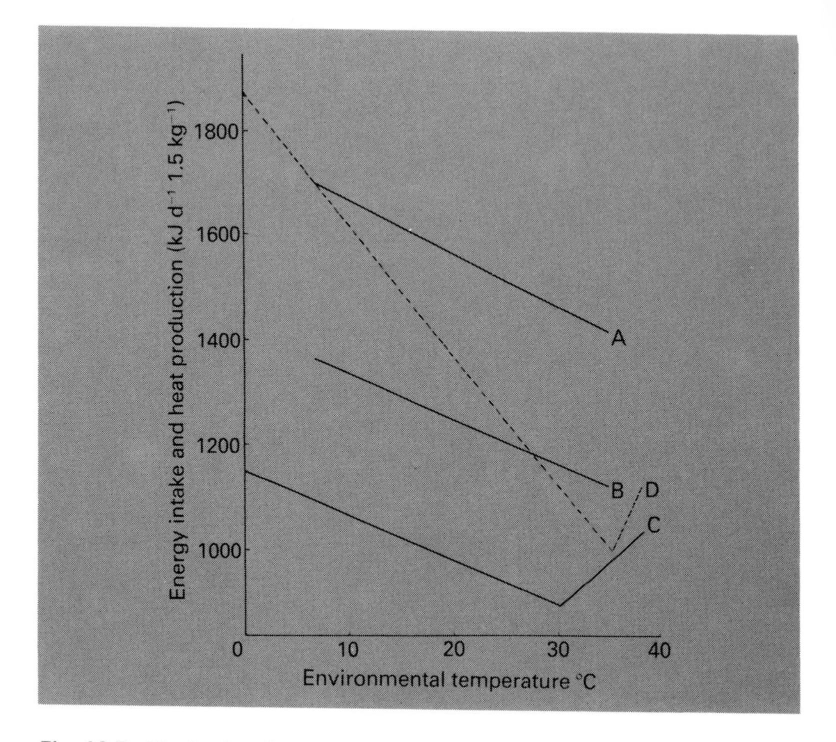

Fig. 12.7 The intake of metabolizable energy and heat production in relation to environmental temperature. Line A, energy intake, and Line B, heat production, are from 3-week energy balances (Davis, Hassan and Sykes, 1973); Line C, heat production of well-feathered hens, and Line D, heat production of poorly-feathered hens, are from hourly rates of oxygen consumption. All values calculated to 1.5 kg bodyweight (Sykes, 1977).

provided that laying birds receive about 15 g protein per day (Fig. 12.9); since feed intake is reduced as the temperature rises, this requires a high protein diet if the protein intake is to be maintained. The rate of lay is probably at a maximum at 18–21°C. Poorly-feathered birds eat more at a given temperature and perform less satisfactorily at lower temperatures.[103, 488]

Adaptation to heat and cold

At high temperatures, birds adopt a wing-spreading posture, and pant; growth rate is reduced and the combs and wattles are enlarged.

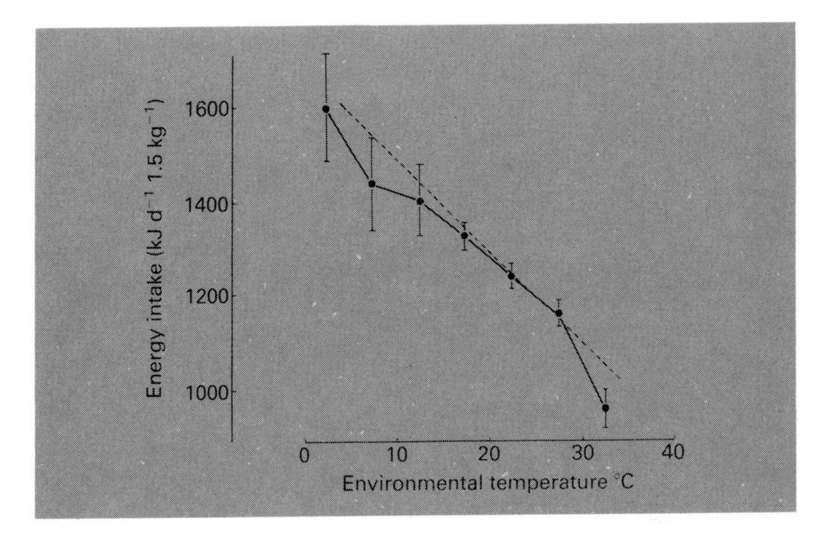

Fig. 12.8 The intake of metabolizable energy (ME) in relation to environmental temperature; mean results from 9 trials at 5°C intervals expressed as kJ d^{-1} for a bird of 1.5 kg bodyweight. Calculated regression (------). ME = 1690–20.1°C (r = 0.81) (Sykes, 1977).

At high temperatures, relative humidity is very important; for 8-week old male chickens, death from heat prostration can occur if the relative humidity is above 30% at an environmental temperature of 41°C.[431] Genetic adaptation to warm conditions is suggested by the probability that the domestic chicken is derived from the jungle fowl of south-east Asia.

In the cold, there is ruffling of the feathers and the birds huddle, a social form of thermoregulation that considerably reduces heat loss[302] (cf. huddling in pigs in the cold, see Chapter 8).

Broiler chicken

It is recommended that young chicken should be initially at an environmental temperature of 29–32°C, falling to 21°C at 2–3 weeks, with allowance for any effects due to wind and radiation (Table 12.2; see Chapter 5). The maximum weight gain occurs in the temperature range 18 to 24°C. For turkeys the optimum temperature is probably 16–18°C. Behavioural indicators are useful in assessing the thermal environment: conditions are satisfactory when the birds move freely in

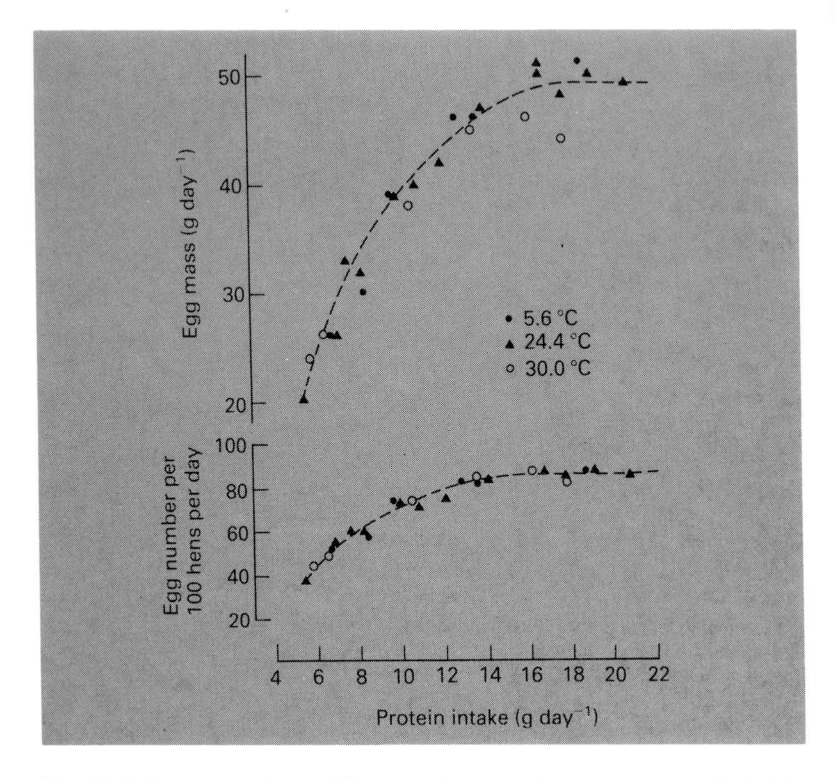

Fig. 12.9 Egg production at different environmental temperatures as affected by protein intake (Sykes, 1977, adapted from Bray and Gesell, 1961).

and out of groups; if there is wing-spreading, or the birds are huddled, conditions are not satisfactory.[103, 151]

Ventilation of poultry houses

The ventilation required can be calculated in terms of the amount of food eaten per day: 7–13 $m^3 h^{-1}$ per kg food per day, up to a maximum in the summer of 100 $m^3 h^{-1}$ per kg per day. At low ventilation rates, ammonia may accumulate; at 10 ppm this is unpleasant, at 25 ppm dangerous to the birds, and at 30 ppm intolerable and dangerous for man. For temperature control in laying houses in the winter low levels of ventilation can be used, down to 900 $m^3 h^{-1}$ per 1000 birds (Table 12.3).

Table 12.2 Temperatures likely to result in similar rates of body heat loss from poultry at different air speeds (Charles and Spencer, 1976, reproduced with the permission of the Controller of Her Majesty's Stationery Office).

	Brooder temperature (°C)	House temperature (°C)	Air speed ($m\,s^{-1}$)	
			Sleeping birds	Active birds
Day old	32	16	—	0.2
	32	21	—	0.4
	32	24	—	0.5
4 weeks old		16	0.3	0.3
		21	0.5	0.6
		24	0.8	1.0
10 weeks old		16	0.3	0.4
		21	0.8	1.0
		24	—	2.5+

Table 12.3 The requirements of poultry for ventilation (Charles and Spencer, 1976, reproduced with the permission of the Controller of Her Majesty's Stationery Office).

Stock	Weight (kg)	Maximum requirement $m^3\,h^{-1}$ per bird	Minimum requirement $m^3\,h^{-1}$ per bird
Pullets and hens	1.2	10	0.8–1.3
	2.0	12	1.3
	2.5	14	1.5
	3.0	14	1.7
	3.5	15	2.0
Broilers	0.05		0.1
	0.4		0.5
	0.9		0.8
	1.4		0.9
	1.8	10	1.3
	2.2	14	1.7
Turkeys	0.5	6	0.7
	2.0	12	1.2
	5.0	15	1.5
	7.0	20	2.0
	11.0	27	2.7

Egg production

The effects of high and low temperatures on egg production appear to be mediated through effects on feed intake; the depression of egg production brought about by high temperatures is probably due to the ensuing reduction in feed intake. Egg laying is hastened by low temperatures and retarded by high temperatures, and starvation leads to cessation of egg production. High temperatures reduce egg size, egg shell thickness and yolk weight. The optimum mean temperature for egg production has been given as the range 10 to 16°C, with production being most efficient between 20 and 30°C.[545] Emmans and Charles[151] remark that rate of lay is unaffected up to about 24°C, with the optimum temperature for economic egg production in the United Kingdom being close to 22°C.

Daylength and seasonality

The domestic fowl shows marked differences in egg production at different times of the year in relation to season. In the northern hemisphere, maximum production occurs in March, with a minimum in September, whereas in the southern hemisphere the maximum is in September and the minimum in March.

The activity of birds generally is considerably affected by variations in light throughout the year. The breeding season is also affected: the uniformity throughout the year at the equator is in marked contrast to the seasonal effects at higher latitudes.[21] There is evidence that birds, as well as mammals, have endogenous rhythms of reproduction, although almost all phases of the cycle may be light-regulated. There is also evidence for both insulative and metabolic adaptation in birds in different seasons.[206]

In the domestic chicken, increasing daylength advances sexual maturity and decreasing daylength delays it; rate of lay is stimulated by increasing daylength and depressed by decreasing daylength. A well established husbandry pattern involves short day rearing on 6 hours light per day, followed by an increase of 20 min per week from the point of lay (which occurs at 20 weeks of age) until there is 17 hours light and 7 hours dark. An alternative is to provide a long constant day, particularly if there are birds of different ages in the group.

The husbandry practice with broiler chickens involves a long daylight period of 23.5 hours throughout life. They are exposed initially to bright lighting (10–20 lux) to encourage feeding and drinking; this is reduced to 1–2 lux by 3–4 weeks. 0.4 lux is estimated as threshold lighting for chickens.[103]

Appendix 1

Appendix 1

Units: equivalents in Système Internationale (S.I.), c.g.s. and the British System

	S.I.	c.g.s.	British
Length, area and volume	1 m	100 cm	3.281 ft
	0.3048 m	30.48 cm	1 ft
	1 km		0.6214 mile
	1.609 km		1 mile
	1 m^2	10^4 cm^2	10.76 ft^2
	0.09290 m^2	929.0 cm^2	1 ft^2
	1 m^3	10^6 cm^3	35.31 ft^3
	0.02832 m^3	28.32×10^3 cm^3	1 ft^3
Velocity	1 m s^{-1}	100 cm s^{-1}	2.237 miles h^{-1} = 196.9 ft min^{-1} = 3.281 ft s^{-1}
	0.4470 m s^{-1}	44.70 cm s^{-1}	1 mile h^{-1} = 88 ft min^{-1} = 1.467 ft s^{-1}
Volume flow rate	1 m^3 s^{-1}	10^6 cm^3 s^{-1}	2119 ft^3 min^{-1} (cfm)
	0.4719×10^{-3} m^3 s^{-1}	471.9 cm^3 s^{-1}	1 ft^3 min^{-1}
	1 m^3 h^{-1}	277.8 cm^3 s^{-1}	0.5886 ft^3 min^{-1}
	3.600 m^3 $h^{-1} \times 10^{-3}$	1 cm^3 s^{-1}	2.119 ft^3 $min^{-1} \times 10^{-3}$
	1.699 m^3 h^{-1}	471.9 cm^3 s^{-1}	1 ft^3 min^{-1}
Mass	1 kg	10^3 g	2.205 lb
	0.4536 kg	453.6 g	1 lb
Density	1 kg m^{-3}	10^{-3} g cm^{-3}	62.4×10^{-3} lb ft^{-3}
	16.02 kg m^{-3}	16.02×10^{-3} g cm^{-3}	1 lb ft^{-3}
Force	1 Newton (N)	10^5 g cm s^{-2} = 10^5 dynes	0.2248 lb force
	4.448 N	444.8×10^3 dynes	1 lb force
Pressure	1 Pascal (Pa) = 1 N m^{-2}	10 g cm^{-1} s^{-2} = 10^{-2} mbar = 7.501×10^{-3} mm Hg	0.02089 lb force ft^{-2}
	47.88 Pa	478.8 g cm^{-1} s^{-2} = 0.4788 mbar = 0.3591 mm Hg	1 lb force ft^{-2}
	1 atmosphere = 101.3 kPa	1013 mbar = 760.0 mm Hg = 760.0 Torr	2116 lb force ft^{-2} = 14.69 lb force in^{-2}

	S.I.	**c.g.s.**	**British**
Temperature	1 °C (or K)	1 °C (or K)	1.8 °F
	0.5556 °C	0.5556 °C	1 °F
Heat	1 J	0.2390 calorie (cal)	0.9485 × 10^{-3} BTU
	4.184 J	1 cal (thermochemical)	3.968 × 10^{-3} BTU
	1.054 kJ	0.2520 kcal	1 BTU
Metabolic rate	1 W = 1 J s^{-1}	0.8604 kcal h^{-1}	3.414 BTU h^{-1}
	1.162 W	1 kcal h^{-1}	3.968 BTU h^{-1}
	0.2929 W	0.2520 kcal h^{-1}	1 BTU h^{-1}
	1 MJ day^{-1} = 11.57 W	0.2390 Mcal day^{-1}	0.9485 × 10^{3} BTU day^{-1}
	4.184 MJ day^{-1} = 48.43 W	1 Mcal day^{-1}	3.968 × 10^{3} BTU day^{-1}
	1.054 kJ day^{-1}	0.2520 kcal day^{-1}	1 BTU day^{-1}
Thermal conductance	1 W m^{-2} °C^{-1}	0.8604 kcal m^{-2} h^{-1} °C^{-1}	0.1762 BTU ft^{-2} h^{-1} °F^{-1}
	1.162 W m^{-2} °C^{-1}	1 kcal m^{-2} h^{-1} °C^{-1}	0.2048 BTU ft^{-2} h^{-1} °F^{-1}
	5.674 W m^{-2} °C^{-1}	4.882 kcal m^{-2} h^{-1} °C^{-1}	1 BTU ft^{-2} h^{-1} °F^{-1}
	1 MJ m^{-2} day^{-1} °C^{-1}	0.2390 Mcal m^{-2} day^{-1} °C^{-1}	2.040 BTU ft^{-2} h^{-1} °F^{-1}
	4.184 MJ m^{-2} day^{-1} °C^{-1}	1 Mcal m^{-2} day^{-1} °C^{-1}	8.534 BTU ft^{-2} h^{-1} °F^{-1}
	0.4903 MJ m^{-2} day^{-1} °C^{-1}	0.1172 Mcal m^{-2} day^{-1} °C^{-1}	1 BTU ft^{-2} h^{-1} °F^{-1}
Thermal insulation	1 °C m^{2} W^{-1}	1.162 °C m^{2} h kcal^{-1}	5.674 °F ft^{2} h BTU^{-1}
	0.8604 °C m^{2} W^{-1}	1 °C m^{2} h kcal^{-1}	4.882 °F ft^{2} h BTU^{-1}
	0.1762 °C m^{2} W^{-1}	0.2048 °C m^{2} h kcal^{-1}	1 °F ft^{2} h BTU^{-1}
	1 °C m^{2} day MJ^{-1} = 0.0864 °C m^{2} W^{-1}	4.184 °C m^{2} day Mcal^{-1}	0.4903 °F ft^{2} h BTU^{-1}
	0.2390 °C m^{2} day MJ^{-1}	1 °C m^{2} day Mcal^{-1}	0.1172 °F ft^{2} h BTU^{-1}
	2.040 °C m^{2} day MJ^{-1}	8.534 °C m^{2} day Mcal^{-1}	1 °F ft^{2} h BTU^{-1}

Note: although care has been taken in the calculation and presentation of these values, the author and publisher do not accept liability for any loss or damage suffered as a result of their application and use.

Appendix 2

Referring to Fig. 2.1 the line for the non-evaporative part of the heat loss cuts the temperature axis at a temperature that falls below the deep body temperature by the small quantity, $H_e\ I_t$, where H_e = rate of evaporative heat loss per unit area of body surface and I_t = the internal thermal insulation per unit area of the animal between the core and the skin surface.[227] This is because the amount of heat, H_e, is transmitted internally to the animal's skin or respiratory tract as non-evaporative heat but from the skin to the environment as evaporative heat. Externally it is then no longer part of the non-evaporative heat flow, and consequently when non-evaporative heat loss is zero the body core temperature is higher than the skin temperature by the amount $H_e\ I_t$. This quantity, which is dimensionally equal to a temperature difference, arises from the relation for non-evaporative heat flow: $H = \Delta T/I$, analogous to Ohm's Law for the flow of electricity through a resistance (see Chapter 4).

This relation can be used to derive the quantity $H_e\ I_t$:

$$H = H_n + H_e \quad \text{(the total heat loss)}$$

$$T_{re} - T_s = H\,I_t \quad \text{(internally, H is all non-evaporative heat)}$$

$$T'_{re} - T_s = H_n\,I_t \quad \text{(when only } H_n \text{ is used to predict body temperature)}$$

$$\therefore\ T_{re} - T'_{re} = (H - H_n)I_t$$

$$= H_e I_t$$

where H_n = non-evaporative heat loss from the skin surface, $W\,m^{-2}$; T_{re} = true rectal (core) temperatures, °C; T'_{re} = apparent core temperature from intercept of sensible heat loss line on temperature axis, °C, and T_s = skin temperature, °C.

As a result of the very large blood flow in the skin that occurs under hot conditions, I_t becomes very small and consequently $H_e\ I_t$ is also small. In the newborn baby, for example, non-evaporative heat loss approaches zero when the environmental temperature is 0.2–0.5°C below the core temperature (Fig. A.1).

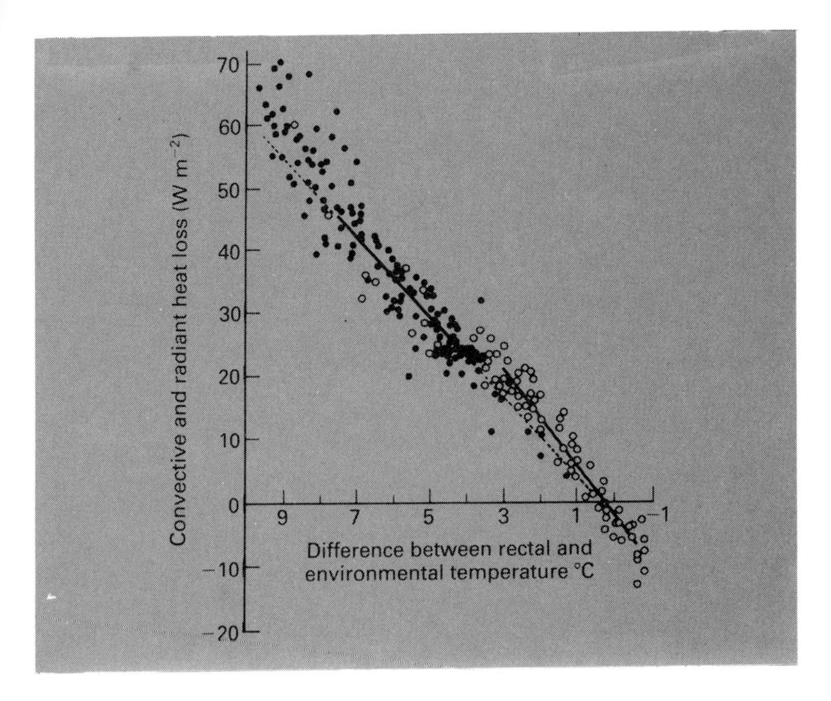

Fig. A.1 The relation between the rectal-environment temperature difference and total heat loss by convection and radiation in 12 babies 0–14 days old and weighing between 1.8 and 2.2 kg. Results obtained when rectal temperature (T_{re}) is 37.2°C or more are indicated by (○); they correspond to section DE of the non-evaporative heat loss line in Fig. 2.1. The lower line is the best-fit relation for all the data obtained when T_{re} is less than 37.2°C (mean 36.9°C, corresponding to section BC in Fig. 2.1) and the operative environmental temperature (T_e) is between 29.5°C and 33°C. The relation has a slope that is equivalent to an insulation of 0.164°C m^2 W^{-1} and an intercept on the x-axis of 0.26°C. The results obtained when T_e is below 29.5°C nearly all lie above the dotted extension of this line, probably because most of the babies were physically active in an environment as cool as this. The upper line is the best-fit relation for all the data obtained when T_{re} is 37.2°C or more (mean 37.4°C) and T_e is between 34°C and 37.5°C. This relation has a slope that is equivalent to an insulation of 0.130 °C m^2 W^{-1} and an intercept of 0.22°C on the x-axis. This intercept differs significantly from zero ($P < 0.05$). Air speed: 0.04–0.05 m s^{-1} (Hey, Katz and O'Connell, 1970, by permission of Journal of Physiology).

Appendix 3

In the following table, the surface area is calculated from 0.1 $W^{0.67}$, where W is body weight in kg. This is the Meeh[360] formula, in which the coefficient, here taken as a mean value of 0.1, ranges from 0.087 for the pig to 0.125 for the rabbit; Lusk[328] lists coefficients for several species. The values of surface area in the table are therefore only approximations.

W, kg	Metabolic body size, $kg^{0.75}$	Surface area, m^2
0.01	0.032	0.0046
0.02	0.053	0.0073
0.05	0.11	0.013
1.00	1.0	0.10
2	1.7	0.16
5	3.3	0.29
10	5.6	0.47
20	9.5	0.74
50	19	1.4
100	32	2.2
200	53	3.5
500	110	6.4
1000	180	10
32		1
89		2
160		3
250		4
350		5
470		6
590		7

Bibliography: suggested further reading

BLAXTER, K. L. (1967). *The Energy Metabolism of Ruminants.* 2nd edition. 329 pp. Charles C. Thomas, Springfield, Illinois.

BUDYKO, M. I. (1974). *Climate and Life*, ed. D. H. Miller. 508 pp. Academic Press, New York.

BURTON, A. C. and EDHOLM, O. G. (1955). *Man in a Cold Environment.* 273 pp. Edward Arnold, London.

DILL, D. B. (editor) (1964). *Adaptation to the Environment.* 1056 pp. American Physiological Society.

FOLK, G. E. (1974). *Textbook of Environmental Physiology.* 308 pp. Lea and Febiger, Philadelphia.

HAFEZ, E. S. E. (editor) (1968). *Adaptation of Domestic Animals.* 415 pp. Lea and Febiger, Philadelphia.

HARDY, J. D., GAGGE, A. P. and STOLWIJK, J. A. J. (editors) (1970). *Physiological and Behavioral Temperature Regulation.* 944 pp. Charles C. Thomas, Springfield, Illinois.

HARDY, R. N. (1972). *Temperature and Animal Life.* 60 pp. Edward Arnold, London.

HARESIGN, W., SWAN, H. and LEWIS, D. (editors) (1977). *Nutrition and the Climatic Environment.* 200 pp. Butterworths, London.

INGRAM, D. L. and MOUNT, L. E. (1975). *Man and Animals in Hot Environments.* 185 pp. Springer Verlag, New York.

JOHNSON, H. D. (editor) (1976). *Progress in Animal Biometeorology*, Vol. 1, Part 1. 603 pp. Swets and Zeitlinger, Amsterdam.

KERSLAKE, D. MCK. (1972). *The Stress of Hot Environments.* 316 pp. Cambridge University Press, Cambridge and London.

KLEIBER, M. (1975). *The Fire of Life.* 2nd edition. 460 pp. John Wiley, New York.

LEBLANC, J. (1975). *Man in the Cold.* Charles C. Thomas, Springfield, Illinois.

MALOIY, G. M. O. (editor) (1972). *Comparative Physiology of Desert Animals.* 413 pp. Zoological Society of London, Academic Press, London.

MONTEITH, J. L. (1973). *Principles of Environmental Physics.* 241 pp. Edward Arnold, London.

MONTEITH, J. L. and MOUNT, L. E. (editors) (1974). *Heat loss from animals and man: assessment and control.* 457 pp. Butterworths, London.

MOUNT, L. E. (1968). *The Climatic Physiology of the Pig*. 271 pp. Edward Arnold, London.

NEWBURGH, L. H. (editor) (1949). *Physiology of Heat Regulation and the Science of Clothing*. 457 pp. Saunders, Philadelphia.

PRECHT, H., CHRISTOPHERSON, J., HENSEL, H. and LARCHER, W. (1973). *Temperature and Life*. 779 pp. Springer-Verlag.

ROBERTSHAW, D. (editor) (1974). *Environmental Physiology*. 326 pp. Butterworths, London.

ROSE, A. H. (editor) (1967). *Thermobiology*. 653 pp. Academic Press, New York.

SCHMIDT-NIELSEN, K. (1964). *Desert Animals: Physiological Problems of Heat and Water*. 277 pp. Oxford University Press, Oxford.

WHITTOW, G. C. (editor). *Comparative Physiology of Thermoregulation:* Vol. I Invertebrates and non mammalian vertebrates (1970), 333 pp.; Vol. II Mammals (1971), 410 pp.; Vol. III Special aspects of thermoregulation (1973), 278 pp. Academic Press, New York and London.

References

1. ADAMS, T. (1971). Carnivores. In *Comparative Physiology of Thermoregulation*, Vol. II, ed. G. Causey Whittow, pp. 151–89. Academic Press, New York.
2. ADOLPH, E. F. (1947). *Physiology of Man in the Desert*. Interscience, New York.
3. ADOLPH, E. F. (1949). Desert. In *Physiology of Heat Regulation and the Science of Clothing*, ed. L. H. Newburgh, pp. 330–8. Saunders, Philadelphia.
4. AGRICULTURAL RESEARCH COUNCIL (1967). The nutrient requirements of farm livestock. No. 3 *Pigs: technical reviews and summaries*. Her Majesty's Stationery Office, London.
5. ALEXANDER, G. (1961). Temperature regulation in the new-born lamb II. A climatic respiration chamber for the study of thermoregulation. *Aust. J. agric. Res.*, **12**, 1139–51.
6. ALEXANDER, G. (1962). Temperature regulation in the new-born lamb. V. Summit metabolism. *Aust. J. agric. Res.*, **13**, 100–21.
7. ALEXANDER, G. (1974). Heat loss from sheep. In *Heat Loss from Animals and Man: Assessment and Control*, eds. J. L. Monteith and L. E. Mount, pp. 173–203. Butterworths, London.
8. ALEXANDER, G. (1975). Body temperature control in mammalian young. *Br. Med. Bull.*, **31**, 62–8.
9. ALEXANDER, G. and BELL, A. W. (1975). Quantity and calculated oxygen consumption during summit metabolism of brown adipose tissue in new-born lambs. *Biol. neonat.*, **26**, 214–20.
10. ALEXANDER, G. and WILLIAMS, D. (1962). Temperature regulation in the newborn lamb. VI. Heat exchanges in lambs in a hot environment. *Aust. J. agric. Res.*, **13**, 122–43.
11. ALLEN, T. E., BENNET, J. W., DONEGAN, S. M. and HUTCHINSON, J. C. D. (1970). Moisture, its accumulation and site of evaporation in the coats of sweating cattle. *J. agric. Sci., Camb.*, **74**, 247–58.
12. ALLEN, T. E. and BLIGH, J. (1969). A comparative study of the temporal patterns of cutaneous water vapour loss from some domesticated mammals with epitrichial sweat glands. *Comp. Biochem. Physiol.*, **31**, 347–63.
13. ANTHONY, D. J. and LEWIS, E. F. (1961). *Diseases of the pig*, 5th edition. Baillière, Tindall and Cox, London.

14. ARMSBY, H. P. and FRIES, J. A. (1903). The available energy of 'Timothy' Hay. *U.S. Dept. Agriculture. Bull.*, **51**,
15. ARNOLD, G. W. and BIRRELL, H. A. (1977). Food intake and grazing behaviour of sheep varying in body condition. *Anim. Prod.*, **24**, 343–53.
16. ASCHOFF, J., BIEBACH, A., HEISE, A. and ŚCHMIDT, T. (1974). Day-night variation in heat balance. In *Heat Loss from Animals and Man: Assessment and Control*, eds. J. L. Monteith and L. E. Mount, pp. 147–72. Butterworths, London.
17. ASCHOFF, J. and WEVER, R. (1958). Kern und Schale im Wärmehaushalt des Menschen. *Naturwissenschaften*, **45**, 477–85.
18. ATWATER, W. O. and BENEDICT, F. G. (1905). A respiration calorimeter with appliances for the direct measurement of oxygen. *Carnegie Inst. Publ. No. 42.*
19. AUGUET, A. and LEFÈVRE, J. (1929). Nouvelle chambre calorimetrique du laboratoire de bioenergitique. *C. r. hebd. Seanc. Acad. Sci., Paris*, **100**, 251–3.
20. AULIE, A. (1977). The effect of intermittent cold exposure on the thermoregulatory capacity of bantam chicks, *Gallus domesticus. Comp. Biochem. Physiol.*, **56A**, 545–9.
21. BAKER, J. R. (1938). The relation between latitude and breeding seasons in birds. *Proc. Zool. Soc., Lond.*, **108**, 557–82.
22. BAKER, M. A. and HAYWARD, J. N. (1968). The influence of the nasal mucosa and the carotud rete hypothalamic temperature in sheep. *J. Physiol., Lond.*, **198**, 561–79.
23. BALDWIN, B. A. (1974). Behavioural thermoregulation. In *Heat Loss from Animals and Man: Assessment and Control*, eds. J. L. Monteith and L. E. Mount, pp. 97–117. Butterworths, London.
24. BALDWIN, B. A. and INGRAM, D. L. (1967). Behavioural thermoregulation in pigs. *Physiol. Behav.*, **2**, 15–21.
25. BALDWIN, B. A. and INGRAM, D. L. (1968). The effects of food intake and acclimatization to temperature on behavioural thermoregulation in pigs and mice. *Physiol. Behav.*, **3**, 395–400.
26. BANNERMAN, M. M. and BLAXTER, K. L. (1969). *The husbanding of red deer.* Aberdeen University Press.
27. BARNETT, S. A. (1959). The skin and hair of mice living at a low environmental temperature. *Q. Jl exp. Physiol.*, **44**, 35–42.
28. BARNETT, S. A. (1965). Adaptation of mice to cold. *Biol. Rev.*, **40**, 5–51.
29. BARNETT, S. A. and MOUNT, L. E. (1967). Resistance to cold in mammals. In *Thermobiology*, ed. A. H. Rose, pp. 411–471. Academic Press, London.
30. BARROTT, H. G. and PRINGLE, EMMA M. (1947). Effect of environment on growth and feed and water consumption of chickens. I. The effect of temperature of environment during the first nine days after hatch. *J. Nutr.*, **34**, 53–8.
31. BARROTT, H. G. and PRINGLE, EMMA M. (1949). Effect of environment on growth and feed and water consumption of chickens. II. The effect of temperature and humidity of environment during the first eighteen days after hatch. *J. Nutr.*, **37**, 153–61.
32. BARROTT, H. G. and PRINGLE, EMMA M. (1950). Effect of environment on growth and feed and water consumption of chickens. III. The effect of temperature of environment during the period from 18 to 32 days of age. *J. Nutr.*, **41**, 25–30.
33. BARTHOLOMEW, G. A. and EPTING, R. J. (1975). Rates of post-flight cooling in sphinx moths. In *Perspectives of Biophysical Ecology*, eds. D. M. Gates and R. B. Schmerl, pp. 405–15. Springer-Verlag, New York.
34. BASS, D. E. (1963). Thermoregulatory and circulatory adjustments during acclimatization to heat in man. In *Temperature, Its Measurement and Control in Science and Industry*, Vol. 3, Pt. 3, ed. J. D. Hardy, pp. 299–305. Reinhold, New York.

35. BAUDINETTE, R. V., LOVERIDGE, J. P., WILSON, K. J., MILLS, C. D. and SCHMIDT-NIELSEN, K. (1976). Heat loss from feet of herring gulls at rest and during flight. *Am. J. Physiol.*, **230**, 920–4.

36. BAUM, E., BRÜCK, K. and SCHWENNIKE, H. P. (1976). Adaptive modifications in the thermoregulatory system of long-distance runners. *J. appl. Physiol.*, **40**, 404–10.

37. BAZETT, H. C., LOVE, L., NEWTON, M., EISENBERG, L., DAY, R. and FORSTER, II, R. (1948). Temperature changes in blood flowing in arteries and veins in man. *J. appl. Physiol.*, **1**, 3–19.

38. BEDFORD, T. (1946). Environmental warmth and its measurement. *Medical Research Council.* Her Majesty's Stationery Office, London.

39. BELDING, H. S. (1970). The search for a universal heat stress index. In *Physiological and Behavioral Temperature Regulation*, eds. J. D. Hardy, A. P. Gagge and J. A. J. Stolwijk, pp. 193–202. Charles C. Thomas, Springfield, Illinois.

40. BELL, A. W., HILDITCH, J. E., HORTON, P. W. and THOMPSON, G. E. (1976). The distribution of blood flow between individual muscles and non-muscular tissues in the hind limb of the young ox (*Bos taurus*): values at thermoneutrality and during exposure to cold. *J. Physiol., Lond.*, **257**, 299–43.

41. BENEDICT, F. G. (1938). Vital energetics. *Carnegie Inst. Publ. 503*, Washington.

42. BENNETT, J. W. (1972). The maximum metabolic response of sheep to cold: effects of rectal temperature, shearing, feed consumption, body posture, and body weight. *Aust. J. agric. Res.*, **23**, 1045–58.

43. BENNETT, J. W. and HUTCHINSON, J. C. D. (1964). Thermal insulation of short lengths of Merino fleece. *Aust. J. agric. Res.*, **15**, 427–45.

44. BENZINGER, T. H. (1971). Peripheral cold reception and central warm reception, sensory mechanisms of behavioral and autonomic thermostasis. In *Physiological and Behavioral Temperature Regulation*, eds. J. D. Hardy, A. P. Gagge and J. A. J. Stolwijk, Chap. 56. Charles C. Thomas, Springfield, Illinois.

45. BENZINGER, T. H. and KITZINGER, C. (1949). Direct calorimetry by means of the gradient principle. *Rev. Sci. Instrum.*, **20**, 849–60.

46. BERMAN, A. (1971). Thermoregulation in intensively lactating cows in near-natural conditions. *J. Physiol., Lond.*, **215**, 477–89.

47. BERRY, I. L. and SHANKLIN, M. D. (1961). Environmental physiology and shelter engineering. LXIV. Physical factors affecting thermal insulation of livestock hair coats. *Res. Bull. 802, Mo. agric. Exp. Sta.*

48. BIANCA, W. (1965). Cattle in a hot environment. *J. Dairy Res.*, **32**, 291–345.

49. BIANCA, W. (1968). Thermoregulation. In *Adaptation of Domestic Animals*, ed. E. S. E. Hafez, pp. 97–118. Lea and Febiger, Philadelphia.

50. BIANCA, W. and FINDLAY, J. D. (1962). The effect of thermally-induced hyperpnea on the acid-base status of the blood of calves. *Res. vet. Sci.*, **3**, 38–49.

51. BIANCA, W. and KUNZ, P. (1978). Physiological reactions of three breeds of goats to cold, heat and high altitude. *Livestock production Science*, **5**, 57–69.

52. BIRKEBAK, R. C. (1966). Heat transfer in biological systems. *Intern. Rev. gen. exptl. Zool.*, **2**, 269–344.

53. BLAGDEN, C. (1774). Experiments and observations in an heated room. *Phil. Trans. Roy. Soc.*, **65**, 111–23.

54. BLAGDEN, C. (1775). Further experiments and observations in an heated room. *Phil. Trans. Roy. Soc.*, **65**, 484–94.

55. BLAXTER, K. L. (1964). The effect of outdoor climate in Scotland on sheep and cattle. *Vet. Rec.*, **76**, 1445–55.

56. BLAXTER, K. L. (1965a). *Energy Metabolism.* Academic Press, London.

57. BLAXTER, K. L. (1965b). Animal production in adverse environments. *J. Univ. Newcastle-upon-Tyne Agric. Soc.*, **19**, 10–15.

58. BLAXTER, K. L. (1967). *The Energy Metabolism of Ruminants*, 2nd ed. Charles C. Thomas, Springfield, Illinois.

59. BLAXTER, K. L. (1971). Deer farming. *Scottish Agriculture*, **51**, 225–30.

60. BLAXTER, K. L. (1977). Environmental factors and their influence on the nutrition of farm livestock. In *Nutrition and the Climatic Environment*, eds. W. Haresign, H. Swan and D. Lewis, pp. 1–16. Butterworths, London.

61. BLAXTER, K. L., BROCKWAY, J. M. and BOYNE, A. W. (1972). A new method for estimating the heat production of animals. *Q. Jl exp. Physiol.*, **57**, 60–72.

62. BLAXTER, K. L., CLAPPERTON, J. L. and WAINMAN, F. W. (1966). The extent of differences between six British breeds of sheep in their metabolism, feed intake and utilization, and resistance to climatic stress. *Br. J. Nutr.*, **20**, 283–94.

63. BLAXTER, K. L., GRAHAM, N. MCC., WAINMAN, F. W. and ARMSTRONG, D. G. (1959). Environmental temperature, energy metabolism and heat regulation in sheep. II. The partition of heat losses in closely clipped sheep. *J. agric. Sci., Camb.*, **52**, 25–40.

64. BLAXTER, K. L. and WAINMAN, F. W. (1961). Environmental temperature and the energy metabolism and heat emission of steers. *J. agric. Sci., Camb.*, **56**, 81–90.

65. BLAXTER, K. L. and WAINMAN, F. W. (1964). The effect of increased air movement on the heat production and emission of steers. *J. agric. Sci., Camb.*, **62**, 207–14.

66. BLAXTER, K. L. and WAINMAN, F. W. (1966). The fasting metabolism of cattle. *Br. J. Nutr.*, **20**, 103–11.

67. BLIGH, J. (1961). The synchronous discharge of apocrine glands of the Welsh Mountain sheep. *Nature, Lond.*, **189**, 582–3.

68. BLIGH, J. (1966). The thermosensitivity of the hypothalamus and thermoregulation in mammals. *Biol. Rev.*, **41**, 317–67.

69. BLIGH, J. (1973). *Temperature Regulation in Mammals and Other Vertebrates.* North Holland, Amsterdam and London.

70. BLIGH, J. (1976). Temperature regulation in cattle. In *Progress in Animal Biometeorology*, Vol. I, Part I, ed. H. D. Johnson, pp. 74–80. Swets and Zeitlinger, Amsterdam.

70a. BLIGH, J. and ROBINSON, S. G. (1965). Radiotelemetry in a veterinary research project. *Med. biol. Illust.*, **15**, 94–9.

71. BOND, J., WINCHESTER, C. F., CAMPBELL, L. E. and WEBB, J. C. (1963). Effects of loud sounds on the physiology and behavior of swine. *U.S. Dept. Agric., Tech. Bull. 1280*. Washington, D.C.

72. BOND, T. E., KELLY, C. F. and HEITMAN, JR., H. (1952). Heat and moisture loss from swine. *Agr. Engrg, St. Joseph, Mich.*, **33**, 148–52.

73. BOND, T. E., KELLY, C. F., MORRISON, S. R. and PEREIRA, N. (1967). Solar, atmospheric and terrestrial radiation received by shaded and unshaded animals. *Trans. Am. Soc. agric. Engrs.*, **10**, 622–25, 627.

74. BRAY, D. J. and GESELL, J. A. (1961). Studies with corn-soya laying diets. 4. Environmental temperature—a factor affecting performance of pullets fed diets suboptimal in protein. *Poultry Science*, **40**, 1328–35.

75. BROCKWAY, J. M., MCDONALD, J. D. and PULLAR, J. D. (1965). Evaporative heat-loss mechanisms in sheep. *J. Physiol., Lond.*, **179**, 554–68.

76. BROCKWAY, J. M. and MALOIY, G. M. O. (1968). Energy metabolism of the red deer. *J. Physiol., Lond.*, **194**, 22–23P.

77. BRODY, S. (1945). *Bioenergetics and Growth.* Reinhold, New York.

77a. BROOK, A. H. and SHORT, B. F. (1960). Sweating in sheep. *Aust. J. agric. Res.*, **11**, 557–69.

78. BROUWER, E. (1965). Report of the sub-committee on constants and factors. In *Energy metabolism*, ed. K. L. Blaxter, pp. 441–3. Academic Press, London.

79. BROWN, G. D. and HUTCHINSON, J. C. D. (1973). Climate and animal production. In *The Pastoral Industries of Australia*, eds. G. Alexander and O. B. Williams, pp. 336–70. Sydney University Press, Australia.

80. BRUCE, J. M. (1977). Farm Buildings Research and Development Studies, *Scottish Farm Buildings Investigation Unit*, **8**, 9.
81. BRÜCK, K. (1961). Temperature regulation in the newborn infant. *Biologia Neonat.*, **3**, 65–119.
82. BRÜCK, K. (1970). Heat production and temperature regulation. In *Physiology and the Perinatal Period.* Parts I and II, ed. U. Stave, pp. 493–557. Appleton Century Crofts.
83. BRÜCK, K., BAUM, E. and SCHWENNIKE, H. P. (1976). Cold-adaptive modifications in man induced by repeated short-term cold-exposures and during a 10-day and -night cold-exposure. *Pflügers Arch.*, **363**, 125–33.
84. BUDYKO, M. I. (1974). *Climate and Life*, ed. D. H. Miller. Academic Press, New York and London.
85. BURTON, A. C. and EDHOLM, O. G. (1955). *Man in a Cold Environment.* Edward Arnold, London; reprinted 1969, Hafner Publishing Company, New York and London.

86. CABANAC, M. (1969). Plaisir ou Déplaisir de la Sensation Thermique et Homeothermie. *Physiol. Behav.*, **4**, 359–64.
87. CAIRNIE, A. B. and PULLAR, J. D. (1959). An investigation into the efficient use of time in the calorimetric measurement of heat output. *Br. J. Nutr.*, **13**, 431–9.
88. CALDER, W. A. and KING, J. R. (1974). Thermal and caloric relations of birds. In *Avian Biology*, Vol. IV, eds. D. S. Farner and J. R. King, pp. 259–413. Academic Press, London.
89. CALDER, W. A. and SCHMIDT–NIELSEN, K. (1966). Evaporative cooling and respiratory alkalosis in the pigeon. *Proc. Nat. Acad. Sci.*, **55**, 750–6.
90. CANNON, P. and KEATINGE, W. R. (1960). The metabolic rate and heat loss of fat and thin men in cold and warm water. *J. Physiol., Lond.*, **154**, 329–44.
91. CAPSTICK, J. W. (1921). A calorimeter for use with large animals. *J. agric. Sci., Camb.*, **11**, 408–31.
92. CAPSTICK, J. W. and WOOD, T. B. (1922). The effect of change of temperature on the basal metabolism of swine. *J. agric. Sci., Camb.*, **12**, 257–68.
93. CARPENTER, G. A. (1974). Ventilation of buildings for intensively housed livestock. In *Heat Loss from Animals and Man: Assessment and Control*, eds. J. L. Monteith and L. E. Mount, pp. 389–403. Butterworths, London.
94. CARPENTER, G. A. and RANDALL, J. M. (1975). The interpretation of daily-temperature records to optimise the insulation of intensive livestock buildings. *Agricultural Meteorology*, **15**, 245–55.
95. CENA, K. (1974). Radiative heat loss from animals and man. In *Heat Loss from Animals and Man: Assessment and Control*, eds. J. L. Monteith and L. E. Mount, pp. 33–58. Butterworths, London.
96. CENA, K. and CLARK, J. A. (1973). Thermal radiation from animal coats: coat structure and measurements of radiative temperature. *Phys. Med. Biol.*, **18**, 432–43.
97. CENA, K. and MONTEITH, J. L. (1975a). Transfer processes in animal coats. I. Radiative transfer. *Proc. R. Soc. Lond. B.*, **188**, 377–93.
98. CENA, K. and MONTEITH, J. L. (1975b). Transfer processes in animal coats. II. Conduction and convection. *Proc. R. Soc. Lond. B.*, **188**, 395–411.
99. CENA, K. and MONTEITH, J. L. (1975c). Transfer processes in animal coats. III. Water vapour diffusion. *Proc. R. Soc. Lond. B.*, **188**, 413–23.
100. CENA, K. and MONTEITH, J. L. (1976). Heat transfer through animal coats. In *Progress in Animal Biometeorology*, Vol. I, Part I, ed. H. D. Johnson, pp. 343–51. Swets and Zeitlinger, Amsterdam.
101. CHAFFEE, R. R. J. and ROBERTS, J. C. (1971). Temperature acclimation in birds and mammals. *Ann. Rev. Physiol.*, **33**, 155–202.

102. CHAMBERS, A. B. (1970). A psychrometric chart for physiological research. *J. appl. Physiol.*, **29**, 406–12.

103. CHARLES, D. R. and SPENCER, P. G. (1976). The climatic environment of poultry houses. *Ministry of Agriculture, Fisheries and Food Bulletin 212*. Her Majesty's Stationery Office.

104. CHOSHNIAK, I. and SHKOLNIK, A. (1977). Rapid rehydration in the Black Bedouin goats: red blood cells fragility and role of the rumen. *Comp. Biochem. Physiol.*, **56 A**, 581–3.

105. CLARK, J. A., CENA, K. and MONTEITH, J. L. (1973). Measurements of the local heat balance of animal coats and human clothing. *J. appl. Physiol.*, **35**, 751–4.

106. CLARK, R. P. and COX, R. N. (1974). An application of aeronautical techniques to physiology. 2. Particle transport within the human microenvironment. *Med. Biol. Engrg, May 1974*, 275–9.

107. CLARK, R. P., GOFF, M. R. and MULLAN, B. J. (1977). Heat-loss studies beneath hovering helicopters. *J. Physiol., Lond.*, **267**, 6–7P.

108. CLARK, R. P., MULLAN, B. J. and PUGH, L. G. C. E. (1977). Skin temperature during running—a study using infra-red colour thermography. *J. Physiol., Lond.*, **267**, 53–62.

109. CLARK, R. P. and TOY, N. (1975a). Natural convection around the human head. *J. Physiol., Lond.*, **244**, 283–93.

110. CLARK, R. P. and TOY, N. (1975b). Forced convection around the human head. *J. Physiol., Lond.*, **244**, 295–302.

111. CLOSE, W. H. (1978). The effects of plane of nutrition and environmental temperature on the energy metabolism of the growing pig. 3. The efficiency of energy utilization for maintenance and growth. *Br. J. Nutr.*, **40**, 433–38.

112. CLOSE, W. H. and MOUNT, L. E. (1975). The rate of heat loss during fasting in the growing pig. *Br. J. Nutr.*, **34**, 279–90.

113. CLOSE, W. H. and MOUNT, L. E. (1978). The effects of plane of nutrition and environmental temperature on the energy metabolism of the growing pig. 1. Heat loss and critical temperature. *Br. J. Nutr.*, **40**, 413–21.

114. CLOSE, W. H., MOUNT, L. E. and BROWN, D. (1978). The effects of plane of nutrition and environmental temperature on the energy metabolism of the growing pig. 2. Growth rate, including protein and fat deposition. *Br. J. Nutr.*, **40**, 423–31.

115. CLOUDSLEY-THOMPSON, J. L. (1970). Terrestrial invertebrates. In *Comparative Physiology of Thermoregulation*, ed. G. C. Whittow, pp. 15–70. Academic Press, New York and London.

116. COLIN, J. and HOUDAS, Y. (1967). Experimental determination of coefficient of heat exchanges by convection of human body. *J. appl. Physiol.*, **22**, 31–8.

117. COLIN, J., TIMBAL, J., GUIEU, J., BOUTELIER, C. and HOUDAS, Y. (1970). Combined effect of radiation and convection. In *Physiological and Behavioral Temperature Regulation*, eds. J. D. Hardy, A. P. Gagge and J. A. J. Stolwijk, pp. 81–96. Charles C. Thomas, Springfield, Illinois.

118. COSSINS, A. R. (1977). Adaptation of biological membranes to temperature. The effect of temperature acclimation of goldfish upon the viscosity of synaptosomal membranes, *Biochim. biophy. Acta*, **470**, 395–411.

119. CRANDALL, L. S. (1964). *The Management of Wild Mammals in Captivity*. University of Chicago Press.

120. CRAWFORD, A. (1779). *Experiments and observations on animal heat and the inflammation of combustible bodies, being an attempt to resolve these phenomena into a general law of nature*. Murray and Sewell, London.

121. CRAWFORD, A. (1788). *Experiments and observations on animal heat and the inflammation of combustible bodies*. 2nd edition. Johnson, London.

122. CRESSWELL, E. and THOMSON, W. (1964). An introductory study of air flow and temperature at grazing-sheep heights. *Emp. J. exp. Agr.*, **32**, 131–5.

123. CURTIS, S. E. (1970). Environmental-thermoregulatory interactions and neonatal piglet survival. *J. Anim. Sci.*, **31**, 576–87.

124. CURTIS, S. E., CHRISTISON, G. I. and ROBERTSON, W. D. (1970). Effects of acute cold exposure and age on respiratory quotients in piglets. *Proc. Soc. Exp. Biol. N.Y.*, **134**, 188–91.

125. CURTIS, S. E., HEIDENREICH, C. J. and HARRINGTON, R. B. (1967). Age dependent changes of thermostability in neonatal pigs. *Am. J. vet. Res.*, **28**, 1887–90.

126. CURTIS, S. E., HEIDENREICH, C. J. and MARTIN, T. G. (1967). Relationship between body weight and chemical composition of pigs at birth. *J. Anim. Sci.*, **26**, 749–51.

127. DAUNCEY, M. J. and JAMES, W. P. T. (1979). Assessment of the heart-rate method for determining energy expenditure in man, using a whole-body calorimeter. *Br. J. Nutr.*, in press.

128. DAUNCEY, M. J., MURGATROYD, P. R. and COLE, T. J. (1978). A human calorimeter for the direct and indirect measurement of 24 h energy expenditure. *Br. J. Nutr.*, **39**, 557–66.

129. DAVIDSON, H. R. (1954). *The Production and Marketing of Pigs.* Longmans, Green, London.

130. DAVIS, R. H., HASSAN, O. E. M. and SYKES, A. H. (1973). Energy utilization in the laying hen in relation to ambient temperature. *J. agric. Sci., Camb.*, **81**, 173–7.

131. DAWES, G. S., JACOBSON, H. N., MOTT, J. C. and SHELLEY, H. J. (1960). Some observations on foetal and new-born rhesus monkeys. *J. Physiol., Lond.*, **152**, 271–98.

132. DAWSON, T. J. (1972). Thermoregulation in Australian desert kangaroos. In *Comparative Physiology of Desert Animals*, ed. G. M. O. Maloiy, pp. 133–46. Zoological Society of London, Academic Press.

133. DAWSON, T. J. (1977). Kangaroos. *Scient. Am.*, **237**, 78–89.

134. DAWSON, W. R. (1973). 'Primitive' mammals. In *Comparative Physiology of Thermoregulation*, Vol. III, ed. G. Causey Whittow, pp. 1–46. Academic Press, New York.

135. DAWSON, W. R. (1975). On the physiological significance of the preferred body temperature of reptiles. In *Perspectives of Biophysical Ecology*, eds. D. M. Gates and R. B. Schmerl, pp. 443–73. Springer-Verlag, New York.

136. DAWSON, W. R. and HUDSON, J. W. (1970). Birds. In *Comparative Physiology of Thermoregulation*, Vol. I, Invertebrates and nonmammalian vertebrates, ed. G. Causey Whittow, pp. 223–310. Academic Press, New York.

137. DAWSON, W. R. and SCHMIDT-NIELSEN, K. (1964). Terrestrial animals in dry heat: desert birds. In *Adaptation to the Environment*, ed. D. B. Dill, pp. 481–92. American Physiological Society, Washington, D.C.

138. DAY, R. and HARDY. J. D. (1942). Repiratory metabolism in infancy and in childhood. 26. A calorimeter for measuring the heat loss of premature infants. *Am. J. dis. Child.*, **63**, 1086–95.

139. DEIGHTON, T. (1937). A study of the fasting metabolism of various breeds of hog. III. Metabolism and surface area measurements. *J. agric. Sci., Camb.*, **27**, 317–31.

140. DE WITT, C. B. (1971). Postural mechanisms in the behavioural thermoregulation of a desert lizard, *Dipsosaurus dorsalis*. *J. Physiol. (Paris)*, **63**, 242–5.

141. DILL, D. B. (1938). *Life, Heat and Altitude.* Harvard University Press.

142. DILL, D. B., BOCK, A. V. and EDWARDS, H. T. (1933). Mechanisms for dissipating heat in man and dog. *Am. J. Physiol.*, **104**, 36–43.

143. DONNELLY, J. B., LYNCH, J. J. and WEBSTER, M. E. D. (1974). Climatic adaptation in recently shorn Merino sheep. *Int. J. Biometeor.*, **18**, 233–47.

144. DRENT, R. (1975). Incubation. In *Avian Biology*, Vol. V, eds. D. S. Farner and J. R. King, pp. 333–420. Academic Press, New York and London.

145. DURRER, J. L. and HANNON, J. P. (1962). Seasonal variations in caloric intake of dogs living in an arctic environment. *Am. J. Physiol.*, **202**, 375–8.

146. EDE, A. J. (1967). *An Introduction to Heat Transfer*. Pergamon Press, Oxford.

147. EDHOLM, O. G. (1977). Energy balance in man. *J. Human Nutr.*, **31**, 413–31.

148. EDHOLM, O. G. and GUNDERSON, E. K. E. (eds.) (1973). *Polar Human Biology*. William Heinemann Medical Books Ltd., London.

149. EDHOLM, O. G. and LEWIS, H. E. (1964). Terrestrial animals in cold: man in polar regions. In *Adaptation to the Environment*, ed. D. B. Dill, pp. 435–46. American Physiological Society.

150. ELSNER, R. W. (1963). Comparison of Australian Aborigines, Alacaluf Indians and Andean Indians. *Federation Proceedings*, **22**, 840–2.

151. EMMANS, G. C. and CHARLES, D. R. (1977). Climatic environment and poultry feeding in practice. In *Nutrition and the Climatic Environment*, eds. W. Haresign, H. Swan and D. Lewis, pp. 31–49. Butterworths, London.

152. FANGER, P. O. (1970). Conditions for thermal comfort: introduction of a general comfort equation. In *Physiological and Behavioral Temperature Regulation*, eds. J. D. Hardy, A. P. Gagge and J. A. J. Stolwijk, pp. 152–76. Charles C. Thomas, Springfield, Illinois.

153. FINCH, V. A. (1972). Energy exchanges with the environment of two East African antelopes, the eland and the hartebeest. In *Comparative Physiology of Desert Animals*, ed. G. M. O. Maloiy, pp. 315–26. Zoological Society of London, Academic Press.

154. FINDLAY, J. D. (1950). The effects of temperature, humidity, air movement and solar radiation on the behaviour and physiology of cattle and other farm animals. *Hannah Dairy Res. Inst. Bull.*, **9**.

155. FINDLAY, J. D. (1960). Reactions of cattle to heat and cold. *Agric. Progress*, **34**, 74–7.

156. FOLK, G. E. (1974). *Textbook of Environmental Physiology*. Lea and Febiger, Philadelphia.

157. FOOD AND AGRICULTURE ORGANIZATION, UNITED NATIONS (1975). *Production Yearbook, Vol. 29*. Rome.

158. FOSTER, K. G., HEY, E. N. and KATZ, G. (1969). The response of the sweat glands of the new-born baby to thermal stimuli and to intradermal acetylcholine. *J. Physiol., Lond.*, **203**, 13–29.

159. FOX, R. H. (1974). Heat acclimatization and the sweating response. In *Heat Loss from Animals and Man: Assessment and Control*, eds. J. L. Monteith and L. E. Mount, pp. 277–301. Butterworths, London.

160. FOX, R. H., WOODWARD, P. M., EXTON-SMITH, A. N., GREEN, M. F., DONNISON, D. V. and WICKS, M. H. (1973). Body temperatures in the elderly; a national study of physiological, social, and environmental conditions. *Br. med. J.*, **1**, 200–6.

161. FRANCIS, J. (1965). Definition and use of zebu, Brahman or *Bos indicus* cattle. *Nature, Lond.*, **207**, 13–16.

162. FRASER, A. F. (1968). *Reproductive Behaviour in Ungulates*, p. 1. Academic Press, London.

163. FREEMAN, B. M. (1966). The effects of cold, noradrenaline and adrenaline upon the oxygen consumption and carbohydrate metabolism of the young fowl (*Gallus domesticus*). *Comp. Biochem. Physiol.*, **18**, 369–82.

164. FREEMAN, B. M. (1970). Thermoregulatory mechanisms of the neonate fowl. *Comp. Biochem. Physiol.*, **33**, 219–30.

165. FREEMAN, B. M. (1976). Thermoregulation in the young fowl (*Gallus domesticus*). *Comp. Biochem. Physiol.*, **54A**, 141–4.

166. FRISCH, J. E. and VERCOE, J. E. (1976). Maintenance requirement, fasting metabolism and body composition in different cattle breeds. In *Energy Metabolism of Farm Animals*, ed. M. Vermorel, pp. 209–12. I.N.R.A., Clermont-Ferrand, France.

167. FRY, F. E. J. and HOCHACHKA, P. W. (1970). Fish. In *Comparative Physiology of Thermoregulation*, Vol. I, Invertebrates and nonmammalian vertebrates, ed. G. Causey Whittow, pp. 79–134. Academic Press, New York.

168. FULLER, M. F. (1965). The effect of environmental temperature on the nitrogen metabolism and growth of the young pig. *Br. J. Nutr.*, **19**, 531–46.

169. FULLER, M. F., DUNCAN, W. R. H. and BOYNE, A. W. (1974). Effect of environmental temperature on the degree of unsaturation of depot fats of pigs given different amounts of food. *J. Sci. Fd Agric.*, **25**, 205–10.

170. FUNK, J. P. (1959). Improved polythene-shielded net radiometer. *J. Sci. Instrum.*, **36**, 267–70.

171. GAGGE, A. P. (1936). The linearity criterion as applied to partitional calorimetry. *Am. J. Physiol.*, **116**, 656–68.

172. GAGGE, A. P. (1940). Standard operative temperature, a generalized temperature scale, applicable to direct and partitional calorimetry. *Am. J. Physiol.*, **131**, 93–103.

173. GAGGE, A. P. (1965). Operative temperature, a physical measure of thermal comfort and thermal equilibrium. *Internat. Physiol. Congr.* Tokyo.

174. GAGGE, A. P. (1970). Effective radiant flux, an independent variable that describes thermal radiation on man. In *Physiological and Behavioral Temperature Regulation*, eds. J. D. Hardy, A. P. Gagge and J. A. J. Stolwijk, pp. 34–45. Charles C. Thomas, Springfield, Illinois.

175. GAGGE, A. P., HERRINGTON, L. P. and WINSLOW, C.-E. A. (1937). Thermal interchanges between the human body and its atmospheric environment. *Am. J. Hyg.*, **26**, 84–102.

176. GAGGE, A. P., STOLWIJK, J. A. J., and HARDY, J. D. (1965). A novel approach to measurement of man's heat exchange with a complex radiant environment. *Aerospace Medicine*, **36**, 432–5.

177. GAGGE, A. P., WINSLOW, C.-E. A. and HERRINGTON, L. P. (1938). The influence of clothing on the physiological reactions of the human body to varying environmental temperatures. *Am. J. Physiol.*, **124**, 30–50.

178. GATENBY, R. M. (1977). Conduction of heat from sheep to ground. *Agricultural Meteorology*, **18**, 387–400.

179. GATES, D. M. (1962). *Energy Exchange in the Biosphere*. Harper and Ross, New York.

180. GATES, D. M. (1968). Physical environment. In *Adaptation of Domestic Animals*, ed. E. S. E. Hafez, pp. 46–60. Lea and Febiger, Philadelphia.

181. GELINEO, S. (1954). Le développement ontogénique de la thermoregulation chez le Chien. *C. r. Seanc. Soc. Biol.*, **148**, 1483–5.

182. GHOBRIAL, L. I. (1970). The water relations of the desert antelope *Gazella dorcas dorcas*. *Physiol. Zoöl.*, **43**, 249–56.

183. GOLDSMITH, R. (1974). Acclimatization to cold in man—fact or fiction? In *Heat Loss from Animals and Man: Assessment and Control*, eds. J. L. Monteith and L. E. Mount, pp. 311–19. Butterworths, London.

184. GONZALES-JIMENEZ, E. and BLAXTER, K. L. (1962). The metabolism and thermal regulation of calves in the first month of life. *Br. J. Nutr.*, **16**, 199–212.

185. GRAHAM, N. MCC. (1964). Influence of ambient temperature on the heat production of pregnant ewes. *Aust. J. agric. Res.*, **15**, 982–8.

186. GRAHAM, N. MCC., WAINMAN, F. W., BLAXTER, K. L. and ARMSTRONG, D. G. (1959). Environmental temperature, energy metabolism and heat regulation in sheep. I. Energy metabolism in closely clipped sheep. *J. agric. Sci., Camb.*, **52**, 13–24.

187. GRIDGEMAN, N. T. and HÉROUX, O. (1965). Relation of oxygen uptake to body weight of rats at different environmental temperatures. *Can. J. Physiol. Pharmacol.*, **43**, 351–7.

188. HAFEZ, E. S. E. (1964). Behavioral thermoregulation in mammals and birds (a review). *Int. J. Biometeorol.*, **7**, 231–45.

189. HAFEZ, E. S. E. (1968a). Behavioral adaptation. In *Adaptation of Domestic Animals*, ed. E. S. E. Hafez, pp. 202–14. Lea and Febiger, Philadelphia.

190. HAFEZ, E. S. E. (1968b). Environmental effects on animal productivity. In *Adaptation of Domestic Animals*, ed. E. S. E. Hafez, pp. 74–93. Lea and Febiger, Philadelphia.

191. HAHN, P. (1956). Effect of environmental temperatures on the development of thermoregulatory mechanisms in infant rats. *Nature, Lond.*, **178**, 96–7.

192. HALDANE, J. S. (1889). *The Methods of Investigating Quantitatively the Heat Production and the Respiratory Exchange of Material in Animals.* Thesis, University of Edinburgh.

193. HALDANE, J. S. (1905). The influence of high air temperatures. *J. Hyg. Camb.*, **5**, 494–513.

194. HALES, J. R. S. (1967). The partition of respiratory ventilation of the panting ox. *J. Physiol., Lond.*, **188**, 45–8P.

195. HALES, J. R. S. (1976). Interaction between respiratory and thermoregulatory systems of domestic animals in hot environments. In *Progress in Animal Biometeorology*, Vol. I, Part I, ed. H. D. Johnson, pp. 123–31. Swets and Zeitlinger, Amsterdam.

196. HALES, J. R. S. and WEBSTER, M. E. D. (1967). Respiratory function during thermal tachypnoea in sheep. *J. Physiol., Lond.*, **190**, 241–60.

197. HAMMEL, H. T. (1964). Terrestrial animals in cold: recent studies of primitive man. In *Adaptation to the Environment*, ed. D. B. Dill, pp. 413–34. American Physiological Society.

198. HAMMEL, H. T. and HARDY, J. D. (1963). A gradient type of calorimeter for measurement of thermoregulatory responses in the dog. In *Temperature, its Measurement and Control in Science and Industry.* Vol. 3, Part 3, ed. J. D. Hardy, pp. 31–42. Reinhold, New York.

199. HAMILTON, W. J. (1974). *Farming the Red Deer.* Her Majesty's Stationery Office, Edinburgh.

199a. HANAWALT, V. M. and SAMPSON, J. (1947). Studies on baby pig mortality. V. Relationship between age and time of onset of acute hypoglycaemia in fasting new-born pigs. *Am. J. vet. Res.*, **8**, 235–43.

200. HARDY, J. D. (1961). Physiology of temperature regulation. *Physiol. Rev.*, **41**, 521–606.

201. HARDY, J. D., STOLWIJK, J. A. J. and GAGGE, A. P. (1971). Man. In *Comparative Physiology of Thermoregulation*, Vol. II, Mammals, ed. G. Causey Whittow, pp. 328–80. Academic Press, New York.

202. HARLAN, J. R. (1976). The plants and animals that nourish man. *Scient. Amer.*, **235**, 89–97.

203. HARRISON, G. A. (1963). Temperature adaptation as evidenced by growth of mice. *Federation Proceedings*, **22**, 691–8.

204. HART, J. S. (1962). Seasonal acclimatization in four species of small wild birds. *Physiol. Zoöl.*, **13**, 224–36.

205. HART, J. S. (1963). Physiological responses to cold in non-hibernating homeotherms. In *Temperature, its Measurement and Control in Science and Industry*, Vol. 3, Part 3, ed. J. D. Hardy, pp. 373–406. Reinhold, New York.

206. HART, J. S. (1964). Geography and season: mammals and birds. In *Adaptation to the Environment*, ed. D. B. Dill, pp. 295–322. American Physiological Society.

207. HART, J. S. (1971). Rodents. In *Comparative Physiology of Thermoregulation*, Vol. II, Mammals, ed. G. Causey Whittow, pp. 1–149. Academic Press, New York.

208. HART, J. S., HEROUX, O., COTTLE, W. H. and MILLS, C. A. (1961). The influence of climate on metabolic and thermal responses of infant caribou. *Can. J. Zool.*, **39**, 845–56.

209. HART, J. S. and ROY, O. Z. (1966). Respiratory and cardiac responses to flight in pigeons. *Physiol. Zool.*, **39**, 291–306.

210. HART, J. S. and ROY, O. Z. (1967). Temperature regulation during flight in pigeons. *Am. J. Physiol.*, **213**, 1311–6.

211. HAYWARD, J. S. (1965). Microclimate temperature and its adaptive significance in six geographic races of Peromyscus. *Can. J. Zool.*, **43**, 341–50.

212. HEATH, J. E. (1970). Behavioral regulation of body temperature in poikilotherms. *Physiologist*, **13**, 399–410.

213. HEINRICH, B. (1976). Heat exchange in relation to blood flow between thorax and abdomen in bumblebees. *J. exp. Biol.*, **64**, 561–85.

214. HEITMAN, H. and HUGHES, E. H. (1949). The effects of air temperature and relative humidity on the physiological well-being of swine. *J. Anim. Sci.*, **8**, 171–81.

214a. HEITMAN, H., HAHN, L., BOND, T. E. and KELLY, C. F. (1962). The effects of modified summer environment on behaviour. *Anim. Behaviour*, **10**, 15–9.

215. HELDMAIER, G. (1971). Zitterfreie Wärmebildung und Körpergrösse bei Säugetieren. *Z. vergl. Physiologie*, **73**, 222–48.

216. HELLON, R. (1970) Hypothalmic neurons responding to changes in hypothalmic and ambient temperature. In *Physiological and Behavioral Temperature Regulation*, eds. J. D. Hardy, A. P. Gagge and J. A. J. Stolwijk, pp. 463–71. Charles C. Thomas, Springfield, Illinois.

217. HEMMINGSEN, A. M. (1960). Energy metabolism as related to body size and respiratory surfaces, and its evolution. *Rep. Steno Hosp., Copenhagen*, **9**, part 2, 7–110.

218. HENSEL, H. (1968). Adaptation to cold. In *Adaptation of Domestic Animals*, ed. E. S. E. Hafez, pp. 183–93. Lea and Febiger, Philadelphia.

219. HENSEL, H. (1973). Neural processes in thermoregulation. *Physiol. Rev.*, **53**, 948–1017.

220. HENSEL, H., BRÜCK, K. and RATHS, P. (1973). Homeothermic organisms. In *Temperature and Life*, by H. Precht, J. Christopherson, H. Hensel and W. Larcher, pp. 503–761. Springer-Verlag.

221. HEROUX, O. (1963). Patterns of morphological, physiological and endocrinological adjustments under different environmental conditions of cold. *Federation Proceedings*, **22**, 789–94.

222. HEROUX. O. (1969). Catecholamines, corticosteroids and thryoid hormones in non-shivering thermogenesis under different environmental conditions. In *Physiology and Pathology of Adaptation Mechanisms*, ed. E. Bajusz. Pergamon Press, Oxford.

222a. HERTZMAN, A. B., RANDALL, W. C., PEISS, C. N. and SECKENDORF, R. J. (1952). Regional rates of evaporation from the skin at various environmental temperatures. *J. appl. Physiol.*, **5**, 153–6.

223. HEY, E. N. (1968). Small globe thermometers. *J. Scient. Instrum. Series 2, Vol. 1*, pp. 955–7, 1260.

224. HEY, E. N. (1974). Physiological control over body temperature. In *Heat Loss from Animals and Man: Assessment and C ontrol*, eds. J. L. Monteith and L. E. Mount, pp. 77–95. Butterwoths, London.

225. HEY, E. N. and KATZ, G. (1969). Evaporative water loss in the new-born baby. *J. Physiol., Lond.*, **200**, 605–19.

226. HEY, E. N. and KATZ, G. (1970). The range of thermal insulation in the tissues of the new-born baby. *J. Physiol., Lond.*, **207**, 667–81.

227. HEY, E. N., KATZ, G. and O'CONNELL, B. (1970). The total thermal insulation of the new-born baby. *J. Physiol., Lond.*, **207**, 683–98.

228. HEY, E. N. and MOUNT, L. E. (1966). Temperature control in incubators. *Lancet*, **ii**, 202–3.

229. HEY, E. N. and MOUNT, L. E. (1967). Heat losses from babies in incubators. *Arch. dis. Childh.*, **42**, 75–84.

230. HEY, E. N. and O'CONNELL, B. (1970). Oxygen consumption and heat balance in the cot-nursed baby. *Arch. dis. Childh.*, **45**, 335–43.

231. HICKS, C. S. (1964). Terrestrial animals in cold: exploratory studies of primitive man. In *Adaptation to the Environment*, ed. D. B. Dill, pp. 405–12. American Physiological Society.

232. HILDEBRANDT, G. (1967). The time factor in adaptation. In *Biometeorology*, Vol. 2, Part 1, eds. S. W. Tromp and W. H. Weihe, pp. 258–75. Pergamon Press, Oxford.

233. HILL, A. V. and HILL, A. M. (1914). A self-recording calorimeter for large animals. *J. Physiol., Lond.*, **48**, xiii.

234. HILL, JUNE R. and RAHIMTULLA, K. A. (1965). Heat balance and the metabolic rate of new-born babies in relation to environmental temperature; and the effect of age and of weight on basal metabolic rate. *J. Physiol., Lond.*, **180**, 239–65.

235. HOFFMAN, R. A. (1964). Terrestrial animals in cold: hibernators. In *Adaptation to the Environment*, D. B. Dill, pp. 379–403. American Physiological Society.

236. HOFMEYR, H. S., GUIDRY, A. J. and WALTZ, F. A. (1969). Effects of temperature and wool length on surface and respiratory evaporative losses of sheep. *J. appl. Physiol.*, **26**, 517–23.

237. HOLMES, C. W. (1971). Local cooling of the mammary gland and milk production in the cow. *J. Dairy Res.*, **38**, 3–7.

238. HOLMES, C. W. and CLOSE, W. H. (1977). The influence of climatic variables on energy metabolism and associated aspects of productivity in the pig. In *Nutrition and the Climatic Environment*, eds. W. Haresign, H. Swan and D. Lewis, pp. 51–73. Butterworths, London.

239. HOLMES, C. W. and MOUNT, L. E. (1967). Heat loss from groups of growing pigs under various conditions of environmental temperature and air movement. *Anim. Prod.*, **9**, 435–52.

240. HOLMES, C. W., STEPHENS, D. B. and TONER, J. N. (1976). Heart rate as a possible indicator of the energy metabolism of calves kept out-of-doors. *Livestock Prod. Sci.*, **3**, 333–41.

241. HONG, S. K. (1963). Comparison of diving and nondiving women of Korea. *Federation Proceedings*, **22**, 831–3.

242. HONIG, W. K. (1966). *Operant behavior*. Appleton-Century-Crofts, New York.

243. HUDSON, J. W. (1973). Torpidity in mammals. In *Comparative Physiology of Thermoregulation*, Vol. III, Special aspects of thermoregulation, ed. G. Causey Whittow, pp. 98–165. Academic Press, New York.

244. HULL, D. (1966). The structure and function of brown adipose tissue. *Br. med. Bull.*, **22**, 92–6.

245. HULL, D. (1973). Thermoregulation in young mammals. In *Comparative Physiology of Thermoregulation*, Vol. III, Special aspects of thermoregulation, ed. G. Causey Whittow, pp. 167–200. Academic Press, New York.

246. HUTCHINSON, J. C. D. and BROWN, G. D. (1969). Penetrance of cattle coats by radiation. *J. appl. Physiol.*, **26**, 454–64.

247. IGGO, A. (1969). Cutaneous thermoreceptors in primates and sub-primates. *J. Physiol., Lond.*, **200**, 403–30.

248. INGRAM, D. L. (1964). The effect of environmental temperature on heat loss and thermal insulation in the young pig. *Res. vet. Sci.*, **5**, 357–64.

249. INGRAM, D. L. (1965a). The effect of humidity on temperature regulation and cutaneous water loss in the young pig. *Res. vet. Sci.*, **6**, 9–17.

250. INGRAM, D. L. (1965b). Evaporative cooling in the pig. *Nature, Lond.*, **207**, 415–16.

251. INGRAM, D. L. (1974). Heat loss and its control in pigs. In *Heat Loss from Animals and Man: Assessment and Control*, eds. J. L. Monteith and L. E. Mount, pp. 233–54. Butterworths, London.

252. INGRAM, D. L. (1976a). Evaporative temperature regulation in pigs. In *Progress in Animal Biometeorology*, Vol. I, Part I, ed. H. D. Johnson, pp. 148–57. Swets and Zeitlinger, Amsterdam.

253. INGRAM, D. L. (1976b). Adaptability and thermoregulatory behaviour of pigs. In *Progress in Animal Biometeorology*, Vol. I, Part I, ed. H. D. Johnson, pp. 442–51. Swets and Zeitlinger, Amsterdam.

254. INGRAM, D. L. (1977). Adaptations to ambient temperature in growing pigs. *Pflügers Arch.*, **367**, 257–64.

255. INGRAM, D. L. and KACIUBA-USCILKO, H. (1977). The influence of food intake and ambient temperature on the rate of thyroxine utilization. *J. Physiol., Lond.*, **270**, 431–8.

256. INGRAM, D. L. and LEGGE, K. F. (1970). The thermoregulatory behaviour of young pigs in a natural environment. *Physiol. Behav.*, **5**, 981–7.

257. INGRAM, D. L. and LEGGE, K. F. (1972). The influence of deep body and skin temperature on thermoregulatory responses to heating of the scrotum in pigs. *J. Physiol., Lond.*, **224**, 477–87.

258. INGRAM, D. L. and MOUNT, L. E. (1975a). *Man and Animals in Hot Environments.* Springer-Verlag, New York.

259. INGRAM, D. L. and MOUNT, L. E. (1975b). Solar radiation and heat balance in animals and man. In *Light as an Ecological Factor:* II, eds. G. C. Evans, R. Bainbridge and O. Rackham, pp. 497–509. Blackwell, Oxford.

260. INGRAM, D. L. and SLEBODZINSKI, A. (1965). Oxygen consumption and thyroid gland activity during adaptation to high ambient temperatures in young pigs. *Res. vet. Sci.*, **6**, 522–30.

261. INGRAM, D. L. and WEAVER, M. E. (1969). A quantitative study of blood vessels of the pig's skin and the influence of environmental temperature. *Anat. Rec.*, **163**, 517–24.

262. INGRAM, D. L. and WHITTOW, G. C. (1962). The effects of variations in respiratory activity and in the skin temperatures of the ears on the temperature of the blood in the external jugular vein of the ox (*Bos taurus*). *J. Physiol., Lond.*, **163**, 221–21.

263. IRVING, L. (1956a). Physiological insulation of swine as bar-skinned mammals. *J. appl. Physiol.*, **9**, 414–20.
264. IRVING, L. (1956b). Metabolism and insulation of swine as bare-skinned mammals. *J. appl. Physiol.*, **9**, 421–6.
265. IRVING, L. (1964). Terrestrial animals in cold: birds and mammals. In *Adaptation to the Environment*, ed. D. B. Dill, pp. 361–78. American Physiological Society.
266. IRVING, L. (1973). Aquatic mammals. In *Comparative Physiology of Thermoregulation*, Vol. III, Special aspects of thermoregulation, ed. G. Causey Whittow, pp. 47–96. Academic Press, New York.
266a. IRVING, L. and KROG, J. (1954). Body temperatures of arctic and subarctic birds and mammals. *J. appl. Physiol.*, **6**, 667–80.

267. JACKSON, P. C. and SCHMIDT-NIELSEN, K. (1964). Countercurrent heat exchange in respiratory passages. *Proc. Natl. Acad. Sci. U.S.*, **51**, 1192–7.
268. JACQUEZ, J. A., HUSS, J., MCKEEHAN, W., DIMITROFF, J. M. and KUPPENHEIM, H. F. (1955). Spectral reflectance of human skin in the region 0.7–2.6 μm. *J. appl. Physiol.*, **8**, 297–9.
269. JACQUEZ, J. A., KUPPENHEIM, H. F., DIMITROFF, J. M., MCKEEHAN, W. and HUSS, J. (1955). Spectral reflectance of human skin in the region 235–700 μm. *J. appl. Physiol.*, **8**, 212–9.
270. JAKOB, M. (1949). *Heat Transfer*. Vol. I. Wiley, New York.
271. JAKOB, M. (1957). *Heat Transfer*. Vol. II. Wiley, New York.
272. JANSKY, L. (1973). Non-shivering thermogenesis and its thermoregulatory significance. *Biol. Rev.*, **48**, 85–132.
273. JANSKY, L. (1976). Effects of cold and exercise on energy metabolism of small mammals. In *Progress in Animal Biometeorology*, Vol. I, Part I, ed. H. D. Johnson, pp. 239–58. Swets and Zeitlinger, Amsterdam.
274. JENKINSON, D. MCE. (1967). On the classification of sweat glands and the question of the existence of an apocrine secretory process. *Brit. Vet. J.*, **123**, 311–5.
275. JÉQUIER, E., PITTET, P. and GYGAX, P. H. (1978). Thermic effect of glucose and thermal body insulation in lean and obese subjects: a calorimetric approach. *Proc. Nutr. Soc.*, **37**, 45–53.
276. JESSEN, C. (1977). Interaction of air temperature and core temperatures in thermoregulation of the goat. *J. Physiol., Lond.*, **264**, 585–606.
277. JOHNSON, H. D. (1976). Effects of temperature on lactation of cattle. In *Progress in Animal Biometeorology*, Vol. I, Part I, ed. H. D. Johnson, pp. 358–66. Swets and Zeitlinger, Amsterdam.
278. JOHNSON, H. D., HAHN, L., KIBLER, H. H., SHANKLIN, M. D. and EDMUNDSON, J. D. (1967). Heat and acclimation influences on lactation of Holstein cattle. *Mo. agric. exp. Stat. Res. Bull. No. 916.*
279. JOHNSON, H. D., RAGSDALE, A. C., BERRY, I. L. and SHANKLIN, M. D. (1963). LXVI. Temperature-humidity effects including influence of acclimation in feed and water consumption of Holstein cattle. *Mo. agric. exp. Stat. Res. Bull. No. 846.*
280. JOHNSON, K. G. (1976). Evaporative temperature regulation in sheep. In *Progress in Animal Biometeorology*, Vol. I, Part I, ed. H. D. Johnson, pp. 140–7. Swets and Zeitlinger, Amsterdam.
280a. JOHNSON, K. G., MALOIY, G. M. O. and BLIGH, J. (1972). Sweat gland function in the red deer (*Cervus elaphus*). *Am. J. Physiol.*, **223**, 604–7.
281. JOHNSON, S. F. and GESSAMEN, J. A. (1973). An evaluation of heart rate as an indirect monitor of free-living energy metabolism. In *Ecological Energetics of Homeotherms*, ed. J. A. Gessamen, pp. 44–54. Utah State University Press.

282. JOYCE, J. P. and BLAXTER, K. L. (1964). The effect of air movement, air temperature and infra-red radiation on the energy requirements of sheep. *Br. J. Nutr.*, **18**, 5–27.

283. JOYCE, J. P., BLAXTER, K. L. and PARK, C. (1966). The effect of natural outdoor environments on the energy requirements of sheep. *Res. vet. Sci.*, **7**, 342–59.

284. KACIUBA-USCILKO, H. (1971). The effect of previous thyroxine administration on the metabolic response to adrenaline in new-born pigs. *Biol. Neonat.*, **419**, 220–6.

285. KACIUBA-USCILKO, H. and INGRAM, D. L. (1977). The effect of propanolol on cold-induced thermogenesis in the newborn pig. *Comp. Biochem. Physiol.*, **56C**, 53–5.

286. KAMAL, T. H. (1976). Indices for heat adaptability of domestic animals. In *Progress in Animal Biometeorology*, Vol. I, Part I, ed. H. D. Johnson, pp. 470–5. Swets and Zeitlinger, Amsterdam.

287. KAMAL, T. H., JOHNSON, H. D. and RAGSDALE, A. C. (1962). Metabolic reactions during thermal stress (35° to 95°F) in dairy animals acclimated at 50° and 80°F. *Mo. agric. exp. Stat. Res. Bull. No. 785.*

288. KAMPEN, M. VAN (1976). Evaporative temperature regulation in birds. In *Progress in Animal Biometeorology*, Vol. I, Part I, ed. H. D. Johnson, pp. 158–66. Swets and Zeitlinger, Amsterdam.

289. KAMPEN, M. VAN, MITCHELL, B. W. and SIEGEL, H. S. (1978). Influence of sudden temperature changes on oxygen consumption and heart rate in chickens in light and dark environments. *J. agric. Sci., Camb.*, **90**, 605–9.

290. KAY, R. N. B. (1970). Meat production from wild herbivores. *Proc. Nutr. Soc.*, **29**, 271–8.

291. KAYSER, C. (1961). *The Physiology of Natural Hibernation.* Pergamon Press, Oxford.

292. KEATINGE, W. R. (1960). The effects of subcutaneous fat and of previous exposure to cold on the body temperature, peripheral blood flow and metabolic rate of men in cold water. *J. Physiol., Lond.*, **153**, 166–78.

293. KEATINGE, W. R. (1969). *Survival in Cold Water.* p. 140. Blackwell, Oxford.

294. KELLY, C. F., BOND, T. E. and HEITMAN, H., JR. (1954). The role of thermal radiation in animal ecology. *Ecology*, **35**, 562–9.

295. KELLY, C. F., BOND, T. E. and HEITMAN, H. JR. (1963). Direct 'air' calorimetry for livestock. *Trans. Am. Soc. agric. Engrs*, **6**, 126–8.

296. KERSLAKE, D. MCK. (1972). *The Stress of Hot Environments.* Cambridge University Press, London.

297. KERSLAKE, D. MCK. and BREBNER, D. F. (1970). Maximum sweating at rest. In *Physiological and Behavioral Temperature Regulation.* eds. J. D. Hardy, A. P. Gagge and J. A. J. Stolwijk, pp. 139–51. Charles C. Thomas, Springfield, Illinois.

298. KING, J. R. and FARNER, D. S. (1964). Terrestrial animals in humid heat: birds. In *Adaptation to the Environment*, ed. D. B. Dill, pp. 603–24. American Physiological Society.

299. KLEIBER, M. (1932). Body size and metabolism. *Hilgardia*, **6**, 315–53.

300. KLEIBER, M. (1947). Body size and metabolic rate. *Physiol. Rev.*, **27**, 511–41.

301. KLEIBER, M. (1965). Metabolic body size. In *Energy Metabolism*, ed. K. L. Blaxter, pp. 427–35. Academic Press, London.

302. KLEIBER, M. (1975). *The Fire of Life*, 2nd edition. John Wiley, New York.

303. KROG, H. and MONSON, M. (1954). Notes on the metabolism of a mountain goat. *Am. J. Physiol.*, **178**, 515–6.

304. KUNO, Y. (1956). *The Physiology of Human Perspiration.* 2nd. edition. Churchill, London.

305. LADELL, W. S. S. (1964). Terrestrial animals in humid heat: man. In *Adaptation to the Environment*, ed. D. B. Dill, pp. 625–59. American Physiological Society.

306. LADELL, W. S. S., WATERLOW, J. C. and HUDSON, M. F. (1944). Desert climate: physiological and clinical observations. *Lancet*, **ii**, 491–7.

307. LAMPKIN, G. H. and QUARTERMAIN, J. (1962). Observations on the grazing habits of grade and Zebu cattle. II. Their behaviour under favourable conditions in the tropics. *J. agric. Sci., Camb.*, **58**, 119–23.

308. LANDSBERG, H. E. (1964). Controlled climate (outdoor and indoor). In *Medical Climatology*, ed. S. Licht. E. Licht, New Haven, Connecticut.

309. LANGHAAR, H. L. (1951). *Dimensional Analysis and Theory of Models.* John Wiley and Sons, Inc., New York.

310. LASIEWSKI, R. C. (1963). Oxygen consumption of torpid, resting, active, and flying humming birds. *Physiol. Zoöl.*, **36**, 122–40.

311. LASIEWSKI, R. C. (1964). Body temperatures, heart and breathing rate, and evaporative water loss in hummingbirds. *Physiol. Zoöl.*, **37**, 212–23.

312. LASIEWSKI, R. C. and DAWSON, W. R. (1967). A re-examination of the relation between standard metabolic rate and body weight in birds. *Condor*, **69**, 13–23.

313. LAVOISIER, A. L. and LAPLACE, DE. (1780). Mémoire sur la chaleur. *Histoire de l'Acad. Roy. des Sci.*, **85**, 355.

313a. LEBLANC, J. (1973). Evaluation of adaptation to the polar environment by autonomic nervous system response. In *Polar Human Biology*, eds. O. G. Edholm and E. K. E. Gunderson, pp. 256–64. William Heinemann Medical Books Ltd, London.

314. LEBLANC, J. (1975). *Man in the Cold.* Charles C. Thomas, Springfield, Illinois.

315. LEBLANC, J. and MOUNT, L. E. (1968). The effects of noradrenaline and adrenaline on oxygen consumption rate and arterial blood pressure in the new-born pig. *Nature, Lond.*, **217**, 77–8.

316. LEE, D. H. K. (1964). Terrestrial animals in dry heat: man in the desert. In *Adaptation to the Environment*, ed. D. B. Dill, pp. 551–82. American Physiological Society.

317. LEE, D. H. K. and VAUGHAN, J. A. (1964). Temperature equivalent of solar radiation on man. *Int. J. Biometeor.*, **8**, 61–9.

318. LEITCH, I., HYTTEN, F. E. and BILLEWICZ, W. Z. (1960). The maternal and neonatal weights of some mammalia. *Proc. Zool. Soc. Lond.*, **133**, 11–28.

319. LEITHEAD, C. S. and LIND, A. R. (1964). *Heat Stress and Heat Disorders.* Cassell, London.

320. LE MAHO, Y., DELCLITTE, P. and CHATONNET, J. (1976). Thermoregulation in fasting emperor penguins under natural conditions. *Am. J. Physiol.*, **231**, 913–22.

320a. LEMAIRE, R. (1960). Considérations physiologiques sur la climatisation en milieu désertique. *Journées d'Informations Med. Soc. Sahariennes*, 101–12.

321. LENTZ, C. P. and HART, J. S. (1960). The effect of wind and moisture on heat loss through the fur of new-born caribou. *Can. J. Zool.*, **38**, 679–88.

322. LEUNG, P. B. M. and HOROWITZ, B. A. (1976). Free-feeding patterns of rats in response to changes in environmental temperature. *Am. J. Physiol.*, **231**, 1220–4.

323. LEWIS, H. E., FOSTER, A. R., MULLAN, B. J., COX, R. N. and CLARK, R. P. (1969). Aerodynamics of the human microenvironment. *Lancet*, **i**, 1273–7.

324. LIND, A. R. (1963). Tolerable limits for prolonged and intermittent exposures to heat. In *Temperature, Its Measurement and Control in Science and Industry*, Vol. 3, Part 3, ed. J. D. Hardy, pp. 337–345. Reinhold, New York.

325. LINZELL, J. L. (1972). Milk yield, energy loss in milk, and mammary gland weight in different species. *Dairy Sci. Abstr.*, **34**, 351–60.

326. LOUW, G. N. (1972). The role of advective fog in the water economy of certain Namib Desert animals. In *Comparative Physiology of Desert Animals*, ed. G. M. O. Maloiy, pp. 297–314. Academic Press, Zoological Society of London.

327. LUNDY, H., MACLEOD, M. G. and JEWITT, T. R. (1978). An automated multi-calorimeter system: preliminary experiments on laying hens. *Br. Poult. Sci.*, **19**, 173–86.

328. LUSK, G. (1928). *The Elements of the Science of Nutrition.* 4th edition. Saunders, Philadelphia.

329. LYMAN, C. P. (1948). The oxygen consumption and temperature regulation of hibernating hamsters. *J. exp. Zool.*, **109**, 55–78.

330. LYMAN, C. P. and CHATFIELD, P. O. (1955). Physiology of hibernation in mammals. *Physiol. Rev.*, **35**, 403–25.

331. McCANCE, R. A. (1961). Characteristics of the newly born. In *Somatic Stability in the Newly Born*, eds. G. E. W. Wolstenholme and M. O'Connor, pp. 1–4. Churchill, London.

332. McCANCE, R. A., EL NEIL, H., EL DIN, N., WIDDOWSON, E. M., SOUTHGATE, D. A. T., PASSMORE, R., SHIRLING, D. and WILKINSON, R. T. (1971). The response of normal men and women to changes in their environmental temperatures and ways of life. *Phil. Trans. R. Soc. Ser. B*, **259**, 533–65.

333. McCARRICK, R. B. and DRENNAN, M. J. (1972). Effect of winter environment on growth of young beef cattle. *Anim. Prod.*, **14**, 97–105.

334. McDOWELL, R. E. (1972). *Improvement of Livestock Production in Warm Climates.* W. H. Freeman and Co., San Francisco.

335. MACGRATH, W. S., VANDER NOOT, G. W., GILBREATH, R. L. and FISHER, H. (1968). Influence of environmental temperature and dietary fat on backfat composition of swine. *J. Nutr.*, **96**, 461–6.

336. McGUIRE, J. H. (1953). *D.S.I.R. Fire Research Special Rep. No. 2.* Her Majesty's Stationery Office, London.

337. McINTYRE, D. G. and EDERSTROM, H. E. (1958). Metabolic factors in the development of homeothermy in dogs. *Am. J. Physiol.*, **194**, 293–6.

338. McLEAN, J. A. (1963). The partition of insensible losses of body weight and heat from cattle under various climatic conditions. *J. Physiol., Lond.*, **167**, 427–47.

339. McLEAN, J. A. (1971). A gradient layer calorimeter for large animals. *J. Instn. Heat. Engrs.*, **39**, 1–8.

340. McLEAN, J. A. (1972). On the calculation of heat production from open-circuit calorimetric measurements. *Br. J. Nutr.*, **27**, 597–600.

341. McLEAN, J. A. (1974). Loss of heat by evaporation. In *Heat Loss from Animals and Man: Assessment and Control*, eds. J. L. Monteith and L. E. Mount, pp. 19–31. Butterworths, London.

342. McLEAN, J. A. and CALVERT, D. T. (1972). Influence of air humidity on the partition of heat exchanges of cattle. *J. agric. Sci., Camb.*, **78**, 303–7.

343. McLEAN, J. A. and WATTS, P. R. (1976). Analytical refinements in animal calorimetry. *J. appl. Physiol.*, **40**, 827–31.

344. MACMILLEN, R. E. and LEE, A. K. (1967). Australian desert mice: independence of exogenous water. *Science, N. Y.*, **158**, 383–5.

345. McNAB, B. K. (1966). An analysis of the body temperatures of birds. *The Condor*, **68**, 47–55.

346. MACFARLANE, W. V. (1964). Terrestrial animals in dry heat: ungulates. In *Adaptation to the Environment*, ed. D. B. Dill, pp. 509–39. American Physiological Society.

347. MACFARLANE, W. V. (1968a). Adaptation of ruminants to tropics and deserts. In *Adaptation of Domestic Animals*, ed. E. S. E. Hafez, pp. 164–82. Lea and Febiger, Philadelphia.

348. MACFARLANE, W. V. (1968b). Comparative functions of ruminants in hot environments. In *Adaptation of Domestic Animals*, ed. E. S. E. Hafez, pp. 264–76. Lea and Febiger, Philadelphia.

349. MACFARLANE, W. V. (1976a). Water metabolism in sheep in relation to heat and cold. *Progress in Animal Biometeorology*, Vol. I, Part I, ed. H. D. Johnson, pp. 49–57. Swets and Zeitlinger, Amsterdam.

350. MACFARLANE, W. V. (1976b). Adaptation to climatic zones and the ecophysiology of ruminants. *Progress in Animal Biometeorology*, Vol. I, Part I, ed. H. D. Johnson, pp. 425–33. Swets and Zeitlinger, Amsterdam.

351. MACFARLANE, W. V. (1976c). Ecophysiological hierarchies. *Israel J. med. Sci.*, **12**, 723–31.

352. MALOIY, G. M. O. (1970). Water economy of the Somali donkey. *Am. J. Physiol.*, **219**, 1522–7.

352a. MALOIY, G. M. O. (1971). Temperature regulation in the Somali donkey (*Equus asinus*). *Comp. Biochem. Physiol.*, **39A**, 403–12.

353. MALOIY, G. M. O. (1972). Renal salt and water excretion in the camel (*Camelus dromedarius*). In *Comparative Physiology of Desert Animals*, ed. G. M. O. Maloiy, pp. 243–59. Academic Press, New York.

354. MALOIY, G. M. O. (1976). Water and salt metabolism in the camel. In *Progress in Animal Biometeorology*, Vol. I, Part I, ed. H. D. Johnson, pp. 58–66. Swets and Zeitlinger, Amsterdam.

355. MALOIY, G. M. O. and HOPCRAFT, D. (1971). Thermoregulation and water relations of two East African antelopes: the hartebeest and impala. *Comp. Biochem. Physiol.*, **38A**, 525–34.

356. MANN, T. P. and ELLIOTT, R. I. K. (1957). Neonatal cold injury due to accidental exposure to cold. *Lancet*, **i**, 299–34.

357. MARTIN, C. J. (1903). Thermal adjustments and respiratory exchange in monotremes and marsupials. A study in the development of homeothermism. *Phil. Trans. Roy. Soc. London Ser. B.*, **195**, 1–37.

358. MASON, I. L. (1951). *A World Dictionary of Breeds, Types and Varieties of Livestock.* Commonwealth Agricultural Bureaux.

359. MAWSON, W. F. Y. (1976). Management, including breeding and nutrition of cattle. In *Progress in Animal Biometeorology*, Vol. I, Part I, ed. H. D. Johnson, pp. 504–11. Swets and Zeitlinger, Amsterdam.

360. MEEH, K. (1879). Oberflächenmessungen des menschlichen Körpers. *Ztschr. Biol.*, **15**, 425–58.

361. MENUAM, B. and RICHARDS, S. A. (1975). Observations on the sites of respiratory evaporation in the fowl during thermal panting. *Respiration Physiology*, **25**, 39–52.

362. MESTYAN, G. and VARGA, F. (1960). Chemical thermoregulation of full-term and premature newborn infants. *J. Pediatrics*, **56**, 623–9.

363. MILLER, A. T., JR. and BLYTH, C. S. (1958). Lack of insulating effect of body fat during exposure to internal and external heat loads. *J. appl. Physiol.*, **12**, 17–9.

364. MITCHELL, D. (1969). In *The husbanding of Red Deer*, eds. M. M. Bannerman and K. L. Blaxter, pp. 16–27. Aberdeen University Press, Aberdeen.

365. MITCHELL, D. (1974a). Convective heat transfer from man and other animals. In *Heat Loss from Animals and Man: Assessment and Control*, eds. J. L. Monteith and L. E. Mount, pp. 59–76. Butterworths, London.

366. MITCHELL, D. (1974b). Physical basis of thermoregulation. In *Environmental Physiology*, ed. D. Robertshaw, pp. 1–32. Butterworths, London.

367. MITCHELL, D., ATKINS, A. R. and WYNDHAM, C. H. (1972). Mathematical and physical models of thermoregulation. In *Essays on Temperature Regulation*, eds. J. Bligh and R. E. Moore, pp. 37–54. North Holland, Amsterdam and London.

368. MITCHELL, B., STAINES, B. W. and WELCH, D. (1977). *Ecology of Red Deer: a Research Review Relevant to Their Management in Scotland.* Institute of Terrestrial Ecology, Cambridge.

369. MONTEITH, J. L. (1973). *Principles of Environmental Physics.* Edward Arnold, London.

370. MONTEITH, J. L. (1974). Specification of the environment in relation to heat loss. In *Heat Loss from Animals and Man: Assessment and Control*, eds. J. L. Monteith and L. E. Mount. Butterworths, London.

371. MORRIS, D. (1965). *The Mammals.* Hodder and Stoughton, London.

372. MORRISON, S. R., BOND, T. E. and HEITMAN, H. (1968). Effect of humidity on swine at high temperature. *Trans. Am. Soc. agric. Engrs.*, **11**, 526–8.

373. MORRISON, S. R., HEITMAN, H. and BOND, T. E. (1969). Effect of humidity on swine at temperatures above optimum. *Int. J. Biometeorol.*, **13**, 135–9.

374. MORRISON, S. R., HEITMAN, H., BOND, T. E. and FINN-KELCEY, P. (1966). The influence of humidity on growth rate and feed utilization of swine. *Int. J. Biometeorol.*, **10**, 163–8.

374a. MORRISON, S. R. and MOUNT, L. E. (1971). Adaptation of growing pigs to changes in environmental temperature. *Anim. Prod.*, **13**, 51–7.

375. MOULE, G. R. (1968). World distribution of domestic animals. In *Adaptation of Domestic Animals*, ed. H. S. E. Hafez, pp. 18–33. Lea and Febiger, Philadelphia.

376. MOUNT, L. E. (1959). The metabolic rate of the new-born pig in relation to environmental temperature and to age. *J. Physiol., Lond.*, **147**, 333–45.

377. MOUNT, L. E. (1960). The influence of huddling and body size on the metabolic rate of the young pig. *J. agric. Sci., Camb.*, **55**, 101–5.

378. MOUNT, L. E. (1962). Evaporative heat loss in the new-born pig. *J. Physiol., Lond.*, **164**, 274–81.

379. MOUNT, L. E. (1963a). The environmental temperature preferred by the young pig. *Nature, Lond.*, **199**, 122–23.

380. MOUNT, L. E. (1963b). Responses to thermal environment in new-born pigs. *Fedn Proc. Fedn Am. Socs. exp. Biol.*, **22**, 818–23.

381. MOUNT, L. E. (1964a). Radiant and convective heat loss from the new-born pig. J. Physiol., Lond., **173**, 96–113.

382. MOUNT, L. E. (1964b). The tissue and air components of thermal insulation in the new-born pig. *J. Physiol., Lond.*, **170**, 286–95.

383. MOUNT, L. E. (1965). The young pig and its physical environment. In *Energy Metabolism*, ed. K. L. Blaxter, pp. 379–85. Academic Press, London.

384. MOUNT, L. E. (1966). Thermal and metabolic comparisons between the new-born pig and human infant. In *Swine in Biomedical Research*, eds. L. K. Bustad and R. O. McClellan, pp. 501–9. Battelle Memorial Institute.

385. MOUNT, L. E. (1967). The heat loss from new-born pigs to the floor. *Res. vet. Sci.*, **8**, 175–86.

386. MOUNT, L. E. (1968a). *The Climatic Physiology of the Pig.* Edward Arnold, London.

387. MOUNT, L. E. (1968b). Adaptation of swine. In *Adaptation of Domestic Animals*, ed. E. S. E. Hafez, pp. 277–91. Lea and Febiger, Philadelphia.

388. MOUNT, L. E. (1969). The respiratory quotient in the newborn pig. *Br. J. Nutr.*, **23**, 407–13.

389. MOUNT, L. E. (1971). Metabolic rate and thermal insulation in albino and hairless mice. *J. Physiol., Lond.*, **217**, 315–26.

390. MOUNT, L. E. (1974a). Thermal neutrality. In *Heat Loss from Animals and Man: Assessment and Control*, eds. J. L. Monteith and L. E. Mount, pp. 425–39. Butterworths, London.

391. MOUNT, L. E. (1974b). *Climatic Variables in Animal Production: the Exploitation of Research in Practice.* The Fourth Hannah Lecture, The Hannah Research Institute, Ayr, Scotland.
392. MOUNT, L. E. (1975). The assessment of thermal environment in relation to pig production. *Livestock Production Science*, **2**, 381–92.
393. MOUNT, L. E. (1976a). Heat loss in relation to plane of nutrition and thermal environment. *Proc. Nutr. Soc.*, **35**, 81–6.
394. MOUNT, L. E. (1976b). Effects of heat and cold on energy metabolism of the pig. In *Progress in Animal Biometeorology*, Vol. I, Part I, ed. H. D. Johnson, pp. 227–38. Swets and Zeitlinger, Amsterdam.
395. MOUNT, L. E. (1976c). Energy expenditure during the growing period. In *Early Nutrition and Later Development*, ed. A. W. Wilkinson, pp. 156–63. Pitman, London.
396. MOUNT, L. E. (1977). The use of heat transfer coefficients in estimating sensible heat loss from the pig. *Anim. Prod.*, **25**, 271–9.
397. MOUNT, L. E. (1978). Heat transfer between animal and environment. *Proc. Nutr. Soc.*, **37**, 21–7.
398. MOUNT, L. E., CLOSE, W. H. and VERSTEGEN, M. W. A. (1973). The effective critical temperature in groups of pigs. *Proc. Nutr. Soc.*, **32**, 71A.
399. MOUNT, L. E., HOLMES, C. W., CLOSE, W. H., MORRISON, S. R. and START, I. B. (1971). A note on the consumption of water by the growing pig at several environmental temperatures and levels of feeding. *Anim. Prod.*, **13**, 561–3.
400. MOUNT, L. E., HOLMES, C. W., START, I. B. and LEGGE, A. J. (1967). A direct calorimeter for the continuous recording of heat loss from groups of growing pigs over long periods. *J. agric. Sci., Camb.*, **68**, 47–55.
401. MOUNT, L. E. and INGRAM, D. L. (1965). The effects of ambient temperature and air movement on localized sensible heat-loss from the pig. *Res. vet. Sci.*, **6**, 84–91.
402. MOUNT, L. E. and ROWELL, J. G. (1960). Body size, body temperature and age in relation to the metabolic rate of the pig in the first five weeks after birth. *J. Physiol., Lond.*, **154**, 408–16.
403. MOUNT, L. E. and STEPHENS, D. B. (1970). The relation between body size and maximum and minimum metabolic rates in the new-born pig. *J. Physiol., Lond.*, **207**, 417–27.
404. MOUNT, L. E. and WILLMOTT, JANE V. (1967). The relation between spontaneous activity, metabolic rate and the 24-hour cycle in mice at different environmental temperatures. *J. Physiol., Lond.*, **190**, 371–80.

405. NAY, T. and HAYMAN, R. H. (1956). Sweat glands in Zebu (*Bos indicus L.*) and European (*B. taurus L.*) cattle. 1. Size of individual glands, the denseness of their population and their depth below the skin surface. *Aust. J. agric. Res.*, **7**, 482–94.
406. NEERGAARD, L. and THORBEK, G. (1967). *Variation in Heat Production in Growing Pigs.* 1. Some observations on the relationship between feed intake and heat production in pigs fed barley and skim-milk. In *Energy Metabolism of Farm Animals*, eds. K. L. Blaxter, G. Thorbek and J. Kielanowski. Oriel Press, Newcastle.
407. NEWBURGH, L. H. (editor) (1949). *Physiology of Heat Regulation and Science of Clothing.* Saunders, Philadelphia.
408. NEWLAND, H. W., MCMILLEN, W. N. and REINEKE, E. P. (1952). Temperature adaptation in the baby pig. *J. anim. Sci.*, **11**, 118–33.
409. NIELSEN, M. (1938). Die Regulation der Korpertemperatur bei Muskelarbeit. *Scand. Arch. Physiol.*, **79**, 193–230.

410. NIELSEN, M. and PEDERSEN, L. (1952). Studies on the heat loss by radiation and convection from the clothed human body. *Acta Physiol. Scand.*, **27**, 272–94.

411. NISHI, Y. and GAGGE, A. P. (1970). Direct evaluation of convective heat transfer coefficient by napthalene sublimation. *J. appl. Physiol.*, **29**, 830–8.

412. NORDAN, H. C., COWAN, I. M. and WOOD, A. J. (1970). The feed intake and heat production of the young black-tailed deer (*Odocoileus hemionus columbianus*). *Can. J. Zool.*, **48**, 275–82.

413. NORMAN, J. N. (1965). Cold exposure and patterns of activity at a polar station. *Br. Antarct. Surv. Bull. No.* **6**, 1–13.

414. ØRITSLAND, N. A. (1970). Temperature regulation of the polar bear (*Thalactos maritimus*). *Comp. Biochem. Physiol.*, **37**, 225–33.

415. PANARETTO, B. A., HUTCHINSON, J. C. D. and BENNETT, J. W. (1968). Protection afforded by plastic coats to shorn sheep. *Proc. Aust. Soc. Anim. Prod.*, **7**, 264–9.

416. PARER, J. T. (1963). Wool length and radiant heating effects in sheep. *J. agric. Sci., Camb.*, **60**, 141–4.

417. PAYNE, W. J. A., LAING, W. I. and RAIKOVA, E. N. (1951). Grazing behaviour of dairy cattle in the tropics. *Nature, Lond.*, **167**, 610–11.

418. PEAKER, M. and LINZELL, J. L. (1975). *Salt Glands in Birds and reptiles.* Cambridge University Press.

419. PEARSON, O. P. (1948). Metabolism of small mammals, with remarks on the lower limit of mammalian size. *Science, N. Y.*, **108**, 44.

420. PEPLER, R. D. (1963). Performance and well-being in heat. In *Temperature, Its Measurement and Control in Science and Industry*, Part 3, Vol. 3, ed. J. D. Hardy, pp. 319–36. Reinhold, New York.

421. PICKWELL, G. V. (1968). Energy metabolism in ducks during submergence asphyxia: assessment by a direct method. *Comp. Biochem. Physiol.*, **27**, 455–85.

422. PINSHOW, B., FEDAK, M. A., BATTLES, D. R. and SCHMIDT-NIELSEN, K. (1976). Energy expenditure for thermoregulation and locomotion in emperor penguins. *Am. J. Physiol.*, **231**, 903–12.

423. PRIESTLEY, C. H. B. (1957). The heat balance of sheep standing in the sun. *Aust. J. agric. Res.*, **8**, 271–80.

424. PROSSER, C. L. (1964). Perspectives of adaptation: theoretical aspects. In *Adaptation to Environment*, ed. D. B. Dill, pp. 11–25. American Physiological Society.

425. PROUTY, L. R., BARRETT, M. J. and HARDY, J. D. (1949). A simple calorimeter for the simultaneous determination of heat loss and heat production in animals. *Rev. Sci. Instrum.*, **20**, 357–63.

426. PRYCHODKO, W. (1958). The effect of aggregation of laboratory mice on food intake at different temperatures. *Ecology*, **39**, 500–3.

427. PUGH, L. G. C. and EDHOLM, O. G. (1955). The physiology of Channel swimmers. *Lancet*, **ii**, 761–8.

428. PULLAR, J. D. (1958). *Direct Calorimetry of Animals by the Gradient Layer Principle.* First Symposium on Energy Metabolism, European Association for Animal Production, Rome.

429. RANDALL, J. M. (1975). The prediction of air flow patterns in livestock buildings. *J. agric. Engng. Res.*, **20**, 199–215.

430. RED DEER COMMISSION (1973). *Annual Report for 1972.* Her Majesty's Stationery Office, Edinburgh.

431. REECE, F. N. and DEATON, J. W. (1976). Effects of temperature on the growth of the domestic chicken. In *Progress in Animal Biometeorology*, Vol. I, Part 1, ed. H. D. Johnson, pp. 337–342. Swets and Zeitlinger, Amsterdam.

432. REIGNAULT, V. and REISET, J. (1849). Recherches chimiques sur la respiration des animaux. *Ann. de chim. et de phys. Ser. 3*, **26**, 299–519.

433. RENNIE, D. W. (1963). Comparison of non-acclimatized Americans and Alaskan Eskimos. *Federation Proceedings*, **22**, 828–30.

434. REYNOLDS, G. S. (1968). *A Primer of Operant Conditioning.* Scott, Foresman and Co., Glenview, Illinois.

435. RICHARDS, S. A. (1970). The biology and comparative physiology of thermal panting. *Biol. Rev.*, **45**, 223–64.

436. RICHARDS, S. A. (1971). The significance of changes in the temperature of the skin and body core of the chicken in the regulation of heat loss. *J. Physiol., Lond.*, **216**, 1–10.

437. RICHARDS, S. A. (1974). Aspects of physical thermoregulation in the fowl. In *Heat Loss from Animals and Man: Assessment and Control*, eds. J. L. Monteith and L. E. Mount, pp. 255–75. Butterworths, London.

438. RICHARDS, S. A. (1976). Evaporative water loss in domestic fowls and its partition in relation to ambient temperature. *J. agric. Sci., Camb.*, **87**, 527–32.

439. RICHARDS, S. A. (1977). The influence of loss of plumage on temperature regulation in laying hens. *J. agric. Sci., Camb.*, **89**, 393–8.

440. RICHET, C. (1889). *La Chaleur Animale.* Felix Alcan, Bibliothèque Scientifique International, Paris.

441. RISING, J. D. (1976). Temperature regulation in birds. In *Progress in Animal Biometeorology*, Vol. I, Part I, ed. H. D. Johnson, pp. 103–7. Swets and Zeitlinger, Amsterdam.

442. ROBBINS, C. T., PRIOR, R. L., MOEN, A. N. and VISEK, W. J. (1974). Nitrogen metabolism of white-tailed deer. *J. anim. Sci.*, **38**, 186–91.

443. ROBERTSHAW, D. (1968). The pattern and control of sweating in the sheep and the goat. *J. Physiol., Lond.*, **198**, 531–9.

444. ROBERTSHAW, D. (1976). Evaporative temperature regulation in ruminants. In *Progress in Animal Biometeorology*, Vol. I, Part I, ed. H. D. Johnson, pp. 132–9. Swets and Zeitlinger, Amsterdam.

445. ROBERTSHAW, D. and TAYLOR, C. R. (1969). A comparison of sweat gland activity in eight species of East African bovids. *J. Physiol., Lond.*, **203**, 135–43.

446. ROBINSON, K. and LEE, D. H. K. (1941). Reactions of the pig to hot atmospheres. *Proc. Roy. Soc., Queensland*, **53**, 145–58.

447. ROBINSON, S. (1949). Physiological adjustments to heat. In *Physiology of Heat Regulation and the Science of Clothing*, ed. L. H. Newburgh, pp. 193–239. Saunders, Philadelphia.

448. ROBINSON, S. (1963). Circulatory adjustments of men in hot environments. In *Temperature, Its Measurement and Control in Science and Industry*, Vol. 3, Part 3, eds. C. M. Hertzfeld and J. D. Hardy, pp. 287–97. Reinhold, New York.

449. ROGERS, A. F. (1973). Antarctic climate, clothing and acclimatization. In *Polar Human Biology*, eds. O. G. Edholm and E. K. E. Gunderson, pp. 265–89. Heinemann, London.

450. ROMIJN, C. and VREUGDENHIL, E. L. (1969). Energy balance and heat regulation in the White Leghorn fowl. *Neth. J. vet. Sci.*, **2**, 32–58.

450a. ROSENMANN, M. and MORRISON, P. (1963). The physiological response to heat and dehydration in the guanaco. *Physiol. Zool.*, **36**, 45–51.

451. ROTHSTEIN, A. and TOWBIN, E. J. (1947). Blood circulation and temperature of men dehydrating in the heat. In *Physiology of Man in the Desert*, ed. E. F. Adolph, pp. 172–96. Interscience.

452. RUBNER, M. (1894). Die Wuelle der thierischen Wärme. *Biol.*, **30**, 73–142.

453. SAINSBURY, D. W. B. (1967). *Animal Health and Housing.* Bailliere, Tindall and Cassell, London.

454. SARRUS and RAMEAUX (1839). Mémoire addressé à l'Académie Royale. *Bulletin de l'académie royale de médecine*, **3**, 1094–100.

455. SCHMIDT-NIELSEN, K. (1964). *Desert Animals: Physiological Problems of Heat and Water.* Oxford University Press, Oxford.

456. SCHMIDT-NIELSEN, K. (1977). 1. Problems of scaling: locomotion and physiological correlates. In *Scale Effects in Animal Locomotion*, ed. T. J. Pedley, pp. 1–21. Academic Press, New York and London.

457. SCHMIDT-NIELSEN, K., HAINSWORTH, F. R. and MURRISH, D. E. (1970). Countercurrent heat exchange in the respiratory passages: effect on water and heat balance. *Resp. Physiol.*, **9**, 263–76.

458. SCHMIDT-NIELSEN, K., JARNUM, S. A. and HOUPT, T. R. (1957). Body temperature of the camel and its relation to water economy. *Am. J. Physiol.*, **188**, 103–12.

459. SCHMIDT-NIELSEN, K., TAYLOR, C. R. and SHKOLNIK, A. (1972). *Desert Snails: Problems of Survival.* Academic Press, London.

460. SCHOLANDER, P. F. (1955). Evolution of climatic adaptation in homeotherms. *Evolution*, **9**, 15–26.

461. SCHOLANDER, P. F. (1958). Counter current exchange; a principle in biology. *Hvalrådets Skrifter, No. 44. Oslo.*

462. SCHOLANDER, P. F., HAMMEL, H. T., HART, J. S., LE MESSURIER, D. H. and STEEN, J. (1958). Cold adaptation in Australian Aborigines. *J. appl. Physiol.*, **13**, 211–8.

463. SCHOLANDER, P. F., WALTERS, V., HOCK, R. and IRVING, L. (1950). Body insulation of some Arctic and tropical mammals and birds. *Biol. Bull.*, **99**, 225–36.

464. SHANKLIN, M. D. and STEWART, R. E. (1958). Relief of thermally induced stress in dairy cattle by radiation cooling. *Univ. Miss. Agr. Expt. Sta. Res. Bull. 670.*

465. SHEPHERD, J. T. and WEBB-PEPLOE, M. M. (1970). Cardiac output and blood flow distribution during work in heat. In *Physiological and Behavioral Temperature Regulation*, eds. J. D. Hardy, A. P. Gagge and J. A. H. Stolwijk, pp. 237–53. Charles C. Thomas, Springfield, Illinois.

466. SHKOLNIK, A., BORUT, A. and CHOSHNIAK, J. (1972). Water economy of the Beduin goat. In *Comparative Physiology of Desert Animals*, ed. G. M. O. Maloiy, pp. 229–42. Zoological Society of London, Academic Press, London.

467. SIEGEL, H. S. (1968). Adaptation of poultry. In *Adaptation of Domestic Animals*, ed. H. S. E. Hafez, pp. 292–309. Lea and Febiger, Philadelphia.

468. SIEGEL, H. S. (1976a). Effects of cold on energy metabolism in birds. In *Progress in Animal Biometeorology*, Vol. I, Part I, ed. H. D. Johnson, pp. 259–65. Swets and Zeitlinger, Amsterdam.

469. SIEGEL, H. S. (1976b). Effects of heat on energy metabolism in birds. In *Progress in Animal Biometeorology*, Vol. I, Part I, ed. H. D. Johnson, pp. 204–9. Swets and Zeitlinger, Amsterdam.

470. SIMPSON, A. M. (1976). *A Study of the Energy Metabolism and Seasonal Cycles of Captive Red Deer.* Ph.D. Thesis, University of Aberdeen.

471. SIPLE, P. A. (1949). Clothing and climate. In *Physiology of Heat Regulation and the Science of Clothing*, ed. L. H. Newburgh, pp. 389–442. Saunders, Philadelphia.

472. SKINNER, B. F. (1938). *The Behaviour of Organisms: an Experimental Analysis.* Appleton-Century, New York.

473. SLEE, J. (1974). The retention of cold acclimatization in sheep. *Anim. Prod.*, **19**, 201–10.

474. SMITH, C. V. (1964). A quantitative relationship between environment, comfort and animal productivity. *Agricultural Meteorology*, **1**, 249–70.

475. SMITH, C. V. (1974). Farm buildings. In *Heat Loss from Animals and Man: Assessment and Control*, eds. J. L. Monteith and L. E. Mount, pp. 345–65. Butterworths, London.
476. SOHOLT, L. F., DILL, D. B. and ODDERSHEBE, I. (1977). Evaporative cooling in desert heat: sun and shade; rest and exercise. *Comp. Biochem. Physiol.*, **57A**, 369–71.
477. SØRENSEN, P. H. (1962). Influence of climatic environment on pig performance. In *Nutrition of Pigs and Poultry*, eds. J. T. Morgan and D. Lewis, pp. 88–103. Butterworths, London.
478. SPEDDING, C. R. W. (1965). *Sheep Production and Grazing Management*. Baillière, Tindall and Cox, London.
479. STANIER, M. W. (1974). Effect of body weight, ambient temperature and huddling on oxygen consumption and body temperature of young mice. *Comp. Biochem. Physiol.*, **51A**, 79–82.
480. STEEN, I. and STEEN, J. B. (1965a). The importance of the legs in the thermoregulation of birds. *Acta physiol. scand.*, **63**, 285–91.
481. STEEN, I. and STEEN, J. B. (1965b). Thermoregulatory importance of the beaver's tail. *Comp. Biochem. Physiol.*, **15**, 267–70.
482. STEINBACH, J. (1976a). Effects of temperature on the growth of pigs. In *Progress in Animal Biometeorology*, Vol. I, Part I, ed. H. D. Johnson, pp. 320–7. Swets and Zeitlinger, Amsterdam.
483. STEINBACH, J. (1976b). Effect of heat on reproduction of swine. In *Progress in Animal Biometeorology*, Vol. I, Part I, ed. H. D. Johnson, pp. 393–8. Swets and Zeitlinger, Amsterdam.
484. STEPHENS, D. B. (1971). The metabolic rates of newborn pigs in relation to floor insulation and ambient temperature. *Anim. Prod.*, **13**, 303–13.
485. STOMBAUGH, D. P., ROLLER, W. L., ADAMS, T. and TEAGUE, H. S. (1973). Temperature regulation in neonatal piglets during mild cold and severe heat stress. *Am. J. Physiol.*, **225**, 1192–8.
486. STOTHERS, J. K. and WARNER, R. M. (1978). Oxygen consumption and neonatal sleep states. *J. Physiol., Lond.*, **278**, 435–40.
487. STURKIE, P. D. (1965). *Avian Physiology*. 2nd edition, pp. 200–1. Cornell University Press.
488. SYKES, A. H. (1977). Nutrition-environment interactions in poultry. In *Nutrition and the Climatic Environment*, eds. W. Haresign, H. Swan and D. Lewis, pp. 17–29. Butterworths, London.
489. SYKES, A. R., GRIFFITHS, R. G. and SLEE, J. (1976). Influence of breed, birth weight and weather on the body temperature of newborn lambs. *Anim. Prod.*, **22**, 395–402.

489a. TAYLOR, C. R. (1969). The eland and the oryx. *Sci. Amer.*, **220**, 88–95.
490. TAYLOR, C. R. (1970). Dehydration and heat: effects on temperature regulation of East African ungulates. *Am. J. Physiol.*, **219**, 1136–9.
491. TAYLOR, C. R. (1972). The desert gazelle: a paradox resolved. In *Comparative Physiology of Desert Animals*, ed. G. M. O. Maloiy, pp. 215–27. Zoological Society of London, Academic Press, London.
492. TAYLOR, C. R. (1974). Exercise and thermoregulation. In *Environmental Physiology*, ed. D. Robertshaw, pp. 163–84. Butterworths, London.
493. TAYLOR, C. R. and LYMAN, C. P. (1972). Heat storage in running antelopes: independence of brain and body temperatures. *Am. J. Physiol.*, **222**, 114–7.
494. TAYLOR, C. R., ROBERSHAW, D. and HOFMANN, R. (1969). Thermal panting: a comparison of wildebeest and zebu cattle. *Am. J. Physiol.*, **217**, 907–10.
494a. TAYLOR, C. R., SPINAGE, C. A. and PLYMAN, C. (1969). Water relations of the water buck, an East African antelope. *Am. J. Physiol.*, **217**, 630–4.

495. TAYLOR, P. M. (1960). Oxygen consumption in new-born rats. *J. Physiol., Lond.*, **154**, 153–68.

496. TEMPLETON, J. R. (1970). Reptiles. In *Comparative Physiology of Thermoregulation* Vol. 1, ed. G. C. Whittow, pp. 167–221. Academic Press, New York.

497. TERRILL, C. E. (1968). Adaptation of sheep and goats. In *Adaptation of Domestic Animals*, ed. E. S. E. Hafez, pp. 246–63. Lea and Febiger, Philadelphia.

498. THAUER, R. (1961). Mécanismes périphériques et centraux de la régulation de la température. *Archs Sci. physiol.*, **15**, 95–123.

499. THAUER, R. (1970). Thermosensitivity of the spinal cord. In *Physiological and Behavioral Temperature Regulation*, eds. J. D. Hardy, A. P. Gagge and J. A. J. Stolwijk, pp. 472–92. Charles C. Thomas, Springfield, Illinois.

500. THOMPSON, G. E. (1973). Climatic physiology of cattle. *J. Dairy Res.*, **40**, 441–73.

501. THOMPSON, G. E. (1976). Effects of cold on energy metabolism of cattle. In *Progress in Animal Biometeorology*, Vol. I, Part I, ed. H. D. Johnson, pp. 210–17. Swets and Zeitlinger, Amsterdam.

502. THOMPSON, G. E., GARDNER, J. W. and BELL, A. W. (1975). The oxygen consumption, fatty acid and glycerol uptake of the liver in fed and fasted sheep during cold exposure. *Q. Jl exp. Physiol.*, **60**, 107–21.

503. THOMPSON, G. E. and THOMSON, E. M. (1972). Effect of cold exposure on mammary circulation, oxygen consumption and milk secretion in the goat. *J. Physiol., Lond.*, **272**, 187–96.

504. THORBEK, G. (1969). Studies on the energy metabolism of growing pigs. In *Energy Metabolism of Farm Animals*, eds. K. L. Blaxter, J. Kielanowski and G. Thorbek, pp. 281–9. Oriel Press, Newcastle.

505. THORBEK, G. (1974). Energy metabolism in fasting pigs at different live weight as influenced by temperature. In *Energy Metabolism of Farm Animals*, eds. K. H. Menke, H. -J. Lantzsch and J. R. Reichl, pp. 147–50. University of Hohenheim.

506. THORBEK, G. (1977). The energetics of protein deposition during growth. *Nutr. Metab.*, **21**, 105–18.

507. TREGEAR, R. T. (1965). Hair density, wind speed and heat loss in mammals. *J. appl. Physiol.*, **20**, 796–801.

508. TUCKER, V. A. (1968). Respiratory exchange and evaporative water loss in the flying budgerigar. *J. exp. Biol.*, **48**, 67–87.

509. TUCKER, V. A. (1970). Energetic cost of locomotion in mammals. *Comp. Biochem. Physiol.*, **34**, 841–6.

510. TUCKER, V. A. (1971). Flight energetics in birds. *Amer. Zool.*, **11**, 115–24.

511. UNDERWOOD, C. R. and WARD, E. J. (1966). The solar radiation area of man. *Ergonomics*, **9**, 155–68.

512. VERCOE, J. E. (1976). Adaptability of different breeds of cattle to heat. In *Progress in Animal Biometeorology*, Vol. I, Part I, ed. H. D. Johnson, pp. 434–41. Swets and Zeitlinger, Amsterdam.

513. VERCOE, J. E. and FRISCH, J. E. (1974). Fasting metabolism, liveweight and voluntary feed intake of different breeds of cattle. In *Energy Metabolism of Farm Animals*, eds. K. H Menke, H. -J. Lantzsch and J. R. Reichl, pp. 131–4. University of Hohenheim.

514. VERSTEGEN, M. W. A., CLOSE, W. H., START, I. B. and MOUNT, L. E. (1973). The effects of environmental temperature and plane of nutrition on heat loss, energy retention and deposition of protein and fat in groups of growing pigs. *Br. J. Nutr.*, **30**, 21–35.

515. VERSTEGEN, M. W. A. and VAN DER HEL, W. (1974). The effects of temperature and type of floor on metabolic rate and effective critical temperature in groups of growing pigs. *Anim. Prod.*, **18**, 1–11.

516. WAITES, G. M. H. (1962). The effect of heating the scrotum of the ram on respiration and body temperature. *Q. Jl exp. Physiol.*, **47**, 314–23.

517. WAITES, G. M. H. (1970). Temperature regulation and the testis. In *The Testis*, eds. A. D. Johnson, W. R. Gomes and N. L. Vandemark, pp. 241–79. Academic Press, New York.

518. WAITES, G. M. H. and MOULE, G. R. (1961). Relation of vascular heat exchange to temperature regulation in the testis of the ram. *J. Reprod. Fert.*, **2**, 213–24.

519. WAITES, G. M. H. and VOGLMAYR, J. K. (1963). The functional activity and control of the apocrine sweat glands of the scrotum of the ram. *Aust. J. agric. Res.*, **6**, 839–51.

520. WALKER, J. E. C., WELLS, R. E., JR. and MERRILL, E. W. (1961). Heat and water exchange in the respiratory tract. *Am. J. Med.*, **30**, 259–67.

521. WEAVER, M. E. and INGRAM, D. L. (1969). Morphological changes in swine associated with environmental temperature. *Ecology*, **50**, 710–13.

522. WEBSTER, A. J. F. (1967). Continuous measurement of heart rate as an indicator of the energy expenditure of sheep. *Br. J. Nutr.*, **21**, 769–85.

523. WEBSTER, A. J. F. (1971). Prediction of heat losses from cattle exposed to cold outdoor environments. *J. appl. Physiol.*, **30**, 684–90.

524. WEBSTER, A. J. F. (1974). Heat loss from cattle with particular emphasis on the effects of cold. In *Heat Loss from Animals and Man: Assessment and Control*, eds. J. L. Monteith and L. E. Mount, pp. 205–31. Butterworths, London.

525. WEBSTER, A. J. F. (1976a). Effects of cold on endocrine responses of ruminants and other domestic animals. In *Progress in Animal Biometeorology*, Vol. I, Part I, ed. H. D. Johnson, pp. 33–41. Swets and Zeitlinger, Amsterdam.

526. WEBSTER, A. J. F. (1976b). Effects of cold on energy metabolism of sheep. In *Progress in Animal Biometeorology*, Vol. I, Part I, ed. H. D. Johnson, pp. 218–26. Swets and Zeitlinger, Amsterdam.

527. WEBSTER, A. J. F. and GORDON, J. G. (1977). Air temperature and heat losses from calves in the first weeks of life. *Anim. Prod.*, **24**, 142.

528. WEBSTER, A. J. F., LOBLEY, G., REEDS, P. J. and PULLAR, J. D. (1978). Protein mass, protein synthesis and heat loss in the Zucker rat. *Proc. Nutr. Soc.*, **37**, 21A.

529. WEBSTER, A. J. F., OSUJI, P. O., WHITE, F. and INGRAM, J. F. (1975). The influence of food intake on portal blood flow and heat production in the digestive tract of the sheep. *Br. J. Nutr.*, **34**, 125–39.

530. WEINER, J. S. (1966). Aspects of adaptation. *Advmt Sci., Lond.*, **23**, 370–8.

531. WEIR, J. B. DE V. (1949). New methods for calculating metabolic rate with special reference to protein metabolism. *J. Physiol., Lond.*, **109**, 1–9.

532. WEISS, B. and LATIES, V. G. (1961). Behavioral thermoregulation. *Science, N.Y.*, **133**, 1338–44.

533. WEST, C. D., TRAEGER, J. and KAPLAN, S. A. (1955). A comparison of the relative effectiveness of hydropenia and pitressin in producing a concentrated urine. *J. Clin. Invest.*, **34**, 887–98.

534. WEST, G. C. (1965). Shivering and heat production in wild birds. *Physiol. Zoöl.*, **38**, 111–20.

535. WHITTOW, G. C. (1962). The significance of the extremities of the ox (*Bos taurus*) in thermoregulation. *J. agric. Sci., Camb.*, **58**, 109–20.

536. WHITTOW, G. C. (1965). Regulation of body temperature. In *Avian Physiology*, ed. P. D. Sturkie, Chapter 8. Comstock Press, Ithaca.

537. WHITTOW, G. C. (1971). Ungulates. In *Comparative Physiology of Thermoregulation*, Vol. II, Mammals, ed. G. Causey Whittow, pp. 192–281. Academic Press, New York.

538. WHITTOW, G. C. (1973). Evolution of thermoregulation. In *Comparative Physiology of Thermoregulation*, Vol. III, Special aspects of thermoregulation, ed. G. Causey Whittow, pp. 202–58. Academic Press, New York.

539. WHITTOW, G. C. (1976). Temperature regulation in marine mammals. In *Progress in Animal Biometeorology*, Vol. I, Part I, ed. H. D. Johnson, pp. 88–94. Swets and Zeitlinger, Amsterdam.

540. WHITTOW, G. C. and FINDLAY, J. D. (1968). Oxygen cost of thermal panting. *Am. J. Physiol.*, **214**, 94–9.

541. WHITTOW, G. C., SCAMMELL, C. A., LEONG, MARGARET and RAND, D. (1977). Temperature regulation in the smallest ungulate, the Lesser Mouse Deer (*Tragulus Javanicus*). *Comp. Biochem. Physiol.*, **56A**, 23–6.

542. WIDDOWSON, E. M. (1950). Chemical composition of newly born mammals. *Nature, Lond.*, **166**, 626–8.

543. WILKIE, D. R. (1960). Thermodynamics and the interpretation of biological heat measurements. *Progress in Biophysics*, **10**, 260–98.

544. WILKIE, D. R. (1977). 2. Metabolism and body size. In *Scale Effects in Animal Locomotion*, ed. T. J. Pedley, pp. 23–36. Academic Press, New York.

545. WILSON, W. O. (1976). Effects of temperature on oviposition and egg formation in poultry. In *Progress in Animal Biometeorology*, Vol. I, Part I, ed. H. D. Johnson, pp. 411–16. Swets and Zeitlinger, Amsterdam.

546. WILSON, W. O., HILLERMAN, J. P. and EDWARDS, W. H. (1952). The relation of high environmental temperature to feather and skin temperatures of laying pullets. *Poultry Science*, **31**, 843–50.

546a. WINSLOW, C. -E. A. and HERRINGTON, L. P. (1949). *Temperature and Human Life.* Princeton University Press, Princeton, New Jersey.

547. WINSLOW, C. -E. A., and GAGGE, A. P. (1941). Influence of physical work on physiological reactions to the thermal environment. *Am. J. Physiol.*, **134**, 664–81.

548. WINSLOW, C. -E. A., GAGGE, A. P. and HERRINGTON, L. P. (1939). The influence of air movement upon heat losses from the clothed human body. *Am. J. Physiol.*, **127**, 505–18.

549. WINSLOW, C. -E. A., GAGGE, A. P. and HERRINGTON, L. P. (1940). Heat exchange and regulation in radiant environments above and below air temperature. *Am. J. Physiol.*, **131**, 79–92.

550. WINSLOW, C. -E. A., HERRINGTON, L. P. and GAGGE, A. P. (1936a). A new method of partitional calorimetry. *Am. J. Physiol.*, **116**, 641–55.

551. WINSLOW, C. -E. A., HERRINGTON, L. P. and GAGGE, A. P. (1936b). The determination of radiation and convection exchanges by partitional calorimetry. *Am. J. Physiol.*, **116**, 669–84.

552. WINSLOW, C. -E. A., HERRINGTON, L. P. and GAGGE, A. P. (1937). Physiological reactions of the human body to varying environmental temperatures. *Am. J. Physiol.*, **120**, 1–22.

553. WINSLOW, C. -E. A., HERRINGTON, L. P. and GAGGE, A. P. (1938). The relative influence of radiation and convection upon the temperature regulation of the clothed body. *Am. J. Physiol.*, **124**, 51–61.

554. WOLFE, J. L. and BARNETT, S. A. (1977). Effects of cold on nest-building by wild and domestic mice, *Mus musculus* L. *Biol. J. Linn. Soc.*, **9**, 73–85.

555. WOOLEY, J. B. and OWEN, R. B. (1977). Metabolic rates and heart rate-metabolism relationships in the black duck (*Anas rubripes*). *Comp. Biochem. Physiol.*, **57A**, 363–7.

556. WORSTELL, D. M. and BRODY, S. (1953). Environmental physiology and shelter engineering with special reference to domestic animals. XX. Comparative physiological reactions of European and Indian cattle to changing temperature. *Mo. Agric. Exp. Sta. Res. Bull. No. 515.*

557. WYNDHAM, C. H. (1970). The problem of heat intolerance in man. In *Physiological and Behavioral Temperature Regulation*, eds. J. D. Hardy, A. P. Gagge and J. A. J. Stolwijk, pp. 324–41. Charles C. Thomas, Springfield, Illinois.

558. WYNDHAM, C. H. and STRYDOM, N. B. (1969). Acclimatizing men to heat in climatic rooms on mines. *J. South African Inst. Mining and Metallurgy*, October 1969, pp. 60–4.

559. YAGLOU, C. P. (1927). Temperature, humidity and air movement in industries. The effective temperature index. *J. Indust. Hyg.*, **9**, 297–309.

560. YAGLOU, C. P. (1949). Indices of comfort. In *Physiology of Heat Regulation and the Science of Clothing*, ed. L. H. Newburgh. Saunders, Philadelphia.

561. YAGLOU, C. P. and MILLER, W. E. (1925). Effective temperature with clothing. *Trans. Am. Soc. Heat. Vent. Engrs.*, **31**, 89–99.

562. YANG, A. T. S. (1977). *Water in the Ecophysiology of Pigs.* Ph.D. Thesis, University of Adelaide.

563. YOSHIMURA, H. (1960). Acclimatization to heat and cold. In *Essential Problems in Climatic Physiology*, eds. H. Yoshimura, K. Ogata and S. Itoh. Nankodo, Kyoto.

564. YOUSEF, M. K. (1976). Thermoregulation in wild ungulates. In *Progress in Animal Biometeorology*, Vol. I, Part I, ed. H. D. Johnson, pp. 108–22. Swets and Zeitlinger, Amsterdam.

564a. YOUSEF, M. K. and DILL, D. B. (1969). Resting energy metabolism and cardiorespiratory activity in the burro, *Equus asinus. J. appl. Physiol.*, **27**, 229–32.

565. YOUSEF, M. K., HAHN, L. and JOHNSON, H. D. (1968). Adaptation of cattle. In *Adaptation of Domestic Animals*, ed. E. S. E. Hafez, pp. 233–45. Lea and Febiger, Philadelphia.

566. YOUSEF, M. K. and LUICK, J. R. (1975). Responses of reindeer, *Rangifer tarandus*, to heat stress. In *First International Reindeer/Caribou Symposium*, Inst. arct. Biol., Univ. Alaska.

567. ZERVANOS, S. M. and HADLEY, N. F. (1973). Adaptational biology and energy relationships of the Collared peccary (*Tayassu tajacu*). *Ecology*, **54**, 759–74.

Index

Bold page numbers refer to major entries.

absolute humidity, 43
absolute temperature, 54
absorptance, 52, 219
  human skin, 52
  pig skin, 194
absorption of radiation, 59
  coefficient, 60
absorptivity, 51
acclimation, metabolic, 122, 124, 136, 143, 154, 273
acclimatization, 116, 117, 158, 170
  to cold, **135**, 154, 216, 232, 274
  to hot environments, 46, **138**, 217
acid-base balance, 49, 219
active body mass, 25
activity, 10, 12, 30, 32, 73, 81, 143, 144, 146, 151, 155, 167, 172, 174, 175, 178, 179, 193, 248
adaptation, 7, **116**, 135, 145, 286
  behavioural, 116, **117**, 143, 185, 198, 250, 267, 277
  cold, 36, 116, **135**, 153, 154, 216
  genetic, 116, 125, 283
  heat storage, 124
  high temperature, **138**, 235, 250
  hypothermic, 142, 154, 155
  insulative, 122, **124**, 154
  metabolic, **122**, 154
  morphological, 116, 125, 199, 250, 277
  physiological, **120**, 154, 160
aestivation, 1, 141
age, 17, 35
air,
  flow, 64, 69
air—*cont.*
  movement, 45, 64, 80, 84, 106, 149, 197, 285
  still air, 18, 19, 92, 93, 94, 96, 97, 152, 168, 174, 196
  thermal conductivity, 66, 92
  thermal expansion, 68
  velocity, 68, 69, 110, 151, 172
air-ambient insulation, 79, 82, 83, **96**, 213
air sacs in birds, 278, 279
Alacaluf Indians, 154
Alaskan mountain goat, 235
albedo, 101, 102
Allen's Rule, 126
alkalosis, respiratory, 49, 219, 235, 257, 279
alpaca, 127, 128
alveolar ventilation, 219, 221
alveoli, pulmonary, 48
ama, 124
ammonia in poultry houses, 284
amphibia, 2
amphibious mammals, 136
anaerobic metabolism, 38–9
anaesthetics, 38
animal heat, 73
animal productivity, 111, 138, 146, 200, 202, 244, 247, 248
antarctic weather, 105
antelope, 91, 128, 247, 260, 263
antelope ground squirrel, 140
appendages, 81, 87, 90, 138, 213, 250
appetite, 210
aquatic mammals, 91, 131, 133, 136, 138

arctic glaucous gull, 276
arctic mammals, 86, 92
arid regions, 103. 260
arterio-venous anastomoses, 87
Artiodactyla, 183, 228, 247, 249
ass, 247, 251
athletes, 147, 155
atmosphere, 57, 99
Australian Aborigine, 124, 142, 154
Awassi sheep, 237–8
Ayrshire cattle, 94, 212, 218

banteng, 257
basal metabolism, 21, 32, 35
bat, 141
bear, 142
beaver, 138
bedding, 30, 110, 116, 120, 145, 177, 182, 185, 191, 197, 201
beech trees, 24
beetle, 144
behaviour, 4, 28, 57, 79, 115, 121, 199, 210
behavioural thermoregulation, 4, 115, 188
Bergmann's Rule, 126
birds, 1, 2, 48, 50, 75, 138, 121, **269**
  body temperature, 269–70
  brown fat, 273
  heat prostration, 283
  migration, 269
  nasal gland, 280
  poorly-feathered, 272, 275, 281, 282
  torpidity, 141, 276
birth, 130, 178, 179, 241, 243
birth weight, 185, 207
bison, 132
black body, 51, 55, 60, 96
  temperature, 56, 64
black duck, 269
blood,
  circulation, 160, 165, 168, 220–1
  flow, 7, 79, 155, 165–8, 276
  pH, 219
  plasma volume, 267
  pressure, 155–6, 221
  red cells, 243, 250–1, 259, 262, 267
  vessels of skin, 199
  viscosity, 168
  volume, 165–6, 168, 221
boar, wild, 183
body,
  conformation, 116, 199
  profile, 57, 59, 250
  shape, 22, 64, 69
  shell, 2, 154
  size, 10, **21**, 24, 80, 92, 139, 142, 209
  type, 125
body cooling, 146, 154, 193
body surface, 8
body surface—*cont.*
  area, 21–2, 24–5, 50, 58, 72, 91, 118, 146, 218, 292
body temperature, 2–3, 7, 32, 35–6, 40, 46, 48, 79, 90, 92, 116, 138, 141, 159, 170, 194, 276, 277, *see also* deep-body temperature
  mean, 155
  birds, 269–70
body weight, 21, 23–5, 35, 129
  birds, 269, 271, 292
  exponent, 22, 179
  hot climates, 261–2, 267
  cattle, 225
  man, 162, 165, 170, 179
  maternal, 129, 131, 133
  newborn, 129–31, 133
  pig, 185, 190
  sheep, 229
body water, 260
bomb calorimeter, 13
Bos indicus, 209, 218, 238, 248, 259, 264
Bos taurus, 209, 218, 248, 255, 259, 264
boundary layer, 45, 64, 66, 84, 97
Bovidae, 228, 247
Brahman, cattle 209, 224, 238
brain, 4, 90, 166
  temperature, 91, 259
breeds,
  pig, 184
  cattle, 209, 226, 227
  sheep, 228, 242, 250
broiler chicken, 285–6
brown fat, 7, 35–7, 142, 179, 243, 273
budgerigar, 280
buffalo, 128, 132, 247, 251–2, 255, 257
bumblebee, 3
buoyancy force, 64, 71
burro, 251–2, 267
burrow, 138–9, 143, 145

calf, *see* newborn calf
calorimetry, 12, 33, **73**, 146, 148, 198
  bomb, 13
  direct, 75
  gradient layer, 76
  heat sink, 75
  ice, 74
  indirect, 76
  open and shut, 77
  partitional, 76
  respiration, 76
camel, 125, 127–8, 138, 162, 182, 247, 250–4, 257, 260–2, **263**, 266, 268, 280
carbon dioxide, 49, 76–7
  partial pressure, 219
cardiac output, 167
carcass analysis, 76, 245

cardiovascular changes, 160, 165, 168, 220–1
caribou, 95, 127, 130, 246, 251
carotid rete, 89–90, 259
cat, 3, 130, 165
catecholamines, 136, 273
cattle, 13, 17, 24, 35, 46, 49, 56, 59, 80, 93, 95, 110, 127, 128, 137, 208, **209**, 246–7, 250, 255, 257–8, 260–2
  beef, 212, 214–5, 226
  breeds, 94, 209, 221–7, 264
  Bos indicus, 209, 218, 238, 248, 259, 264
  Bos taurus, 209, 218, 248, 255, 259, 264
  dairy, 212, 214, 226
  pneumonia, 1, 24–5
  selection, 226
cattle shades, 220
Cervidae, 228, 248
cetacean, 91
characteristic dimension, 65
cheek temperature, 149–50
cheetah, 11, 38
chemical combustion, 73, 75
chick, newly hatched, 273
chicken, 269, 273, 277, 279–81, 283, 286
chilblains, 153
chipmunk, 142
chloride in sweat, 161
circadian variation, *see* 24-hourly variation
circulatory failure in heat, 165, 170–1
classical conditioning, 117
climate, 154
  micro-climate, 113, 152, 155
  types, 100
  zones, **101**
clo unit, 83
cloaca of birds, 280
clothing, 45, 79, 83, 84, 95, 138, 145, 151, 152, 155, 164, 171–2, 176
  vapour permeability, 84, 175
  zones, 172
coat, 42, 45, 62, 79, 84, 90, 125, 131, 134, 135, 209, 213, 237–8, 258, 264
  absorption coefficient, 62
  colour, 53, 58, 59, 176, 194, 219, 258
  hair density, 58, 62, 93–4, 183
  reflection coefficient, 62
  shedding in cattle, 216
  temperature, 59
  thermal conductivity, 58, 92
  thermal insulation, 36, **92**, 124, 213
coefficients for heat transfer, 42, 76, 81, 96 106, 107, 112
cold, 90, 92, 105, 123, 135, 142, 148, 155, 210, 215, 273
  limit, 7, 14, 16, 19–20, 25, 122, 180, 187, 242
  stress, 232
cold—*cont.*
  susceptibility, 145, 157, 179, 182, 184, 190
  test, 155
  tolerance, 124, 209, 216
cold-induced vasodilatation, 87, 153, 213
cold injury, 153
comfort, thermal, 103, 105, 114, 171, 173
  zone, 16, 174
conditioning, 117
  classical, 117
  operant, 117, 199–202
conductance, thermal, 15, 31, 81, 82, 96, 274
  units, 289
conduction of heat, 9, 40, 64, **72**, 114, 197, 230
conductivity, thermal, 66, 72, 82, 92–3
  air, 66, 92
  coat, 58, 92
conservation of energy, law, 12
controlled hyperthermia, 160
convection, 9, 40, **64**, 240–1
  forced, 64, 96, 110
  natural (free), 64, 93, 97, 110
  mixed, 64
convective heat transfer, 56, **64**, 81, 97, 114, 165, 172, 195
  coefficients, **65**, 96, 105, 151
  cylinder, 69, 70
  human head, 69, 70
convective insulation, 81, 96
conversion efficiency of food, 206
cooling, 91, 160
coprophagy, 140
core temperature, *see* deep-body temperature
cortisol, 136–7
counter-current heat exchange, 8, 48, **90**, 138, 140, 276, 280
cow, 21, 24, 71–2, 92, 110, 129, 132
coypu, 130
Crawford, 73
critical temperature, **7**, 14–5, 17, 19, 26, 27, 30, 47, 89, 92, 110, 122–3
  birds, 274, 276
  effective, 17, 31, 111, 192, 197
  cattle, 212, 214
  deer, 246
  goat, 235
  man, 146, 154
  pig, 185, 187, 190–1, 199, 201, 204, 208
  sheep, 233–4

daylength effect in chicken, 286
dead-space ventilation, 219, 221, 257
deep-body temperature, 1, 2, 7, 9, 11, 14, 15, 17, 32, 36, 40, 48, 72, 83, 87, 122, 124, 134

deep-body temperature—*cont.*
birds, 269, 271, 279
cattle, 211
deer, 246
hot climates, 251–2
man, 171
pig, 186
sheep, 241
deer, 132, **245**, 247
caribou, 95, 127, 130, 246, 251
distribution, 245
elk, 245
Lesser Mouse Deer, 245
moose, 127, 245
red deer, **245**, 251
reindeer, 132, 137, 246, 251
dehydration, 138, 153, 162, 163, 166, 170, 259–62, 266, 280
desert, 102, **103**, 108, 120, 128, 138–9, 143, 153, 161–5, 176, 241, 249, 259, 263–4
animals, 48
arthropods, 144
iguana, 3, 6, 143
journey, 163–4
rodents, 48, 139
snail, 142
wood-louse, 143
desiccation, 103
dewlap, 209, 250
dew point temperature, 43–4, 48, 103, 114
diaferometer, 77
diapause, 144
diffusion resistance, 45, 83–4, 87
digestible energy, 13
dimensionless numbers, 65–6
disease, 79, 210, 226, 248
diuresis, 165
diving mammals and birds, 38
dog, 11, 38, 49, 50, 76, 127, 165
husky, 94–5, 127
donkey, 128, 251, 253, 256, 267–8, 280
draughts of air, 204, 215
drought, 138, 144, 263
drugs, 38
dry bulb temperature, 43
duck, 38, 269
duiker, 255
Duroc pig, 184

ear, 88, 125, 199, 250, 277
East African mammals, 255–6, 261
effective critical temperature, 17, 31, 111, 192, 197
effective radiant flux, 107, 113
effective temperature scale, 106, 169
efficiency of food utilization, 30
egg, hen, 286
hatching, 273
egg, hen—*cont.*
incubation, 273
production, 248, 284–6
eland, 251–3, 255, 260, 263, 267
elderly people, 156
elephant, 38, 132
elk, 245
embryo, 39, 207, 223
emissivity, 51, 96
skin, 52, 53
wall, 56
emotional responses, 38
Emperor penguin, 274
endocrines, 36, 124, 136–7, 220
energy, 11
balance, 12, 32, 73, 148, 282
digestible, 13
equivalents of oxygen consumption, 77
exchange, 146
expenditure, 148, 153, 160, 189
food, 12
gross, 13
metabolizable, 12, 29
retention, **12**, **29**, 111, 201, 203
environment, thermal, 28, **99**, 151
assessment, **105**
micro-environment, 8, 113, 143, 145, 200
specification, 112
standardized, 19, 109
environmental temperature, 19
equatorial regions, 99, 101
equivalent still air temperature, 105
equivalent still shade temperature, 105
equivalent temperature, 113–4
Eskimo, 155–6, 175
eutherian mammal, 2, 141, 269
evaporation, 7, 9, 164, 176
evaporative heat loss, 8, 14, 15, 40, **42**, 72, 107, 114, 138
birds, 272, 277, 290
cattle, 212, 217
deer, 246
hot climates, 250, 253, 258, 268
man, 146, 148, 180
pig, 185, 193
sheep, 235
evaporative heat loss, cutaneous, **46**
birds, 278
cattle, 217–8
hot climates, 253, 255–6
pig, 193
evaporative heat loss, respiratory, 140, 144
birds, 277–8
cattle, 218–9
hot climates, 255–6, 264
pig, 194
evaporative impedance, 82, 84

ewe, 27, 229
exercise, 27, 141, 146, 155, 167, 177, 241, 259
expired air, 42, 48
exponent of body weight, 22, 179

face, 149, 151
faeces, 12, 140
  water content, 261, 263
farm animals, 21
fasting, 22, 186, 215
  metabolism, 24, 191, 211–2, 245
fat, 12, 30, 36–7, 86, 93–4, 232, 281
  brown, 7, 35–7, 142, 179, 243, 273
  composition, 136
  gain in pig, 184, 202
  in newborn pig, 185
  subcutaneous, 8, 79, 85, 130, 135, 138, 145, 156, 182, 185, 245, 281
  tail in sheep, 128, 250
  unsaturated, 199
  white, 36
feeding level, *see* plane of nutrition
feeding, heat increment of, 29, 30, 201
feathers, 275, 281, 283
fertility, 207, 244, 248
fishes, 2
fleece, 110, 228–33, 239, 240
flight, 38, 280
floor, 9, 72–3, 110, 118
  heat loss from pig, 118, 197
  heat loss from sheep, 230
foal, 186
food conversion efficiency, 206
food energy, 12
food intake, 7, 10, 12, 27, 28, 31–2, 79, 123
  birds, 276, 281–2, 285
  cattle, 209, 217, 221–2, 224
  hot climates, 248
  man, 148, 153
  pig, 199, 206
  sheep, 232
foot, 146, 154, 167, 177, 281
forage, 248, 263
forehead temperature, 146, 149, 150
forest, 100, 102
Fourier's Law, 82
fox, 93, 94, 136
fowl, 36, **247**
free energy, 12
frostbite, 87, 153
full radiator, 51

Gaspé fishermen, 124, 155
gastro-intestinal tract, 35
gazelle, 38, 128, 247, 251–5, 259, 262–3
genetic adaptation, 116, 125, 283
gerbil, 140, 144
giraffe, 251–2
glass, transmittance, 54, 58
globe thermometer, 68, 71–2, 106, 169
glomerular filtration rate, 259
glucocorticoids, 136
gluconeogenesis, 136
glycogenolysis, 136
goat, 11, 127, 132, 238, 241, 247, 255–7, 259, 260
  Alaskan mountain, 235
  breeds, 229
  metabolism, 235
  mountain, 235–6
  Saanen, 229
  Sinai, 235–6, 262
  world total, 229
goldfish, 136, 143
goose, 269
gradient layer calorimeter, 76
Grashof number, 66
grazing, 211, 221, 232, 250, 260, 263
greenhouse, 55
gross energy, 12
ground, 57, 72, 113, 115, 230
ground squirrel, 142
group effects: pig, 188–91
growth, 12, 31, 73, 190, 203, 207, 222, 225, 248, 281–2
  rate, postnatal, 16, 185
growth hormone, 136–7
guanaco, 251
guinea pig, 75, 130
gular flutter, 50, 279
gull, 90

habituation, 117
hair density, 58, 62, 93, 94, 183
hamster, 142–3
hands, 149, 153, 155, 167
hare, 127
hartebeest, 251–2
hatching of eggs, 273
head, human,
  convective heat transfer, 69, 70
  protection in desert, 177
heart rate, 38–9, 159, 167, 169, 217, 276
heat balance, **7**, 95, 112, 238
  equations, 76
  sheep, 238
heat acclimatization, 46, **138**, 217
heat casualties, 169
heat of combustion, 13
heat conservation, 5, 6, 8, 130, 175, 276
heat disorders, 170
heat dissipation, 5, 7, 8, 138, 160, 175, 276, 281
heat exchange, 9, 42, 49, 53, 90, 171
  counter-current, 8, 48, **90**, 138, 140, 276, 280
  partition, 76

heat exchanger, 90, 276
heat exhaustion, 170–1, 161
heat flow, 82, 111
heat hyperpyrexia, 171
heat increment of feeding, 29, 30, 201
heat load, 9, 40, 58, 140
  birds, 277
  cattle, 219
  hot climates, 267
  man, 147, 160, 169, 170, 175, 177
  sheep, 229, 239
heat loss, 8, 9, 14, 18, 31, 33, **40**, 75–6, 80, 110
  birds, 285, 291
  man, 148, 151
  pig, 188, 191, 197
  sheep, 230, 237
heat production, **6**, 10, 12, **14**, 48, 58, 76, 90, 92, 118, 124
  birds, 272, 282–3
  cattle, 211–2, 217
  deer, 245
  extra thermoregulatory, 27, 111, 202
  goat, 235
  hot climates, 260
  man, 171, 179–80
  pig, 191, 205
  sheep, 233
  sites in animal, 35
heat prostration in birds, 283
heat sink calorimeter, 75
heat storage, **10**, 14, 32, 267
heat stress, 162, 167, **168**, 207, 210
  indices, 168
heat stroke, 168, 171
heat susceptibility, 190, 216
heat syncope, 170–1
heat tolerance, 115, 157, 167–8, 170, 226, 248
heat transfer, xi, 9, **40**, 105
  conductive, *see* conductive heat transfer
  convective, *see* convective heat transfer
  coefficients, xi, **42**, 76, 81, 96, 106–7, 112
  evaporative, *see* evaporative heat loss
  non-evaporative, *see* non-evaporative heat transfer
  radiative, *see* radiation
  sensible, *see* non-evaporative heat transfer
heat of vaporization, 43, 48, 85
heat of warming ingesta, 73, 198
Heberden, 110
helicopter, 151
Henle, loop of, 258–9
heron, 90
herring gull, 90, 124, 281
hibernation, 1, 36, 141, 143
hidromeiosis, 162
hippopotamus, 136, 182, 251
homeotherm, 1, 8, 11, 16, 24, 32, 92, 118, 130
horse, 38, 95, 127–8, 132, 247, 254, 256, 263
hospital incubator, 54, 56–7, 97, 178
hot limit, 4, 7, 16, 19
house sparrow, 273–4
housing,
  cattle, 225
  pig, 203
huddling, 8, 30, 111, 116, 118, 119, 131, 145, 185, 188–91, 198, 201, 275, 283
human infant, 36, 54, 57, 75, 177, **178**
humidity, 42, 43, 103, 106, 110, 113, 128, 138, 165, 172, 176, 219, 248, 262, 283
humming bird, 1, 92, 141, 271, 276
husky dog, 94, 95, 127
hyperthermia, 7, 11, 14–16, 27, 32, 46, 123, 138, 140, 182, 193, 201, 208, 217, 220, 250, 267, 277, 279, 281
hypoglycaemia of newborn, 186
hypothalamus, 4, 17, 237, 279
hypothermia, 6, 14–17, 32, 129, 142, 156, 181–2, 276, 281
hypoxia, 36

ice, 153
ice calorimeter, 74
immersion foot, 153
impala, 251–2
incubation of eggs, 273
incubator, hospital, 54–57, 97, 178
infra-red radiation, 54, 238
insects, 3, 4, 143, 175
inspired air, 48–9
insulation, thermal, 6, 10, 17, 19–21, 26, 31, 72, **79**, 90, 116, 122, 133, 135, 142
  air-ambient, 79, 82–3, **96**, 213
  birds, 275, 281, 290, 291
  cattle, 212–3, 215
  clothing, 149, 174
  coat, 36, **92**, 124, 213
  convective, 81, 96
  deer, 246
  diffusion resistance, 45, 83–4, 87
  external, **79**
  feathers, 273, 275, 281, 283
  internal (tissue), 79, 85, 87, 125
  man, 152, 155–6, 179
  materials, 93
  non-evaporative (sensible), 82
  overall, 80, 118
  pig, 185, 198
  radiative, 81, 96
  sheep, 229, 231, 234
  specific, 80, 85
  tissue, 79, 85, 87, 125
  units, 83, 289
insulation-wind-decrement, 96
insulin, 136–7

invertebrates, 2
irradiance, 60

jerboa, 140

Kalahari Bushman, 124, 154
kangaroo, 141
kangaroo rat, 48, 140, 266
kidney, 167, 263
  concentrating power, 164, 257, 259, 267

lactation, 129, 207, 223, 241
  in cold, 216
lactic acid, 39, 48
lamb, *see* newborn lamb
Large White pig, 184
latent heat of vaporization, 43, 48, 85
Lavoisier, 74
leaf, 71–2
lean body mass, 25
learning, 117
least thermoregulatory effort, 15, 16
lethal limit,
  dehydration, 163
  temperature, 140, 143, 168, 241, 269
lighting threshold for chickens, 286
lipolysis, 136
litter size, 130
liver, 35, 39, 167
lizard, 3, 4, 6, 143–4
llama, 127–8, 256, 263
long-distance runners, 155
long-distance swimmers, 156
long-wave radiation, 53–4, 60
lungs, 45
  alveoli, 48
  birds, 278–9
lynx, 94

maintenance requirement, 12, 21, 29, 73, 79, 203, 211–2, 214–5, 225, 269
mallard, 272
mammal, 1, 3, 6, 36, 121
  amphibious, 136
  aquatic, 91, 131, 133, 136, 138
  arctic, 86, 92
  diurnal and nocturnal, 1, 32
  diving, 38
  East African, 255–6, 261
  eutherian, 2, 141, 169
  marsupial, 2, 141
  monotreme, 2
mammary blood flow, 235
man, 13, 17–19, 38, 45–6, 48, 57, 60, 64, 72–3, 75–6, 86, 90, 106, 108, 132, 136, 138, 140–1, **145**, 208, 229
marine birds, 280
marsupial, 2, 141
maternal weight, 129, 131, 133

mean body temperature, 10, 12
mean radiant temperature, 40, 106, 108, 110
mean skin temperature, 96, 107
meat production, world, 247
Merino sheep, 50, 94, 229, 237–8, 260, 264
metabolic acclimation, 122, 124, 143, 154, 273
metabolic body size, 22, 24–5, 259, 292
metabolic rate, xi, 9, 10, 12, 14, 18, 21, 23–4, 34, 36, 55, 74, 76–7, 92, 97, 115, 122–3, 125, 130, 138, 141–3, *see also* heat production
  birds, 269, 274, 277, 280–1
  deer, 245–6
  man, 154, 178
  maximum, 19, 20, 34, 83, 122, 131, 170, 179–80, 187, 273
  minimum, 7, 15–6, 27, 31, 34, 122, 146, 187, 202, 235, 269
metabolism, 25, 137
  aerobic and anaerobic, 38–9
  basal, 21, 32, 35
  fasting, 22, 186, 215
  resting, 35, 146, 157
  summit, 230
metabolizable energy, 12, 29
  intake, 205
methane, 12, 77
micro-climate, 113, 152, 155
micro-environment, 8, 113, 143, 145, 200
migration of birds, 269
milk,
  composition, 130, 132
  energy, 130, 132
  yield, 129, 185, 210, 223–4, 248
miner's cramp, 171
mitochondria, 36–7
mixed convection, 64
monkey, 130
Monotreme, 2
moose, 127, 245
morphological adaptation, 116, 125, 199, 250, 277
mouse, 17, 21, 24, 32, 34, 38, 46, 92, 94, 97, 118, 120, 123, 125, 129–30, 133, 141
  deer mouse, 120
  grasshopper mouse, 140
  hairless, 95
  tail length, 125
mouth temperature, 161
mud-bath, 185, 250
multiple choice, 120, 198
muscle, 35
muscular work, 35, 36
musk-ox, 128, 132, 251

nasal gland of birds, 280
nasopharynx, 91, 256

natural convection, 64, 93
neonatal cold injury, 181
nest, 8, 118, 182
newborn, 39
  calf, 38, 93, 96, 130, 186, 212, 214, 223
  foal, 186
  guinea pig, 131
  infant, 17, 19, 20, 36, 54, 57, 75, 83, 130, 177, **178**, 207, 290, 291
  kitten, 17
  lamb, 19, 20, 50, 83, 130–1, 180, 186, 207, **241**
  mammal, 35, 130, 134
  monkey, 130
  mouse, 131
  physiological tolerance, 129
  pig, 17–20, 56, 83, 86, 89, 94, 118–9, 130, 180, 182, **185**, 198
  puppy, 17, 130
  rabbit, 36–7
  rat, 130–1
  weight, 129–31, 133
Newton's Law of Cooling, 83
nocturnal activity, 138, 140
non-evaporative heat transfer, 8, 9, 14–5, **40**, 72, 148, 290
non-shivering thermogenesis, 5, 7, **35**, 135–6, 179, 243, 273
noradrenaline, 35
nose skin temperature, 149
Nusselt number, 66

obesity, 156
Ohm's Law, 82, 290
operant conditioning, 117, 199–202
operative temperature, 106, 108–9, 179–80, 291
optimal temperature, 16
oryx, 251–3, 255, 260, 263
ostrich, 144, 280
ox, 50, 88, 220, 222, 256
oxygen consumption, *see* metabolic rate
  energy equivalents, 77
  methods of measurement, 76
  predictive equation, 77
ozone, 99

pack rat, 140
panting, 5, 8, 11, 45, 49, 50, 138, 141
  birds, 282
  cattle, 219
  hot climates, 253, 256, 262–3
  pig, 208
  sheep, 235, 237, 243
partial efficiency, 30
partitional calorimetry, 76
peccary, 251–2
penguin, 271, 274
per cent wetted area, 42, 84
Perissodactyla, 247, 249
Perspex shield, radiation, 54, 56–8
phlogiston theory, 74
photoperiodicity, 99, 231, 248
photosynthesis, 99
pig, 13, 17–9, 24, 28–9, 35–6, 45, 49, 56, 72–3, 75, 85–6, 88–90, 94–6, 108, 111–2, 117, 120, 123, 125–6, 132, 136, **182**, 229, 247, 250, 255
  breeds, 184
  classification, 183
  farming, 112, 200
  fossil, 183
  gestation period, 183
  housing, 203
  newborn, *see* newborn pig
  world total, 184
pigeon, 269, 273–4
pilo-erection, 5, 6, 8, 92, 116, 212
plane of nutrition, 7, 17, **26**, 123, 174, 190. 192, 201–2, 215, 233, 241
plants, 99, 103, 140, 144, 263, 265
plasma, 168, 263
  volume, 267
  osmotic pressure, 267
plastics, 54
pneumonia in cattle, 225
poikilotherm, 1, 24–5
point of lay, 286
polar bear, 86, 136
polar human biology, 152, 172
polar regions, 99, 105, 146, 274
polyethylene, transmittance, 55, 58
poorly-feathered birds, 272, 275, 281–2
porpoise, 132, 138
posture, 6, 8, 30, 50, 57, 65, 80, 115–9, 196, 282
potential energy, 11
poultry, 13, **269**
preferred temperature, 16, 120, 198
pregnancy, 207, 223, 241
prickly heat, 171
productivity, animal, 111, 138, 146, 200, 202, 244, 247–8
profile, body, 57, 59, 250
protein, 12, 30, 281
  gain, 202
  synthesis, 35
psychrometer constant, 113
psychometric chart, 43–4
pulmonary ventilation, 49, 220–1, 235, 257
  alveolar, 219, 221
  dead space, 219, 221, 257
  minute volume, 42, 50
  'second phase', 219, 235
  tidal volume, 50, 219–20, 279
pulse rate, 108, 166
pyrexia, 171

Q_{10}, 34

rabbit, 3, 50, 95, 130
radiant energy, 51, 220, 239
  atmospheric, 196
  solar, 196
radiant heat exchange, 51, 106, 138, 194, 215, 237
radiating area, 50–1
radiation, 9, 40, **50**, 109, 215, 239–40
  coefficient, 61, 72, 96
  cooling, 56
  full spectrum, 57, 60
  geometry, 58
  infra-red, 54, 238
  long-wave, 54, 60, 93, 175, 194, 238
  penetration, 58
  short-wave, 53, 59, 60, 175, 194
  solar, 238
  ultra-violet, 99
  visible, 54, 58, 99
radiative insulation, 81, 96
radiative interchange factor, 53, 57
radiometer, 56, 108, 220
radiotelemetry, 38, 269
radius of curvature and heat loss, 81
rain, 100, 102, 110, 124, 215, 230
ram, 90
rat, 117, 123, 130, 132, 136, 140, 162
rate of lay, 282, 286
rectal temperature, 10, 108, 250–1, 259, 262, 266–7, *see also* deep-body temperature
  cattle, 216, 223
  man, 156–7, 166, 171
  pig, 186, 194
  sheep, 243
red blood cells, 243, 250–1, 259, 262, 267
red deer, **245**, 251
red fox, 93–4
reference base for metabolism, 24
reflectance, 53, 219, 258
  pig skin, 194–5
reflection, 59
  coefficient, 60, 62
  power, 51
rehydration, 262
reindeer, 132, 137, 246, 251
relative humidity, 43–4, 103, 110
reproduction, 222, 286
  fertility, 207
reptiles, 2, 144
respiration chamber, 56
respiratory alkalosis, 49, 219, 235, 257, 279
respiratory rate, 50, 217–20, 243–4, 251, 254–5, 257, 276, 279
respiratory heat loss, 48, *see also* evaporative heat loss, respiratory
respiratory minute volume, 42, 50
respiratory quotient, 77, 186
respiratory tract, 42, 44–5, 48, 81, 140, 144, 217, 255, 278
respiratory water loss, 243–4, 263
resting metabolism, 35, 146, 157
rete, carotid, 89, 90, 259
Reynolds number, 66
rhinoceros, 132, 247, 251
robe of desert people, 175
rodents, 48
  desert, 48, 139
ruminant, 13, 221, 254, 260, 262–3
running, 38, 155

saliva, 140–1
salt,
  addition to diet in heat, 48, 162, 164
  balance, 164, 280
  loss, 161–2, 170–1
  tolerance, 257
sand rat, 140
saturation deficit, 45
saturation vapour pressure, 43, 114
Schlieren photography, 67, 97
scorpion, 143
scrotal temperature, 222, 235
scrotum, 90, 235
seal, 94, 136, 138
sea lion, 136
season, 102
  dormancy, 277
sea water, 164
'second phase' breathing, 219, 235
sensible heat transfer, 9, 290
set-point temperature, 4, 5, 155
shade, 114, 120, 144, 194, 220, 248, 250, 267
shape of body, 22, 64, 69
shearing, 231, 240
sheep, 13, 18–9, 27, 35, 38, 46, 49, 50, 56, 59, 73, 79, 80, 88, 91, 95–6, 110, 127–8, 130, 132, 137, 208, **228**, 245, 247, 249, 250, 255, 257, 259, 260, 262–4
  Awassi, 237–8
  breeds, 94, 228–9, 238, 241–2, 250
  distribution, 228
  fat tail, 128, 250
  heat balance, 238
  Merino, 50, 94, 229, 237–8, 260, 264
  shorn, 27, 31, 238, 242, 259
  Soay, 242
  world total, 228
shelter, 152, 182, 199, 215, 225
shivering, 5, 7, 35–6, 135–6, 154–5, 179, 243, 273, 276
short-wave radiation, 53, 59, 60, 175, 194
shrew, 92
simulated animals, 106

skin temperature, 6, 10, 28, 36, 54–5, 68, 72, 83–9, 106
birds, 272, 274, 275, 290
cattle, 221–2
deer, 246
man, 155, 167
pig, 191
skunk, 94
sleep, 172, 178
snow, 95, 101–2, 130
social thermoregulation, 38, 182, 283
solar altitude, 59
solar constant, 99
solar heat gain, 108
solar radiation, 51–2, 57, **99**, 103, 108, 114, 128, 141, 143, 225, 229, 248, 266
South African mines, 168
specific heat of body, 10
specific thermal insulation, 81
spermatogenesis, 90, 223, 237
spermatozoa, 207, 222
sphere, 105
profile, 57
spinal cord, 4
sprays, cooling, 46, 182, 185, 194
standardized environment, 19, 109
equivalent temperature, 109, 112, 186, 204
steer, 92, 96
Stefan–Boltzmann Law and constant, 51, 61
still air, 18–9, 92–4, 96–7, 152, 168, 174, 196
still-birth, 207
stratosphere, 99
straw bed, 197, 204
subcutaneous fat, 8, 79, 85, 130, 135, 138, 145, 156, 182, 185, 245, 281
Suidae, members, 183
summit metabolism, 15
sun, 57, 99, 120, 144, 162, 176, 215, 250, 258, 267
sunburn, 177
sunlight, 58, 63, 71–2, 101, 110, 175
surface,
area, 21–2, 24–5, 50, 58, 72, 91, 118, 146, 218, 292
body, 8
temperature, 64, 76
sweat, 45–6, 255
composition, 46, 48, 160–1
sweat glands, 45–6, 181, 193, 217–8, 255, 264
sweating, 5, 8, 42, 49, 108, 125, 138, 141
cattle, 210, 217
hot climates, 253, 255, 266
man, 145, 157, 159, **160**, 163, 170, 176, 181
maximum rate, 161
men and women, 168
sweating—*cont.*
pig, 208
sheep, 236
threshold temperature, 155, 181, 268
swimming and water immersion, 156–8
symbols, xi
syncope in heat, 170–1

tail, 125, 128, 250
telemetry, 38, 269
temperate regions, 99, 102, 172
temperature, *see also* under context, e.g. dew point
black-body, 56, 64
body, *see* body temperature
brain, 91, 259
core, *see* deep-body temperature
critical, *see* critical temperature
deep-body, *see* deep-body temperature
environmental, 83
equivalent, 113, 114
forehead, 146, 149, 150
foot, 146, 154
mean body, 155
mean radiant, 72, 172, 266
mean skin, 155
mouth, 161
preferred, 198, 210
rectal, *see* rectal temperature
scrotum, 222, 235
set-point, 4, 5, 155
skin, *see* skin temperature
wall, 54, 106
temperature receptors, 4, 5, 236
'temperature-humidity index', 224
temperature regulation, **1**, 16, 19, 20, 30, 72, 116, 129, 139, 147, 155, 165, 170–1, 185, 193, 246, 260, 269, 273, 277, 281
testes, 90
thermal assessment, **105**
thermal capacity, 72
thermal circulation index, 87, 89, 185, 275
thermal comfort, 103, 105, 114, 171, 173
thermal conductance, 15, 31, 81–2, 96, 274
units, 289
thermal conductivity, 66, 72, 82, 92–3
thermal demand of environment, 17, 105, 111, 117, 188, 191
thermal diffusivity, 72
thermal insulation, *see* insulation, thermal
thermal neutrality, 7, 9, 14, 16, 25, 30, 110, 118–9, 122
birds, 276
cattle, 212, 217
deer, 246
goat, 235
man, 177–8
pig, 186–7, 191, 199, 204–5

thermal neutrality—*cont.*
  sheep, 233–4
thermal-wind-decrement, 96, 105, 111
thermocline, 120
thermography, 58, 60
thermoregulation, *see* temperature regulation
thirst, 162, 171
thoracic hump, 209
thyroid, 36, 123, 217, 220
thyroxine secretion rate, 136–7
tidal volume, 50, 219–20, 279
torpidity, 1, 92
  birds, 141, 276
transmittance, 55, 58
treadmill, 162
tropics, 100–2, 154, 165, 172, 206, 210, 222, 226, 260, 264
turbinates, 50, 256
turkey, 269, 283, 285
24-hourly variation, 1, 33–4, 103, 170, 192, 250, 252, 266–7, 272

udder, 216
ultra-violet radiation, 99
ungulates, 128, 251–4, 263
  classification, 249
  in hot climates, 259
unicellular organisms, 24, 25
units, xi, 83–4, 288–9
urea, 48, 140
uric acid in birds, 280
urine, 12, 162, 164–5
  concentration, 140, 229, 259–63
  nitrogen, 77

vaporization, 42, 43, 84
vapour permeability of clothing, 84, 175
vapour pressure, 43–4, 114
vascular volume, 166
vasoconstriction, 5, 6, 15, 18, 79, 86, 116, 130, 167, 185, 212–3
vasodilatation, 5, 15, 68, 79, 85, 87, 165–8, 213, 221, 276
  cold-induced, 87, 153, 213
vasomotor tone, 179, 276
vasopressin, 257, 264
vegetation, 100–2, 115, 127
ventilation, poultry, 284
vicuna, 127
viscosity,
  blood, 168
  kinematic, 68
visible spectrum, 54, 58, 99
vole, 97
volumetric specific heat, 84

wallows, 46, 120, 182, 185, 194, 208, 248, 250
walrus, 136
water,
  consumption, 73, 207, 224, 280
  deficit, 163, 166, 170
  dehydration, 138, 153, 162–3, 166, 170, 259–62, 266, 280
  deprivation, 138, 144, 257, 261, 266
  economy, 138, 267, 280
  gut contents, 260
  loss, 47, 140, 171, 256, 263, 267, 277, 280–1
  turnover, 183, 209, 225, 257–60, 266, 268
  vapour pressure, 9, 40
waterbuck, 251, 255
wavelength, 51, 54
wet bulb globe temperature index, 169
wet bulb temperature, 43
wet, dry index, 169
wether sheep, 24
whale, 132, 138
white fat, 36
white fox, 136
Wien's Displacement Law, 54
wild boar, 183
wildebeest, 251, 253, 260
wind, 18, 48, 95, 110, 124, 174, 198, 215, 230, 241
  chill, 105, 111, 149, 151
  speed, 69, 95, 102–3, 105–6, 195, 230–1
wind-break, 114, 120, 145, 154
winter, 124, 225
wolf, 127
wool, 231, 238
  shearing, 231
work, 11, 147, 159, 162, 164, 167–8, 176

yak, 128
Yorkshire pig, 184

zebu cattle, 209, 263